DIGITAL COMMUNICATION SYSTEMS

PEYTON Z. PEEBLES, JR., Ph.D

PROFESSOR OF ELECTRICAL ENGINEERING
THE UNIVERSITY OF FLORIDA

Prentice/Hall International, Inc.

© 1987 by Prentice-Hall, Inc.
A Division of Simon & Schuster
Englewood Cliffs, New Jersey 07632

Printed in the United States of America

10   9   8   7   6   5   4   3   2   1

ISBN   0-13-211962-5   025

PRENTICE-HALL INTERNATIONAL (UK) LIMITED, *London*
PRENTICE-HALL OF AUSTRALIA PTY. LIMITED, *Sydney*
PRENTICE-HALL CANADA INC., *Toronto*
PRENTICE-HALL HISPANOAMERICANA, S.A., *Mexico*
PRENTICE-HALL OF INDIA PRIVATE LIMITED, *New Delhi*
PRENTICE-HALL OF JAPAN, INC., *Tokyo*
PRENTICE-HALL OF SOUTHEAST ASIA PTE. LTD., *Singapore*
EDITORA PRENTICE-HALL DO BRASIL, LTDA., *Rio de Janeiro*
PRENTICE-HALL, INC., *Englewood Cliffs, New Jersey*

To my wife Barbara and sons
Peyton III and Edward

Contents

Appendix B   Review of Random Signal Theory   391

Preface

This book was written to be a textbook for seniors or first-year graduate students interested in electrical communication systems. The book departs from the usual format in two major respects. First, the text does not cover analog or pulse modulation systems. Only digital communication systems are discussed. The clear trend in today's academia is to teach less analog and more digital material. I believe we shall soon see the day when the principal communication courses are all digital. Analog subjects, if taught at all, will be in a course separate from that covering the digital systems. I hope the book will provide good service to the trend to all-digital courses.

The second way the book departs from the usual format is the omission from the main text of the topics of deterministic signals (Fourier series, Fourier transforms, etc.), networks (transfer functions, bandwidth, etc.), and random signals and noise. The reason is that I believe these topics should no longer be a part of a communication course. There are enough modern communication subjects to be covered without having to include these subjects too. However, for students whose backgrounds are a bit rusty or for instructors who still include some of these subjects, I have provided very succinct reviews in Appendixes A and B. The only other student background required to use the book is that typical of senior electrical engineering students.

Content of the book has been selected to be adequate for a one-semester course in communication theory. By selective omission (*M*-ary subjects in Chapter 6, for example) the content can match a course of one quarter length. Both courses presume three hours per week of classroom exposure. A good preview of the book's content can be obtained from either the table of contents or from reading the short SUMMARY AND DISCUSSION sections at the ends of Chapters 2 through 6.

Because the text is intended to be a textbook, a large number of problems is included (over 370 including appendixes). The more-advanced or lengthy problems and text sections are keyed by a star ($\star$). A complete solutions manual is available to instructors from the publisher.

Several people have helped to make this book possible. Dr. Leon W. Couch, II, of the University of Florida and Dr. D. G. Daut of Rutgers University read the full text and made many valuable improvements. Mr. Mike Meesit independently worked many of the problems and suggested improvements. D. I. Starry cheerfully typed the entire manuscript in its several variations, and the University of Florida made her services available. To these I extend warmest thanks. I am also grateful to Addison-Wesley Publishing Company for permission to reproduce freely material from Chapter 7 of my earlier book, *Communication System Principles* (1976). Finally, my wife, Barbara, provided invaluable assistance in proofreading, and her efforts are deeply appreciated.

Gainesville, FL   Peyton Z. Peebles, Jr.

Chapter 1

Introduction

This book is concerned with the transfer of information from one point to another by use of *digital communication* (*digicom*)† systems. Although we subsequently define digicom systems more precisely, we broadly define them as systems that convey information by using only a finite set of discrete symbols. Printed English text is an example of such symbols. Here a finite set of symbols (26 letters, space, numbers 0 to 9, some special characters, and various punctuation symbols) is used to convey information to a reader.

1.1 SOME HISTORY

The use of digital methods to convey information is not new. Some of the concepts we consider to be very modern and state of the art today were actually in existence over 380 years ago [1]. One of the earliest digicom techniques was due to Francis Bacon (1561–1626), an English philosopher. In 1605 Bacon developed a two-letter alphabet that could be used to represent 24 letters‡ of the usual alphabet by five-letter "words" using the two basic letters. Figure 1.1-1 illustrates the representations for the first few alphabet letters. In today's terminology the five-letter "words" would be called *codewords,* and the collection of all these codewords would be called a

---

† It's hoped that most will forgive the author's minor abuse of the language in coining this new word.

‡ Dr. M. New, English Department, University of Florida, has informed the author that letters *j* and *v* were missing in the early English alphabet.

1

$$
\begin{array}{ll}
\textbf{\textit{A}} = a\,a\,a\,a\,a & \textbf{\textit{H}} = a\,a\,b\,b\,b \\
\textbf{\textit{B}} = a\,a\,a\,a\,b & \textbf{\textit{I}} = a\,b\,a\,a\,a \\
\textbf{\textit{C}} = a\,a\,a\,b\,a & \textbf{\textit{K}} = a\,b\,a\,a\,b \\
\textbf{\textit{D}} = a\,a\,a\,b\,b & \textbf{\textit{L}} = a\,b\,a\,b\,a \\
\textbf{\textit{E}} = a\,a\,b\,a\,a & \textbf{\textit{M}} = a\,b\,a\,b\,b \\
\textbf{\textit{F}} = a\,a\,b\,a\,b & \textbf{\textit{N}} = a\,b\,b\,a\,a \\
\textbf{\textit{G}} = a\,a\,b\,b\,a & \textbf{\textit{O}} = a\,b\,b\,a\,b \\
\end{array}
$$

Figure 1.1-1. Francis Bacon's use of two basic letters to represent alphabet letters by five-letter codewords (circa 1605). (Adapted from [1], © 1983 IEEE, with permission.)

code. Since only two basic letters are used, it would be called a *binary code*.

In 1641, shortly after Bacon's death, John Wilkins (1614–1672), a theologian and mathematician, showed how codewords could be made shorter by using more basic letters [1]. Figure 1.1-2 illustrates Wilkins' two-, three- and five-letter codewords. In present-day terminology these three codes would be called M-ary codes (M = 2, 3, and 5) of respective *lengths* 5, 3, and 2. Because the binary code has two basic letters per position and five positions per codeword, it can represent as many as $2^5 = 32$ alphabet letters

Alphabet Letter	Codewords Using Basic Letters Indicated		
	a, b	a, b, c	a, b, c, d, e
A	$a\,a\,a\,a\,a$	$a\,a\,a$	$a\,a$
B	$a\,a\,a\,a\,b$	$a\,a\,b$	$a\,b$
.	.	.	.
.	.	.	.
.	.	.	.
L	$a\,b\,a\,b\,a$	$c\,b\,b$	$c\,a$
M	$a\,b\,a\,b\,b$	$c\,b\,c$	$c\,b$
.	.	.	.
.	.	.	.
.	.	.	.
Y	$b\,a\,b\,b\,a$	$b\,c\,b$	$e\,c$
Z	$b\,a\,b\,b\,b$	$b\,c\,c$	$e\,d$

Figure 1.1-2. Codewords for Wilkins' two-, three-, and five-letter alphabets (circa 1641). (Adapted from [1], © 1983 IEEE, with permission.)

(more than enough for the 24 for which it was intended). The codes using three and five basic letters may represent as many as $3^3 = 27$ and $5^2 = 25$ alphabet letters, respectively.

In an apparently independent development in 1703, Gottfried Wilhelm Leibniz, a German mathematician, described a binary code using only the two numbers 0 and 1 to represent integers, as illustrated in Fig. 1.1-3. Leibniz's binary code seems to be the earliest forerunner of today's natural binary code (Chap. 3), where we use binary digits **0** and **1**. By replacing the ordered letters $A, B, \ldots, Z$ by integers and letters a and b by 0 and 1,† respectively, we see that the Bacon and Leibniz codes are equivalent.

Integer	Codeword
0	0
1	1
2	1 0
3	1 1
4	1 0 0
5	1 0 1
6	1 1 0
7	1 1 1
8	1 0 0 0
.	.
.	.
.	.

Figure 1.1-3. Leibniz's binary codewords representing integer numbers (circa 1703). (Adapted from [1], © 1983 IEEE, with permission.)

The principal practical uses of the digital codes of Bacon, Wilkins, and Leibniz in communication systems involved various forms of optical telegraph links [1]. The first link was in France in 1794 [1]. After the birth of electrical technology around 1800, when Volta discovered the primary battery, digital communication by electrical methods began to evolve. The most important system was the telegraph perfected by Samuel Morse in 1837. The *Morse code* was in reality a binary code where the two "letters," called a *dot* and a *dash,* were transmitted as short or long electrical pulses, respectively. Later on, in 1875, Emile Baudot developed another code used today in international telegraph [2]; it uses binary digits 0 and 1 instead of dots and dashes and has a fixed number (five) of digits per codeword instead of a variable-length codeword, as in the Morse code.

After the theoretical prediction of electromagnetic radiation in 1864 by James Clerk Maxwell (1831–1879), a Scottish physicist, and its experimental verification in 1887 by Heinrich Hertz (1857–1894), a German physicist,

---

† The left-most unnecessary zeros are dropped in the Bacon code or, equivalently, added to the Leibniz code.

the transmission of information by radio telegraph was later achieved by Marconi in 1897.

Not all digicom systems are designed to convey the alphabet as messages. In 1937 Alec Reeves conceived one of the most important digital techniques in use today [3]. It is called *pulse code modulation* (PCM) and it can convey messages such as the audio waveform produced by a microphone. In PCM the message is first periodically sampled. Each sample, which can be any value in a continuum of possible values, is rounded or *quantized* to one of a finite set of discrete amplitudes. The amplitudes are then no different, conceptually, from the finite-sized alphabet or a finite set of integers as encoded by Bacon or Leibniz earlier. Thus the amplitudes are assigned codewords similar to those of Fig. 1.1-3. Finally, the PCM procedure assigns suitable electrical waveforms to represent the codewords. For example, a rectangular video pulse might be used to represent a binary **1**, whereas no pulse might correspond to a binary **0**.

The preceding short historical sketch is by no means complete (see [1] for more details and other references). It does, however, serve to indicate that many of the basic concepts used in modern digicom systems are not new and have their roots in the minds of men who lived over 380 years ago. It also has served to highlight some of the concepts that remain crucial to modern systems, such as sampling, quantization, and coding.

1.2 SOME DEFINITIONS

Prior to development of details on digicom systems, it is helpful to define some quantities that relate to all succeeding work.

Analog Sources and Signals

An *analog source* of information produces an output that can have any one of a continuum of possible values at any given time. Similarly, an *analog signal* is an electrical waveform that can have any one of a continuum of possible amplitudes at any one time. The sound pressure from an orchestra playing music is an example of an analog source, whereas the voltage from a microphone responding to the sound waves represents an analog signal. We shall often refer to an analog signal simply as a *message*.

Even though the principal thrust of this book is the description of digicom systems, we shall also demonstrate in detail how the digital systems can be used to convey analog messages.

In some developments it is helpful to model a message as a *deterministic waveform* (Appendix A), whereas in others we model it as a sample function of a *random process* (Appendix B). Appendixes A and B are included for those readers desiring to review waveform representations.

Digital Sources and Signals

We define a *digital source* as one with an output that can have only one of a finite set of discrete values at any given time. Most sources in nature are analog. However, when these are combined with some manufactured device, a digital source may result. For example, temperature is an analog quantity but when combined with a thermostat with output values of *on* or *off*, the combination can be considered a digital source. A *digital signal* is defined as an electrical waveform having one of a finite set of possible amplitudes at any time. If the thermostat is designed to output a voltage when on and no voltage when off, its output is a digital signal, also referred to as a message.

In general, we shall be a bit broad in our use of the word *message*. It will also be frequently used to refer to source outputs. Thus a digital source can be said to issue one of a finite set of messages at any one time.

Signal Classifications

Analog and digital information signals defined above are assumed to be *baseband* unless otherwise defined. A baseband waveform is one having its largest spectral components clustered in a band of frequencies at (or near) zero frequency. The term *lowpass* is often used to mean the same as baseband. All practical lowpass messages will have a frequency above which their spectral components may be considered negligible. We shall call this frequency, denoted by W_f for a message labeled $f(t)$, the *spectral extent* of the message (the unit is radians/second).

A *bandpass* signal is one with its largest spectral components clustered in a band of frequencies removed by a significant amount from zero. Most information waveforms are baseband; they are rarely bandpass. The bandpass signal is usually the result of one or more messages affecting (modulating) a higher frequency signal called a *carrier*.

Carrier Modulation

Most information sources are characterized by baseband information signals. However, baseband waveforms cannot be efficiently transmitted by radio methods from one point to another. On the other hand, bandpass waveforms may readily be transmitted by radio. One basic purpose of carrier modulation is, therefore, to shift the message to a higher frequency band for better radiation. In some systems carrier modulation can also result in improved performance in noise.

Often modulation involves changing the amplitude, frequency, or phase (or combination of these) of a carrier as some function of the message. When the message is digital we often refer to these operations as *keying*— for example, *phase shift keying* (PSK).

1.3 DIGITAL SYSTEM BLOCK DIAGRAM AND BOOK SURVEY

A typical communication system involves a transmitting station (sender), a receiving station (user), and a connecting medium called a *channel*. These basic functions are adequate for one-way operation. Two-way communication requires each station have both transmitter and receiver. Because operation each way is similar, only one-way operation need be described.

Transmitting Station

Of course, we are interested mainly in digital systems. A block diagram of the principal functions that may be present in a digicom system is illustrated in Fig. 1.3-1. Because the overall system is digital, the transmitting subsystem can accept such signals directly. It can also work with analog signals if they are converted to digital form in the *analog-to-digital* (A/D) *converter*. A/D conversion involves periodically sampling the analog waveform (to be discussed in Chap. 2) and quantizing the samples. Quantization amounts to rounding samples to the nearest of a number of discrete amplitudes. A digital signal results that is compatible with the digital system. However, in the rounding process information is lost that limits the accuracy with which the analog signal can be reconstructed in the receiver. Reconstruction methods are discussed in Chap. 2 while quantization principles are developed in Chap. 3.

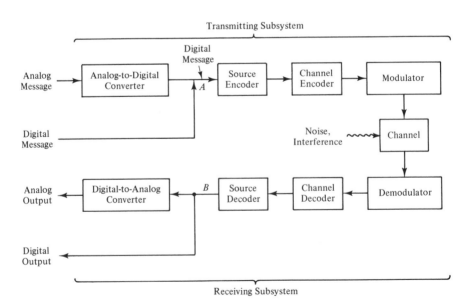

Figure 1.3-1. Functional block diagram of a communication system.

As described earlier, the actual output of the A/D converter at point *A* in Fig. 1.3-1 is a discrete voltage level. For purposes of describing the next function, the source encoder, it is helpful simply to imagine the digital signal, regardless of its origin, to be one of a finite set of discrete symbols (levels) at any given time. The general purpose of the source encoder is to convert effectively each discrete symbol into a suitable digital representation, often binary. For example, let a digital message have five possible symbols (levels), denoted by m_1, m_2, m_3, m_4, and m_5. These levels can be represented by a sequence of *binary digits* 0 and 1, as shown in the middle column of Table 1.3-1. Each sequence has three digits; each digit in the sequence can be 0 or 1. Clearly, there are $2^3 = 8$ possible binary codewords,† so that a three-digit binary representation can handle as many as eight symbols. For comparison, a *ternary* representation using a two-digit sequence is shown in the third column; here each digit can be a 0, a 1, or a 2. There are $3^2 = 9$ possible ternary codewords, so as many as nine symbols could be represented by this ternary code.

TABLE 1.3-1.   Digital Representations for a Five-Symbol Message Source.

Symbols (levels) Available	Symbol Representation	
	Natural Binary	Ternary
m_1	000	00
m_2	001	01
m_3	010	02
m_4	011	10
m_5	100	11

Digital messages are said to possess *redundancy* if their symbols are not equally probable or are not statistically independent [4]. Most practical message sources—the English language, for example—have redundancy; otherwise, *y cld nt undrstnd ths sentce*. A principal purpose of the source encoder is to remove redundancy. The more effective the encoder is, the more redundancy is removed, which allows a smaller average number of binary digits to be used in representing the message. Source encoding is developed in Chap. 3.

In some systems where no channel encoding function is present, the source encoder output is converted directly to a suitable waveform (within the modulation function) for transmission over the channel. Noise and interference added to the waveform cause the receiver's demodulation operation to make errors in its effort to recover (determine) the correct digital

† The *code*, which is the collection of all possible codewords, has *length* 3 and *size* 8.

representation used in the transmitter. By including the channel encoding
function, the effects of channel-caused errors can be reduced. The channel
encoder makes this reduction possible by adding *controlled* redundancy to
the source encoder's digital representation in a known manner such that
errors may be reduced. Channel encoding is discussed in Chap. 3.

Channel

We shall, throughout the book, assume the channel is linear and time
invariant. Unless otherwise defined, it is considered to have infinite bandwidth
with added white Gaussian channel noise of constant power density spectrum
$\mathcal{N}_0/2$ applicable on $-\infty < \omega < \infty$. Thus we study only the additive white
Gaussian noise (AWGN) channel.

Receiving Station

The functions performed in the receiving subsystem merely reflect the
inverse operations of those in the transmitting station. The demodulator
recovers the best possible version of the output that was produced by the
channel encoder at the transmitter. The demodulator's output will contain
occasional errors caused by channel noise. Part of the optimization of
various digital systems centers on minimizing errors made in the demodulator.
Optimum baseband systems are developed in Chaps. 4 and 6, and optimum
bandpass systems are discussed in Chaps. 5 and 6.

The purpose of the channel decoder is to reconstruct, to the best
extent possible, the output that was generated by the source encoder at
the transmitter. It is here that the controlled redundancy inserted by the
channel encoder may be used to identify and *correct* some channel-caused
errors in the demodulator's output. Channel decoding is considered in
Chap. 3.

The source decoder performs the exact inverse of the source encoding
function. For digital messages its output becomes the final receiver output
(point *B* in Fig. 1.3-1). If the original message was analog, the source
decoder output is passed through a *digital-to-analog* (D/A) converter which
reconstructs the original message using the sampling theory described in
Chap. 2.

The reader is cautioned to accept the above discussions with a degree
of open-mindedness. Remember that the discussions center around the
functions that may be present in a digicom system and do not always infer
the actual implementation of a system. Hence practical systems may not
always follow the blocks in Fig. 1.3-1 exactly. For example, a number of
digital systems are developed in Chap. 4 that can each accept an analog
message directly and produce the channel waveform in one operation.
These systems typically use no channel encoding or decoding operations.

In the course of this book the various functions of Fig. 1.3-1 are discussed in detail. In those digital systems that use analog messages through application of A/D and D/A methods, it is helpful to define a simple system against which comparisons of noise performance can be made.

1.4 SIMPLE BASEBAND SYSTEM FOR REFERENCE

The simplest possible analog system is shown in Fig. 1.4-1. A message $f(t)$, with spectral extent W_f, is transmitted directly to a receiver through a channel where white noise of power density $\mathcal{N}_0/2$ is added. The receiver is just a lowpass filter (LPF) with bandwidth W_f. The filter passes the message with negligible distortion but is no wider in bandwidth than necessary in order not to pass excessive noise. If the filter is approximated by an ideal rectangular passband with unity gain, the output signal and noise powers are†

$$S_o = \overline{s_d^2(t)} = \overline{f^2(t)} \tag{1.4-1}$$

$$N_o = \overline{n_d^2(t)} = \frac{1}{2\pi} \int_{-\infty}^{\infty} \mathcal{S}_{n_d}(\omega)\, d\omega = \frac{1}{2\pi} \int_{-W_f}^{W_f} \frac{\mathcal{N}_0}{2}\, d\omega = \frac{\mathcal{N}_0 W_f}{2\pi}, \tag{1.4-2}$$

where $\mathcal{S}_{n_d}(\omega)$ is the power density spectrum of the output noise $n_d(t)$.

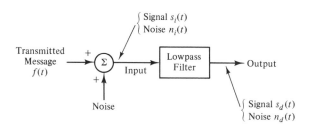

Figure 1.4-1. Simple baseband analog communication system useful in comparing noise performances of other systems [5].

At the input to the filter, signal power S_i is

$$S_i = \overline{f^2(t)} = S_o. \tag{1.4-3}$$

We *define* input noise power N_i as the noise power *in the band of the filter*. Thus

$$N_i = N_o = \frac{\mathcal{N}_0 W_f}{2\pi}. \tag{1.4-4}$$

In systems using analog messages, an excellent measure of noise performance is the output signal-to-noise power ratio. For our elementary

---

† The overbar represents either the infinite-time average for a deterministic signal or the statistical expected (mean) value for a random message (see Appendix B).

baseband system we have

$$\left(\frac{S_o}{N_o}\right)_B = \left(\frac{S_i}{N_i}\right)_B = \frac{2\pi\overline{f^2(t)}}{\mathcal{N}_0 W_f} = \frac{2\pi S_i}{\mathcal{N}_0 W_f}. \tag{1.4-5}$$

These signal-to-noise ratios provide a good basis against which the performance of other, more complicated, systems can be compared.

REFERENCES

[1] Aschoff, V., The Early History of the Binary Code, *IEEE Communications Magazine,* Vol. 21, No. 1, January 1983, pp. 4–10.

[2] Couch, II, L. W., *Digital and Analog Communication Systems,* Macmillan Publishing Co., Inc., New York, 1983.

[3] Carlson, A. B., *Communication Systems, An Introduction to Signals and Noise in Electrical Communication,* second edition, McGraw-Hill Book Co., Inc., New York, 1975. (See also third edition, 1986.)

[4] Sklar, B., A Structured Overview of Digital Communications—A Tutorial Review—Part I, *IEEE Communications Magazine,* Vol. 21, No. 5, August 1983, pp. 4–17. Part II of same title in October 1983 issue, pp. 6–21.

[5] Peebles, Jr., P. Z., *Communication System Principles,* Addison-Wesley Publishing Co., Inc., Reading, Massachusetts, 1976. (Figure 1.4-1 has been adapted.)

Chapter 2

Sampling Principles

2.0 INTRODUCTION

When an analog message is conveyed over an analog communication system, the full message is typically used at all times. To send the same analog signal over a digicom system requires that only its *samples* be transmitted at periodic intervals. Because the receiver can, therefore, receive only samples of the message, it must attempt to reconstruct the original message *at all times* from only its samples. Methods exist whereby this desired end can be accomplished. These methods involve the theory of sampling which is described in this chapter. For the most part we shall need only the theory related to lowpass waveforms (Secs. 2.1–2.5) to continue profitably through the book. However, there are several other interesting topics in sampling theory that are also included for those readers wishing to delve deeper into the subject.

At first glance it may seem astonishing that only samples of a message and not the entire waveform can adequately describe all the information in a signal. However, as remarkable as it may seem, we shall find that under some reasonable conditions, a message can be recovered *exactly* from its samples, even at times in between the samples. To accomplish this purpose we shall introduce some sampling principles (theorems) that apply to either deterministic or random signals (noise). These principles form the very foundation that makes possible most of the digital systems to be studied in the following chapters. A good review of the literature on sampling theory has been given by Jerri [1].

One of the richest rewards to be realized from sampling is that it becomes possible to interlace samples from many different information signals in time. Thus we have a process of *time-division multiplexing* analogous to frequency multiplexing. With the availability of modern high-speed switching circuits, the practicality of time multiplexing is well established and, in many cases, may be preferred over frequency multiplexing.

2.1 SAMPLING THEOREMS FOR LOWPASS NONRANDOM SIGNALS

In this section we shall discuss two sampling theorems. First, we consider sampling a nonrandom lowpass waveform. After developing the applicable sampling theorem, we show how the original unsampled waveform can be recovered from its samples. A simple way of interpreting the sampling theorem using networks is introduced to show this fact. The second theorem is a sort of dual to the first; we show that the spectrum of a time-limited waveform can be determined completely from samples of its spectrum.

Time Domain Sampling Theorem

The lowpass sampling theorem may be stated as follows:

A lowpass signal $f(t)$, bandlimited such that it has no frequency components above W_f rad/s, is uniquely determined by its values at equally spaced points in time separated by $T_s \leq \pi/W_f$ seconds.†

This theorem is sometimes called the *uniform sampling theorem* owing to the equally spaced nature of the instantaneously taken samples.‡ It allows us to completely reconstruct a bandlimited signal from instantaneous samples taken at a rate $\omega_s = 2\pi/T_s$ of at least $2W_f$, which is twice the highest frequency present in the waveform. The *minimum* rate $2W_f$ is called the *Nyquist rate*.

The sampling theorem has been known to mathematicians since at least 1915 [4]. Its use by engineers stems mainly from the work of Shannon [5]. Proof of the theorem begins by assuming $f(t)$ to be an arbitrary waveform, except that its Fourier transform $F(\omega)$ exists and is bandlimited such that its nonzero values exist only for $-W_f \leq \omega \leq W_f$. Such a signal is illustrated in Fig. 2.1-1(a). From the theory of Fourier series, the spectrum may be represented by a Fourier series developed by assuming a periodic spectrum $Q(\omega)$ as shown in (b). If the "period" of the repetition is $\omega_s \geq 2W_f$, it is clear that no overlap of spectral components will occur. With no overlap

---

† If a spectral impulse exists at $\omega = \pm W_f$, the equality is to be excluded [2].

‡ The theorem can be generalized to nonuniform samples taken one per interval anywhere within contiguous intervals $T < \pi/W_f$ in duration [3].

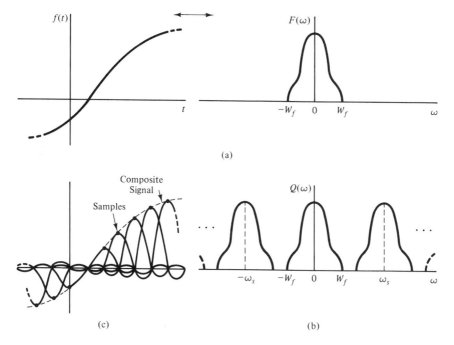

Figure 2.1-1. Waveforms and spectrums related to sampling. (a) A signal and its spectrum, (b) periodic representation for spectrum of (a), and (c) the signal reconstructed from its samples [6].

the Fourier series giving $Q(\omega)$ will also equal $F(\omega)$ for $|\omega| \leq W_f$. Thus

$$Q(\omega) = F(\omega), \qquad |\omega| \leq W_f. \qquad (2.1\text{-}1)$$

By using the complex form of the Fourier series we get

$$Q(\omega) = \sum_{n=-\infty}^{\infty} C_n e^{jn2\pi\omega/\omega_s}, \qquad (2.1\text{-}2)$$

where the series coefficients are given by

$$C_n = \frac{1}{\omega_s} \int_{-\omega_s/2}^{\omega_s/2} Q(\omega) e^{-jn2\pi\omega/\omega_s} \, d\omega \qquad (2.1\text{-}3)$$

for $n = 0, \pm 1, \pm 2, \ldots$. Since in the central period $Q(\omega) = F(\omega)$, the coefficients evaluate to

$$C_n = \frac{2\pi}{\omega_s} \frac{1}{2\pi} \int_{-W_f}^{W_f} F(\omega) e^{-jn2\pi\omega/\omega_s} \, d\omega = \frac{2\pi}{\omega_s} f\left(\frac{-n2\pi}{\omega_s}\right). \qquad (2.1\text{-}4)$$

We see that the coefficients are proportional to instantaneous samples of $f(t)$ at times $-n2\pi/\omega_s = -nT_s$, where T_s is the time between samples:

$$T_s = 2\pi/\omega_s. \qquad (2.1\text{-}5)$$

From (2.1-2) the periodic spectrum $Q(\omega)$ becomes

$$Q(\omega) = \sum_{n=-\infty}^{\infty} \frac{2\pi}{\omega_s} f\left(\frac{-n2\pi}{\omega_s}\right) e^{jn2\pi\omega/\omega_s}, \tag{2.1-6}$$

which is valid for all ω. Next, we recognize that $f(t)$ must be given by the inverse Fourier transform of its spectrum, as given by (2.1-1) with (2.1-6) substituted. After inverse transformation we have

$$f(t) = \sum_{k=-\infty}^{\infty} \frac{2W_f}{\omega_s} f\left(\frac{k2\pi}{\omega_s}\right) \frac{\sin\left[W_f\left(t - \frac{k2\pi}{\omega_s}\right)\right]}{W_f\left(t - \frac{k2\pi}{\omega_s}\right)}, \tag{2.1-7}$$

where we have let $k = -n$. This equation constitutes a proof of the uniform sampling theorem, since it shows that $f(t)$ is known for all time from a knowledge of its samples. The samples determine the amplitudes of time signals in the sum having known form.

The time signals all have the form $\sin(x)/x$. This form is defined in Appendix A as the *sampling function* Sa(x)—that is,

$$\text{Sa}(x) = \frac{\sin(x)}{x}. \tag{2.1-8}$$

The sampling function is sometimes called an *interpolating function*. In terms of sampling functions, (2.1-7) can be restated as

$$f(t) = \frac{W_f T_s}{\pi} \sum_{k=-\infty}^{\infty} f(kT_s) \, \text{Sa}[W_f(t - kT_s)]. \tag{2.1-9}$$

Recall that these results were derived under the restriction $\omega_s \geq 2W_f$, requiring a sampling interval

$$T_s \leq \frac{\pi}{W_f}, \tag{2.1-10}$$

as stated in the theorem. An important special case of (2.1-9) occurs when sampling is at the maximum (Nyquist) period $T_s = \pi/W_f$:

$$f(t) = \sum_{k=-\infty}^{\infty} f(kT_s) \, \text{Sa}[W_f(t - kT_s)]. \tag{2.1-11}$$

Figure 2.1-1(c) illustrates a possible signal $f(t)$ as being the sum of time-shifted sampling functions as given by (2.1-11).

Two other valuable forms of the sampling theorem as embodied in equation form are given without proofs:

$$f(t - t_0) = \frac{W_f T_s}{\pi} \sum_{k=-\infty}^{\infty} f(kT_s - t_0) \text{Sa}[W_f(t - kT_s)] \tag{2.1-12}$$

$$f(t) = \frac{W_f T_s}{\pi} \sum_{k=-\infty}^{\infty} f(kT_s - t_0) \text{Sa}[W_f(t + t_0 - kT_s)]. \tag{2.1-13}$$

The reader may wish to prove (2.1-12) as an exercise. The procedure is to retrace the above proof, except replace $F(\omega)$ by the spectrum $F(\omega)\exp(-j\omega t_0)$ corresponding to the delayed signal $f(t - t_0)$. As a second exercise, (2.1-13) follows from (2.1-12) by inspection. (How?)

Our main results are (2.1-9), (2.1-12), and (2.1-13); these are all valid for complex as well as real signals. Waveforms are assumed to be nonrandom, however. We develop a sampling theorem for random signals in Sec. 2.2.

Example 2.1-1

As an example relating to sampling functions we show that

$$\int_{-\infty}^{\infty} \text{Sa}\left[\frac{\omega_s}{2}(t - kT_s)\right] \text{Sa}\left[\frac{\omega_s}{2}(t - mT_s)\right] dt = \begin{cases} 0, & m \neq k \\ \dfrac{2\pi}{\omega_s}, & m = k. \end{cases}$$

The sampling functions are said to be *orthogonal* because they satisfy this relationship. We shall use Parseval's theorem, which may be written in the form

$$\int_{-\infty}^{\infty} x(t)y^*(t)\,dt = \frac{1}{2\pi}\int_{-\infty}^{\infty} X(\omega)Y^*(\omega)\,d\omega$$

for two signals $x(t)$ and $y(t)$ having Fourier transforms $X(\omega)$ and $Y(\omega)$, respectively. Here we use the asterisk to represent the complex conjugate operation.

Since

$$\text{Sa}\left[\frac{\omega_s}{2}(t - mT_s)\right] \leftrightarrow \begin{cases} \dfrac{2\pi}{\omega_s}e^{-jm2\pi\omega/\omega_s}, & |\omega| < \omega_s/2 \\ 0, & |\omega| > \omega_s/2, \end{cases}$$

where the double-ended arrow represents a Fourier transform pair, we obtain the desired result by integration in the frequency domain.

$$\int_{-\infty}^{\infty} \text{Sa}\left[\frac{\omega_s}{2}(t - kT_s)\right] \text{Sa}\left[\frac{\omega_s}{2}(t - mT_s)\right] dt$$

$$= \frac{1}{2\pi}\int_{-\omega_s/2}^{\omega_s/2} \left(\frac{2\pi}{\omega_s}\right)^2 e^{-j(m-k)2\pi\omega/\omega_s}\,d\omega$$

$$= \frac{2\pi}{\omega_s}\frac{\sin[(m-k)\pi]}{(m-k)\pi} = \begin{cases} 0, & m \neq k \\ 2\pi/\omega_s, & m = k. \end{cases}$$

A Network View of Sampling Theorem

There is a useful way in which lowpass sampling theorems may be viewed by using networks [7]. Consider the network of Fig. 2.1-2. Imagine that a periodic train of impulses is available. We form the product of $f(t)$ and this pulse train to get

$$f_s(t) = f(t)\sum_{k=-\infty}^{\infty} \delta(t - kT_s) = \sum_{k=-\infty}^{\infty} f(kT_s)\delta(t - kT_s). \qquad (2.1\text{-}14)$$

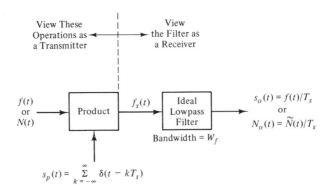

Figure 2.1-2. Block diagram of a network useful in the interpretation of sampling theorems [6].

The product function using impulses is called an *ideal sampler* and $f_s(t)$ is the ideally sampled version of $f(t)$.

Next, assume the filter is ideal with a transfer function

$$H(\omega) = \begin{cases} 1, & |\omega| < W_f \\ 0, & |\omega| > W_f. \end{cases} \tag{2.1-15}$$

Its impulse response, by inverse Fourier transformation of (2.1-15), is

$$h(t) = \frac{W_f}{\pi} \text{Sa}(W_f t). \tag{2.1-16}$$

The filter output response to (2.1-14) now follows easily:

$$s_o(t) = \frac{W_f}{\pi} \sum_{k=-\infty}^{\infty} f(kT_s) \text{Sa}[W_f(t - kT_s)] \tag{2.1-17}$$

$$= \frac{f(t)}{T_s}.$$

The last form in (2.1-17) derives from the sampling theorem, which means that $f(t)$ must be given by (2.1-9).

Let us pause to summarize what has been developed. First, a train of instantaneous samples of $f(t)$ at the product device output has been generated. The sample rate is $\omega_s = 2\pi/T_s \geq 2W_f$. These samples may be viewed as the output of a transmitter. Second, by use of an ideal lowpass filter having a bandwidth equal to the maximum frequency extent W_f of $f(t)$, the output is given by the middle term of (2.1-17). The filter may be viewed as the receiver which must recover $f(t)$. Finally, by application of the sampling theorem, this output equals $f(t)/T_s$. Thus we have shown that, within a constant factor, $f(t)$ is completely reconstructed from its samples by using an ideal lowpass filter. Reconstruction is valid for any sample rate $\omega_s \geq 2W_f$.

The ideal sampler of Fig. 2.1-2 cannot be realized; however, it can be approximated in practice by using a train of very narrow, large amplitude pulses. Fortunately, such measures are not usually necessary, since easily realizable practical techniques exist for sampling, as noted in Sec. 2.3.

Aliasing

Let us examine the spectrum of the ideally sampled signal $f_s(t)$ of (2.1-14). The train of impulses can be replaced by its Fourier series representation to obtain

$$f_s(t) = f(t) \sum_{k=-\infty}^{\infty} \delta(t - kT_s) = \frac{1}{T_s} \sum_{k=-\infty}^{\infty} f(t)e^{jk\omega_s t} \qquad (2.1\text{-}18)$$

The Fourier transform of $f_s(t)$, denoted by $F_s(\omega)$, is

$$F_s(\omega) = \frac{1}{T_s} \sum_{k=-\infty}^{\infty} F(\omega - k\omega_s) \qquad (2.1\text{-}19)$$

from the frequency shifting property of transforms. $F_s(\omega)$ is comprised of scaled (by $1/T_s$) replicas of the message spectrum at all multiples of ω_s. If $F(\omega)$ is bandlimited to W_f and if $\omega_s \geq 2W_f$, these replicas will not overlap, which is required if the filter in Fig. 2.1-2 is to pass an undistorted spectrum $F(\omega)$, the component of $F_s(\omega)$ for $k = 0$.

If, however, $f(t)$ is not bandlimited or if the sampling rate ω_s is not high enough, there can be overlap of spectral components, as illustrated in Fig. 2.1-3. Spectral overlap in the central replica ($k = 0$ component) is called *aliasing*. In Fig. 2.1-3(a) aliasing is due to the message not being bandlimited. This form of aliasing can be minimized by sampling at a high enough rate that replicas become far separated. Another solution might be to prefilter the message to force it to be more bandlimited. In (b) aliasing is due only to sampling at too low a rate; the solution is to raise ω_s. *Spectral folding*, as aliasing is sometimes called, causes higher frequencies to show up as lower frequencies in the recovered message; in voice transmission intelligibility can be seriously degraded.

The *mean-squared aliasing error*, denoted here by E_{a1}, can be defined as the energy folded into the signal's band $|\omega| \leq \omega_s/2$ by the shifted replicas [8]. If ω_s is at least large enough that this energy is mainly due to the adjacent replicas at $\pm\omega_s$, we have†

$$E_{a1} = \frac{2}{2\pi} \int_{\omega_s/2}^{\infty} |F(\omega)|^2 \, d\omega. \qquad (2.1\text{-}20)$$

The ratio of E_{a1} to the undistorted signal's total energy, E_f, will be called

---

† If the message is a power signal, the energy density spectrum $|F(\omega)|^2$ is replaced by the applicable power density spectrum of $f(t)$.

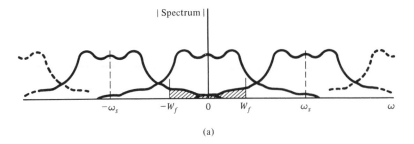

(a)

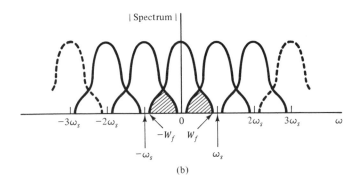

(b)

Figure 2.1-3. (a) Sampled signal spectral overlap due to message not being bandlimited and (b) spectral overlap caused by undersampling of a bandlimited message [6].

the *fractional aliasing error*

$$\frac{E_{a1}}{E_f} = \frac{\displaystyle\int_{\omega_s/2}^{\infty} |F(\omega)|^2 \, d\omega}{\displaystyle\int_{0}^{\infty} |F(\omega)|^2 \, d\omega}. \qquad (2.1\text{-}21)$$

The reciprocal, E_f/E_{a1}, is called the signal-to-distortion ratio by Gagliardi [8]. Other definitions of fractional aliasing error exist. Stremler uses one in which higher frequencies are weighted more heavily than lower ones [9]. In general, (2.1-21) and other definitions serve only as reasonable measures of the intensity of aliasing because the aliasing effects are difficult to measure; they depend on the phases of the overlapping spectral components as well as their amplitudes.

Frequency-Domain Sampling Theorem

A frequency-domain sampling theorem that is analogous to the time-domain theorem may also be stated:

A Fourier spectrum $F(\omega)$, corresponding to a signal $f(t)$, timelimited such that it is nonzero only in the interval $-T_f \leq t \leq T_f$, is uniquely determined by its values at equally spaced points in frequency separated by an amount $W_s \leq \pi/T_f$ rad/s.

By analogy, W_s and T_f here correspond to T_s and W_f, respectively, in the time-domain theorem. The result analogous to (2.1-9) may be found to be

$$F(\omega) = \frac{T_f W_s}{\pi} \sum_{k=-\infty}^{\infty} F(kW_s)\, \text{Sa}[T_f(\omega - kW_s)] \qquad (2.1\text{-}22)$$

for real or complex $f(t)$. This expression can also be proved by repeating the procedure that leads to (2.1-9). Similarly, the forms analogous to (2.1-12) and (2.1-13) are:

$$F(\omega - \omega_0) = \frac{T_f W_s}{\pi} \sum_{k=-\infty}^{\infty} F(kW_s - \omega_0)\text{Sa}[T_f(\omega - kW_s)] \qquad (2.1\text{-}23)$$

$$F(\omega) = \frac{T_f W_s}{\pi} \sum_{k=-\infty}^{\infty} F(kW_s - \omega_0)\text{Sa}[T_f(\omega + \omega_0 - kW_s)]. \qquad (2.1\text{-}24)$$

Aliasing can also occur with frequency-domain sampling. Now, however, replicas are in the time domain and are located at multiples of $T_s = 2\pi/W_s$.

2.2 SAMPLING THEOREM FOR LOWPASS RANDOM SIGNALS

The theory of sampling can also be extended to include random signals or noise [1, 10–16]. Let a lowpass random signal or noise be represented as a sample function of a random process $N(t)$. We assume $N(t)$ to be wide-sense stationary with a power density spectrum $\mathcal{S}_N(\omega)$ bandlimited such that

$$\mathcal{S}_N(\omega) = 0, \qquad |\omega| > W_N, \qquad (2.2\text{-}1)$$

where W_N is the maximum spectral extent of $N(t)$. Because $\mathcal{S}_N(\omega)$ is bandlimited and the noise's autocorrelation function is the inverse transform of $\mathcal{S}_N(\omega)$, the autocorrelation function must have representations of the forms of (2.1-9), (2.1-12), and (2.1-13). They are

$$R_N(\tau) = \frac{W_N T_s}{\pi} \sum_{k=-\infty}^{\infty} R_N(kT_s)\text{Sa}[W_N(\tau - kT_s)] \qquad (2.2\text{-}2)$$

$$R_N(\tau - t_0) = \frac{W_N T_s}{\pi} \sum_{k=-\infty}^{\infty} R_N(kT_s - t_0)\text{Sa}[W_N(\tau - kT_s)] \qquad (2.2\text{-}3)$$

$$R_N(\tau) = \frac{W_N T_s}{\pi} \sum_{k=-\infty}^{\infty} R_N(kT_s - t_0)\text{Sa}[W_N(\tau + t_0 - kT_s)], \qquad (2.2\text{-}4)$$

where $R_N(\tau)$ is the autocorrelation function, T_s is the period between samples of $R_N(\tau)$, and t_0 is an arbitrary constant.

We shall now show that $N(t)$ can be represented by the function

$$\tilde{N}(t) = \frac{W_N T_s}{\pi} \sum_{k=-\infty}^{\infty} N(kT_s)\mathrm{Sa}[W_N(t - kT_s)] \tag{2.2-5}$$

in the sense that the mean-squared difference (or error) between the actual process $N(t)$ and its representation $\tilde{N}(t)$ is zero. In other words, $\tilde{N}(t) = N(t)$ in the sense of zero mean-squared error defined by

$$\overline{\varepsilon^2} = E\{[N(t) - \tilde{N}(t)]^2\} = 0, \tag{2.2-6}$$

where $E\{\cdot\}$ represents the statistical average. By direct expansion (2.2-6) becomes

$$\overline{\varepsilon^2} = R_N(0) - 2E[N(t)\tilde{N}(t)] + E[\tilde{N}^2(t)], \tag{2.2-7}$$

which we show equals zero by finding the various terms.

By substitution of (2.2-5) into the middle right-side term of (2.2-7) we develop the following:

$$E[N(t)\tilde{N}(t)] = \frac{W_N T_s}{\pi} \sum_{k=-\infty}^{\infty} E[N(t)N(kT_s)]\mathrm{Sa}[W_N(t - kT_s)]$$

$$= \frac{W_N T_s}{\pi} \sum_{k=-\infty}^{\infty} R_N(kT_s - t)\mathrm{Sa}[W_N(t - kT_s)] = R_N(0). \tag{2.2-8}$$

The last form of (2.2-8) is the result of applying (2.2-4) with $\tau = 0$ and $t_0 = t$. At this point (2.2-7) becomes

$$\overline{\varepsilon^2} = -R_N(0) + E[\tilde{N}^2(t)]. \tag{2.2-9}$$

By again using (2.2-5)

$$E[\tilde{N}^2(t)] = E\left\{\frac{W_N T_s}{\pi} \sum_{k=-\infty}^{\infty} N(kT_s)\mathrm{Sa}[W_N(t - kT_s)] \right.$$

$$\left. \cdot \frac{W_N T_s}{\pi} \sum_{l=-\infty}^{\infty} N(lT_s)\mathrm{Sa}[W_N(t - lT_s)]\right\}$$

$$= \frac{W_N T_s}{\pi} \sum_{k=-\infty}^{\infty} \mathrm{Sa}[W_N(t - kT_s)]$$

$$\cdot \frac{W_N T_s}{\pi} \sum_{l=-\infty}^{\infty} R_N(lT_s - kT_s)\mathrm{Sa}[W_N(t - lT_s)]. \tag{2.2-10}$$

From (2.2-3) with $\tau = t$ and $t_0 = kT_s$, the second sum in (2.2-10) is recognized as $R_N(t - kT_s) = R_N(kT_s - t)$, so

$$E[\tilde{N}^2(t)] = \frac{W_N T_s}{\pi} \sum_{k=-\infty}^{\infty} R_N(kT_s - t)\mathrm{Sa}[W_N(t - kT_s)] = R_N(0). \tag{2.2-11}$$

Here we used (2.2-4) again with $\tau = 0$ and $t_0 = t$ to obtain the last form of (2.2-11). Finally, $\overline{\varepsilon^2}$ becomes zero when (2.2-11) is used in (2.2-9).

In summary, the above development has shown the following sampling theorem to be valid:

A lowpass wide-sense stationary random process $N(t)$, bandlimited such that its power density has no nonzero frequency components above W_N rad/s or below $-W_N$ rad/s, can be represented by its sample values at equally spaced times separated by $T_s \leq \pi/W_N$ s; the representation is $\tilde{N}(t)$ of (2.2-5), which is valid in the sense that $\tilde{N}(t) = N(t)$ with zero mean-squared error.

Much of our earlier results applicable to nonrandom waveform sampling also apply to random signals. For example, the network interpretation of the lowpass sampling theorem shown in Fig. 2.1-2 applies. When the random process $N(t)$ is ideally sampled in the product device, the sampled version $N_s(t)$ is

$$N_s(t) = \sum_{k=-\infty}^{\infty} N(kT_s)\delta(t - kT_s). \qquad (2.2\text{-}12)$$

Each impulse excites the filter and contributes an output component $N(kT_s)$ times the impulse response (kth impulse). Since the impulse response is (2.1-16), we obtain an output random signal $N_o(t)$ of

$$N_o(t) = \frac{W_N}{\pi} \sum_{k=-\infty}^{\infty} N(kT_s)\mathrm{Sa}[W_N(t - kT_s)] = \frac{\tilde{N}(t)}{T_s} \qquad (2.2\text{-}13)$$

from (2.2-5), if the filter's bandwidth is W_N.

Example 2.2-1
A signal is added to noise; the sum is to be transmitted by ideal sampling. The receiver consists of an ideal filter that acts on the ideally sampled sum. We take two examples, both of which assume the receiver must recover the signal without distortion.

First, suppose the noise is bandlimited to $W_N/2\pi = 2$ kHz, whereas the signal is bandlimited to $W_f/2\pi = 3$ kHz. For no signal distortion, a sample rate of at least 6 kHz is required. Since this rate exceeds $2W_N/2\pi = 4$ kHz, the noise will also be reconstructed without error if the receiver bandwidth is 3 kHz.

Second, suppose the bandwidths are reversed. Signal bandwidth is now 2 kHz and noise bandwidth is 3 kHz. A sample rate of 4 kHz with a receiver bandwidth of 2 kHz leads to perfect message recovery, but the noise is undersampled which produces aliasing. Some power-spectrum sketches will show that the total receiver output noise power is the same as that transmitted if sampling and receiver matched the noise's Nyquist rate, but the form of the power spectrum of the output noise has changed.

2.3 PRACTICAL SAMPLING METHODS

The instantaneous sampling of the preceding sections using impulses has been called *ideal sampling*. It must be approximated in practice by using large, vanishingly narrow pulses (in relation to π/W_f). On the other hand, it may be impractical or even undesirable to use very narrow pulses. Practical transmissions of interest will then use finite duration pulses. In the following paragraphs, we investigate two forms of such practical sampling. In Sec. 2.4 the techniques are generalized.

Natural Sampling

Natural sampling involves a direct product of $f(t)$ and a train of rectangular pulses, as shown in Fig. 2.3-1(a). The spectrum of $f(t)$ is defined as $F(\omega)$; it and $f(t)$ are sketched in (b). The pulse train $s_p(t)$ has amplitude K and has the Fourier series expansion

$$s_p(t) = \frac{K\tau}{T_s} \sum_{k=-\infty}^{\infty} \frac{\sin(k\omega_s\tau/2)}{k\omega_s\tau/2} e^{jk\omega_s t} \tag{2.3-1}$$

with τ being the pulse length and $\omega_s = 2\pi/T_s$ the pulse rate. The train and its spectrum

$$S_p(\omega) = \frac{2\pi K\tau}{T_s} \sum_{k=-\infty}^{\infty} \frac{\sin(k\omega_s\tau/2)}{k\omega_s\tau/2} \delta(\omega - k\omega_s) \tag{2.3-2}$$

are illustrated in (c). The product

$$f_s(t) = s_p(t)f(t) \tag{2.3-3}$$

is the sampled version of $f(t)$.†

The spectrum $F_s(\omega)$ of $f_s(t)$ is helpful in visualizing how $f(t)$ is recovered from its samples. By recalling that a time product of two waveforms has a spectrum given by $1/2\pi$ times the convolution of the two spectrums, we obtain

$$
\begin{aligned}
F_s(\omega) &= \frac{1}{2\pi} F(\omega) * S_p(\omega) \\
&= \frac{K\tau}{T_s} \sum_{k=-\infty}^{\infty} \frac{\sin(k\omega_s\tau/2)}{k\omega_s\tau/2} \int_{-\infty}^{\infty} \delta(x - k\omega_s)F(\omega - x)\, dx \\
&= \frac{K\tau}{T_s} \sum_{k=-\infty}^{\infty} \frac{\sin(k\omega_s\tau/2)}{k\omega_s\tau/2} F(\omega - k\omega_s).
\end{aligned}
\tag{2.3-4}
$$

The product and its spectrum are shown in Fig. 2.3-1(d).

From (2.3-4) it is seen that, so long as $\omega_s \geq 2W_f$, the spectrum of the sampled signal contains nonoverlapping, scaled, and frequency-shifted replicas of the information signal spectrum. By applying $f_s(t)$ to a lowpass filter of

---

† There is an implicit constant involved in the product device having the value 1.0 V^{-1}. The purpose of the constant is to preserve units after the product.

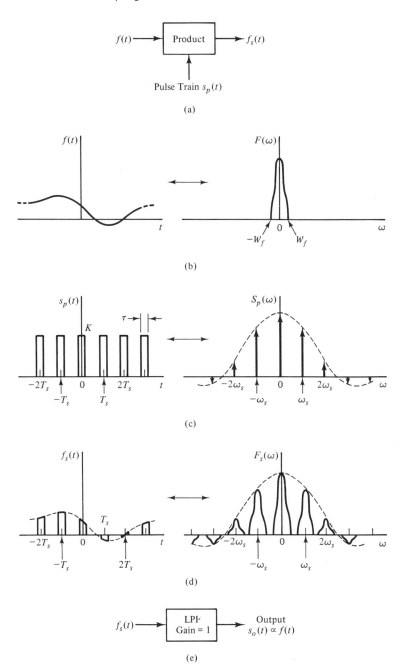

Figure 2.3-1. (a) Method of natural sampling for waveform of (b) using the pulse train of (c) to produce the signal and spectrum of (d). The signal is recovered with the lowpass filter of (e) [6].

bandwidth W_f, as shown in Fig. 2.3-1(e), $f(t)$ is easily recovered without distortion.† The spectrum $S_o(\omega)$ of the output signal $s_o(t)$ from the filter will be

$$S_o(\omega) = \frac{K\tau}{T_s} F(\omega), \tag{2.3-5}$$

while the output signal is

$$s_o(t) = \frac{K\tau}{T_s} f(t). \tag{2.3-6}$$

The foregoing discussion has shown that the product of a message $f(t)$ and a realizable train of rectangular, finite-amplitude pulses forms a realistic sampling method, that of natural sampling. The spectrum of the naturally sampled version $f_s(t)$ of $f(t)$ contains undistorted replicas of the message spectrum $F(\omega)$. The central term is just an amplitude-scaled version of $F(\omega)$ that results in reconstruction of $f(t)$ when selected by a lowpass filter. From (2.3-6) the filter's output is seen to be $f(t)$ scaled by the dc component, $K\tau/T_s$, of the sampling pulse train $s_p(t)$.

The network model of the ideal sampler of Secs. 2.1 and 2.2 hold true for natural sampling. This fact follows because the product device [Fig. 2.3-1(a)] constitutes a practical transmitter of samples of $f(t)$ while the receiver consists of an ideal lowpass filter to recover the message. In fact, if the sample rate exceeds the Nyquist rate ($\omega_s > 2W_f$), there is a "clear" region from $|\omega| = W_f$ to $|\omega| = \omega_s - W_f$ over which a more realistic filter can go from full response to negligible response, as noted earlier.

Flat-Top Sampling

With this type of sampling the amplitude of each pulse in a pulse train is constant during the pulse but is determined by an instantaneous sample of $f(t)$ as illustrated in Fig. 2.3-2(a). The time of the instantaneous sample is chosen to occur at the pulse center for convenience.‡ For comparison, and to illustrate the differences involved, natural sampling is illustrated in (b).

For the assumed ideal rectangular pulses, the flat-top sampled signal is

$$f_s(t) = \sum_{k=-\infty}^{\infty} f(kT_s) K \,\mathrm{rect}\!\left(\frac{t - kT_s}{\tau}\right), \tag{2.3-7}$$

† If $\omega_s > 2W_f$, the filter must uniformly pass all frequencies $|\omega| \le W_f$ but may then roll off in a practical way out to the point $|\omega| = \omega_s - W_f$, where its response must become negligible (ideally zero). As ω_s approaches $2W_f$, the filter must approach the ideal lowpass filter.

‡ In a realizable system the sample time must occur at or before the leading edge of any given sample pulse, since it is impossible to initiate a pulse of the required amplitude before that amplitude is established.

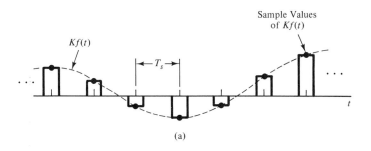

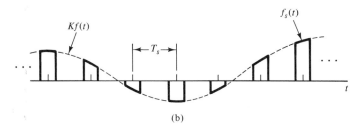

Figure 2.3-2.   (a) Flat-top sampling of a signal $f(t)$ and (b) natural sampling [6].

where K is a scale constant; it is the amplitude of a sampling pulse for a unit input (dc) signal. The function rect $(\cdot)$ is defined by (A.2-1) of Appendix A.

To determine our ability to reconstruct $f(t)$ from the sampled signal (2.3-7), it is helpful to observe that the same expression derives from an ideal sampler followed by an amplifier with gain $K\tau$ and a filter. The filter must have an impulse response

$$q(t) = \frac{1}{\tau}\text{rect}\left(\frac{t}{\tau}\right) \tag{2.3-8}$$

and transfer function

$$Q(\omega) = \frac{\sin(\omega\tau/2)}{\omega\tau/2}. \tag{2.3-9}$$

The network is illustrated in Fig. 2.3-3(a).† It is straightforward to show that the spectrum of $f_s(t)$ is

$$F_s(\omega) = \frac{K\tau}{T_s}\sum_{k=-\infty}^{\infty} \frac{\sin(\omega\tau/2)}{\omega\tau/2} F(\omega - k\omega_s). \tag{2.3-10}$$

$F_s(\omega)$ of (2.3-10) appears on the surface to be similar to (2.3-4) for

---

† This model is only for mathematical and conceptual convenience. Actual implementation in practice may be quite different. The amplifier is implied simply so that $Q(\omega)$ as defined may have a low-frequency gain of unity. The reasoning will become clearer as the reader proceeds to subsequent developments.

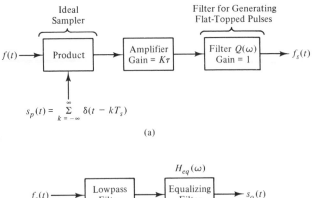

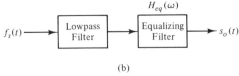

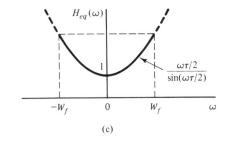

Figure 2.3-3. (a) Generation method for flat-top samples. (b) Signal recovery method requiring the equalization filter with the transfer function of (c) [6].

natural sampling. There is an important difference, however. A lowpass filter operating on (2.3-10) will not give a distortion-free output proportional to $f(t)$. To see this, suppose a lowpass filter alone were used. The output spectrum would be $(K\tau/T_s)F(\omega)\sin(\omega\tau/2)/(\omega\tau/2)$, which is clearly not proportional to $F(\omega)$ as needed. The factor $Q(\omega) = \sin(\omega\tau/2)/(\omega\tau/2)$ represents distortion which may be corrected by adding a second filter, called an *equalizing filter*. It must have a transfer function $H_{eq}(\omega) = 1/Q(\omega)$ for $|\omega| \leq W_f$, that is,

$$H_{eq}(\omega) = \begin{cases} \dfrac{\omega\tau/2}{\sin(\omega\tau/2)}, & |\omega| \leq W_f \\ \text{arbitrary elsewhere.} \end{cases} \qquad (2.3\text{-}11)$$

The equalized output spectrum is

$$S_o(\omega) = \frac{K\tau}{T_s} F(\omega), \qquad (2.3\text{-}12)$$

and the output signal is

$$s_o(t) = \frac{K\tau}{T_s} f(t). \tag{2.3-13}$$

Our analysis has shown, in summary, that flat-top sampling still allows distortion-free reconstruction of the information signal from its samples as long as a proper equalization filter is added to the reconstruction (lowpass) filter path. These operations are given in block diagram form in Fig. 2.3-3(b). The equalizing filter transfer function is shown in (c). As in natural sampling we again find that the output $s_o(t)$ is proportional to $f(t)$ with a proportionality constant equal to the dc component of the sampling pulse train. In the next section this fact will again be evident even when arbitrarily shaped sampling pulses are used.

2.4 PRACTICAL SAMPLING
WITH ARBITRARY PULSE SHAPES

In the real world pulses are never perfectly rectangular as was assumed above for natural and flat-top sampling. To have such pulses would require infinite bandwidth in all circuits, an obviously unrealistic situation. It becomes appropriate to then ask: What can be done in a more realistic system? To answer this question, let us assume an arbitrarily shaped sampling pulse $p(t)$. The sampling pulse train will be a sum of such pulses occurring at the sampling rate ω_s. The spectrum of $p(t)$ is defined as $P(\omega)$.

Consider first a generalization of natural sampling. The sampling pulse train now becomes

$$s_p(t) = \sum_{k=-\infty}^{\infty} p(t - kT_s), \tag{2.4-1}$$

where $p(t)$ is some arbitrary pulse shape as illustrated in Fig. 2.4-1(a). Expanding $s_p(t)$ into its complex Fourier series having coefficients C_k, we develop the sampled signal as

$$f_s(t) = f(t) \sum_{k=-\infty}^{\infty} p(t - kT_s) = \sum_{k=-\infty}^{\infty} C_k f(t) e^{jk\omega_s t}. \tag{2.4-2}$$

The spectrum of $f_s(t)$ becomes

$$F_s(\omega) = \sum_{k=-\infty}^{\infty} C_k F(\omega - k\omega_s). \tag{2.4-3}$$

Both $f_s(t)$ and $F_s(\omega)$ are sketched in Fig. 2.4-1(b).

Again we see that by using a lowpass filter, the signal $f(t)$ may be recovered without distortion. The filter will pass the term of (2.4-3) for $k = 0$. The output is easily found to be

$$s_o(t) = C_0 f(t). \tag{2.4-4}$$

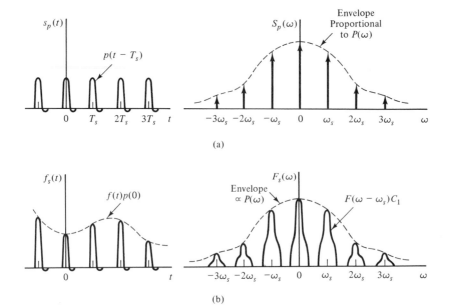

Figure 2.4-1. Spectrums associated with natural sampling using arbitrarily shaped pulses. (a) Train of arbitrary pulses and its spectrum. (b) Sampled signal and its spectrum [6].

Here C_0 is the dc component of the sampling pulse train. We may define parameters K and τ such that $C_0 = K\tau/T_s$, where K is defined as the actual amplitude of $p(t)$ at $t = 0$. We may think of τ as the duration of an *equivalent rectangular pulse* that gives the same pulse train dc level and has the same amplitude at $t = 0$ as the actual pulse. It is easily determined that τ is given by

$$\tau = \frac{1}{p(0)} \int_{-T_s/2}^{T_s/2} p(t)\, dt. \tag{2.4-5}$$

With these definitions (2.4-4) becomes

$$s_o(t) = \frac{K\tau}{T_s} f(t). \tag{2.4-6}$$

The analysis above has shown that the only effect of using an arbitrary pulse shape in the sampling pulse train, as far as signal recovery in natural sampling goes, is to produce a reconstructed signal scaled by a factor equal to the dc component of the pulse train.

Flat-top (instantaneous) sampling may also be generalized. Following the points discussed above we define an arbitrary pulse shape $q(t)$, having a spectrum $Q(\omega)$, which is related to the arbitrary pulse $p(t)$ by

$$p(t) = K\tau q(t). \tag{2.4-7}$$

It is a normalized version of $p(t)$ having amplitude $1/\tau$ at $t = 0$ and unit spectral magnitude† at $\omega = 0$. As before, K is the amplitude of $p(t)$ at $t = 0$ and τ is the equivalent rectangular pulse of $p(t)$ found from (2.4-5). The sampled signal becomes

$$f_s(t) = \sum_{k=-\infty}^{\infty} f(kT_s)K\tau q(t - kT_s). \tag{2.4-8}$$

This is a generalization of (2.3-7). We now recognize that the block diagram of Fig. 2.3-3(a) applies to (2.4-8) if the filter impulse response is $q(t)$. The spectrum at the filter output is now found to be

$$F_s(\omega) = \frac{K\tau}{T_s} \sum_{k=-\infty}^{\infty} Q(\omega)F(\omega - k\omega_s). \tag{2.4-9}$$

Again, in reconstruction of $f(t)$, only the term for $k = 0$ is of interest, since it is the only one passed by the lowpass filter. The output spectrum is $K\tau Q(\omega)F(\omega)/T_s$, and the factor $Q(\omega)$ represents distortion. As in the flat-top case, we may use an equalizing filter to remove the distortion. The required filter transfer function is

$$H_{eq}(\omega) = \frac{1}{Q(\omega)}. \tag{2.4-10}$$

The overall output with equalization becomes

$$S_o(\omega) = \frac{K\tau}{T_s} F(\omega), \tag{2.4-11}$$

$$s_o(t) = \frac{K\tau}{T_s} f(t). \tag{2.4-12}$$

In all our practical sampling methods the output of the reconstruction filter in the receiver is $f(t)$ scaled by the dc component of the unmodulated pulse train.

Example 2.4-1
A flat-top sampling system uses a train of triangular pulses defined by (A.2-3) of Appendix A:

$$p(t) = 2\text{tri}\left(\frac{t}{\tau_0}\right) \leftrightarrow 2\tau_0 \, \text{Sa}^2\left[\frac{\omega\tau_0}{2}\right] = P(\omega)$$

where $\tau_0 = 1 \, \mu s$. The message being sampled is assumed bandlimited to $W_f/2\pi = 5$ kHz and sampling is at the Nyquist rate. We define K, τ, and $q(t)$ for this system and find the receiver's output waveform.

From the definition of K,

$$K = p(0) = 2 \text{ V}.$$

---

† This is seen by calculating the Fourier series coefficient C_0 for the periodic waveforms using pulses of (2.4-7), substituting the Fourier transform of $q(t)$ for $\omega = 0$ and using (2.4-5).

For Nyquist sampling $T_s = \pi/W_f = 1/[2(5)10^3] = 10^{-4}$ s. From (2.4-5)

$$\tau = \frac{1}{2}\int_{-10^{-4}/2}^{10^{-4}/2} 2\,\mathrm{tri}\!\left(\frac{t}{10^{-6}}\right) dt = \int_{-10^{-6}}^{10^{-6}} \mathrm{tri}\!\left(\frac{t}{10^{-6}}\right) dt = 10^{-6}\ \mathrm{s}.$$

Thus

$$q(t) = \frac{p(t)}{K\tau} \leftrightarrow \mathrm{Sa}^2\!\left(\frac{\omega\tau_0}{2}\right) = \mathrm{Sa}^2\!\left[\frac{\omega(10^{-6})}{2}\right] = Q(\omega).$$

Finally, from (2.4-12) the receiver's output is

$$s_o(t) = \frac{K\tau}{T_s} f(t) = 0.02 f(t).$$

Observe that the recovered message has a relatively small amplitude (0.02 scale factor). In the next section, ways of increasing the output are given.

2.5 PRACTICAL SIGNAL RECOVERY METHODS

Practical ways of sampling messages were developed in the last two sections. The transmitted train of samples usually takes the form of either (2.4-8) for flat-top sampling or (2.4-2) for natural sampling. In both cases the sampling pulses could have arbitrary shape. The only message recovery method discussed was the lowpass filter with appropriate equalization, as needed. In most cases these filters produce a low level response because the factor τ/T_s common to all systems is often much less than unity.

In this section we introduce two practical message recovery methods that increase the receiver's output level compared to using a lowpass filter.

Signal Recovery by Zero-Order Sample-Hold

Because of the factor τ/T_s the output of the receiver using the lowpass filter reconstruction method is not as large as we might like. A simple sample-hold circuit may be used to increase the output by a factor T_s/τ. The circuit is shown in Fig. 2.5-1(a). The amplifier gain is arbitrary and assumed to equal unity; it only needs to provide a very low output impedance when driving the capacitor.

For purposes of description, assume the input to the sample-hold circuit is a flat-top signal as illustrated in (b). At the sample points (shown as heavy dots), the switch instantaneously† closes and the output level equals the input sample amplitude. While the switch is open the capacitor holds the voltage, as shown dashed, until the next closure. The output still looks like a flat-top sampled signal, but its pulse width is now T_s instead of τ seconds.

---

† In practice it would close just after the start of a sample pulse and open just before it ends.

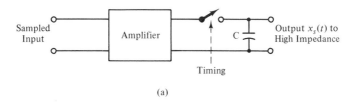

(a)

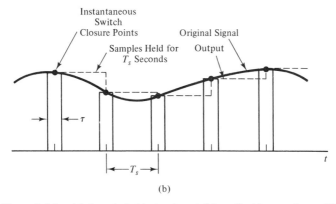

(b)

Figure 2.5-1. (a) Sample-hold circuit and (b) applicable waveforms [6].

By using (2.4-8) as the sampled input, the sample amplitudes are $Kf(kT_s)$, since $q(kT_s) = 1/\tau$. The sampled and held signal becomes

$$x_s(t) = K \sum_{k=-\infty}^{\infty} f(kT_s)h(t - kT_s) \tag{2.5-1}$$

where

$$h(t) = \begin{cases} 1, & 0 < t < T_s \\ 0, & \text{elsewhere} \end{cases} \tag{2.5-2}$$

represents the action of the holding circuit; it is its impulse response.

If $H(\omega)$ is the Fourier transform of $h(t)$, we find the spectrum $X_s(\omega)$ of the output of the sample-hold circuit to be

$$X_s(\omega) = \frac{K}{T_s} \sum_{k=-\infty}^{\infty} H(\omega)F(\omega - k\omega_s), \tag{2.5-3}$$

where, by Fourier transformation of $h(t)$,

$$H(\omega) = T_s \frac{\sin(\omega T_s/2)}{\omega T_s/2} \exp\left(\frac{-j\omega T_s}{2}\right). \tag{2.5-4}$$

From (2.5-3) it is clear that a lowpass filter, to remove components of the spectrum at multiples of ω_s, and an equalizer filter are necessary to recover $f(t)$. The equalizer transfer function required is

$$H_{eq}(\omega) = \begin{cases} T_s/H(\omega), & |\omega| \leq W_f \\ \text{arbitrary}, & \text{elsewhere}. \end{cases} \tag{2.5-5}$$

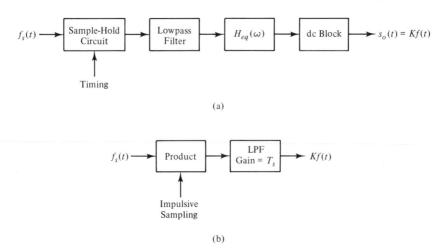

Figure 2.5-2. (a) Receiver using a sample-hold circuit for signal recovery.
b) Equivalent receiver [6].

The equalizer response is the final output

$$s_o(t) = Kf(t). \qquad (2.5\text{-}6)$$

The above operations are illustrated in Fig. 2.5-2(a). A dc block is
shown because some messages contain a dc component that is often not
required (or even desired) in the output. In the present discussion the dc
block can be ignored. From the standpoint of message reconstruction the
overall sample-hold receiving system can be replaced by the equivalent
system illustrated in (b). It is made up of an ideal sampler followed by an
ideal filter having a gain T_s and bandwidth W_f.

The sample-hold circuit described above is called zero-order because
the held voltage may be described by a polynomial of order zero.

First-Order Sample-Hold

Figure 2.5-3 illustrates the process of first-order sampling and holding.
At a particular instant (say kT_s), a sample of the signal is held until the
next sample. Now, however, rather than being a constant level held,
the level between samples varies linearly. The slope is determined by the
present sample (at time kT_s) and the immediately past sample [at $(k - 1)T_s$].

The output of the sample-hold circuit again has the spectrum of
(2.5-3) where the transfer function $H(\omega)$ of the sample-hold circuit is now
[17]

$$H(\omega) = T_s(1 + j\omega T_s)\left[\frac{\sin(\omega T_s/2)}{\omega T_s/2}\right]^2 \exp(-j\omega T_s). \qquad (2.5\text{-}7)$$

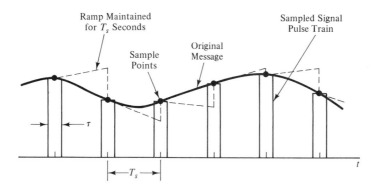

Figure 2.5-3. Waveforms applicable to first-order sample-hold [6].

The block diagram of Fig. 2.5-2 also applies to first-order sampling and holding. The equalization filter transfer function to be used is

$$H_{eq}(\omega) = \begin{cases} T_s/H(\omega), & |\omega| \leq W_f \\ \text{arbitrary}, & \text{elsewhere.} \end{cases} \qquad (2.5\text{-}8)$$

The first-order sample-hold operation derives its name from the fact that its held voltage is described by a polynomial of order one.

Higher-Order Sample-Holds

Sample-hold operations in general may be fractional or higher integer-order. These are discussed in [17]. A zero-order operation is capable of reproducing a constant (zero-order polynomial) signal $f(t)$ perfectly. A first-order sample-hold operation can reproduce exactly a constant or ramp (first-order polynomial) signal $f(t)$. Thus, an nth-order sample-hold can reproduce exactly a polynomial signal of order n. Sample-hold circuits above first-order are rarely used in practice for various reasons including economy. Fractional-order data holds are sometimes preferred in control systems [17].

★2.6 SAMPLING THEOREMS FOR BANDPASS SIGNALS

All our preceding discussions of sampling have dealt only with lowpass waveforms. In this section we consider sampling of bandpass waveforms. Generally, the theory involved in sampling theorems for bandpass signals is complicated. The complication arises from the fact that *two* spectral bands are involved in the bandpass case (one at $+\omega_0$ and one at $-\omega_0$, ω_0 being the carrier frequency), as opposed to only one in the lowpass case (from $-W_f$ to $+W_f$). Sampling again produces shifted replicas of the message spectrum; these are now more difficult to control in order to avoid aliasing. Since the choice of sampling rate, ω_s, is the only control available over the positions of the replicas, there is less freedom in choosing values of ω_s in bandpass sampling.

Direct Sampling of Bandpass Signals

Recall that the lowpass sampling theorem allows a signal $f(t)$ to have the expansion

$$f(t) = \sum_{k=-\infty}^{\infty} f(kT_s)h(t - kT_s), \tag{2.6-1}$$

where T_s is the period between samples taken periodically at times kT_s, $f(kT_s)$ are the samples, and $h(t)$ is a suitable function. The function $h(t)$ was defined by either (2.1-15) or (2.1-16) for lowpass sampling. We found it helpful to note that (2.6-1) could be interpreted as the response of the network of Fig. 2.1-2.† In direct sampling of a bandpass waveform it is again possible to find a function $h(t)$ such that (2.6-1) applies [18]. Thus, in essence, (2.6-1) is the sampling theorem for bandpass signals and its form guarantees that $f(t)$ can be reconstructed from the ideally sampled version of $f(t)$ using the network of Fig. 2.1-2.‡

Let $f(t)$ be a bandpass signal having spectral components only in the range $W_0 \leq |\omega| \leq W_0 + W_f$. It results that the Nyquist (minimum) sampling rate $2W_f$ can now be realized only for certain discrete values of $(W_0 + W_f)/W_f$. For other ratios the minimum rate will be larger than $2W_f$ but never larger than $4W_f$.

The correct function $h(t)$ to be used in (2.6-1) is known to be [18]

$$h(t) = \frac{T_s}{\pi t} \{\sin[(W_0 + W_f)t] - \sin(W_0 t)\}. \tag{2.6-2}$$

With some minor trigonometric manipulation, (2.6-2) becomes

$$h(t) = \frac{W_f T_s}{\pi} \, \text{Sa}\left[\frac{W_f t}{2}\right] \cos(\omega_0 t) \tag{2.6-3}$$

where the carrier frequency of the bandpass waveform is defined as

$$\omega_0 = W_0 + \frac{W_f}{2}. \tag{2.6-4}$$

The waveform becomes

$$f(t) = \frac{W_f T_s}{\pi} \sum_{k=-\infty}^{\infty} f(kT_s) \text{Sa}\left[\frac{W_f(t - kT_s)}{2}\right] \cos[\omega_0(t - kT_s)] \tag{2.6-5}$$

from (2.6-1).

---

† Network output was $f(t)/T_s$ for $h(t)$ defined by (2.1-16). To get exact correspondence—that is, get an output $f(t)$—the filter would have to be assigned a gain of T_s.

‡ The ideally sampled signal and its spectrum are again given by (2.1-14) and (2.1-19), respectively.

Only certain values of sample rate $\omega_s = 2\pi/T_s$ are allowed. By assuming $W_0 \neq 0$ and using results of Kohlenberg [18], the minimum allowable sample rate is

$$\omega_{s(\min)} = \frac{2}{M+1}\left(1 + \frac{W_0}{W_f}\right)W_f, \qquad (2.6\text{-}6)$$

where M is the largest nonnegative integer satisfying

$$M \leq \frac{W_0}{W_f}. \qquad (2.6\text{-}7)$$

A plot of (2.6-6) is given in Fig. 2.6-1. The values of the peaks are

$$\omega_{s(\text{peaks})} = \frac{2(M+2)}{M+1}\,W_f, \qquad M = 0, 1, 2, \ldots . \qquad (2.6\text{-}8)$$

We see that only when W_0/W_f equals an integer do we realize the Nyquist rate.

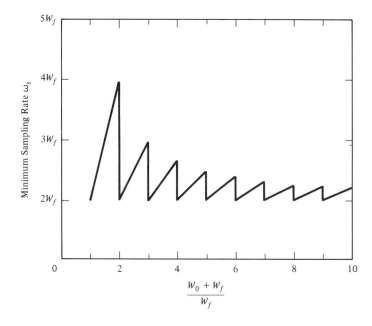

Figure 2.6-1. Minimum sampling rate for first-order sampling of bandpass signals [6].

In general the samples of $f(t)$ are not independent, even when sampling is at the minimum rate given by (2.6-6), except when $W_0/W_f = 0, 1, \ldots$. As $W_0/W_f \to \infty$ they approach independence, however. Thus in narrowband systems where $W_f << \omega_0 = W_0 + (W_f/2)$, samples are approximately independent.

Brown [19] recently discussed direct bandpass waveform sampling, which is also called *first-order sampling*. We shall subsequently define various orders of sampling.

Quadrature Sampling of Bandpass Signals

We continue the discussions on bandpass signal sampling by observing that it is not necessary that a bandpass signal be directly sampled. It is possible to precede sampling by preparatory processing of the waveform [2, 20]. Such an operation will always allow sampling at the Nyquist, or minimum, rate if the signal is bandlimited.

Let

$$f(t) = r(t)\cos[\omega_0 t + \phi(t)] \qquad (2.6\text{-}9)$$

be a bandpass signal having all its spectral components in the band $\omega_0 - (W_f/2) \leq |\omega| \leq \omega_0 + (W_f/2)$. By direct expansion we see that $f(t)$ can be represented by

$$f(t) = f_I(t)\cos(\omega_0 t) - f_Q(t)\sin(\omega_0 t), \qquad (2.6\text{-}10)$$

where $f_I(t)$ and $f_Q(t)$ are the in-phase and quadrature-phase components, respectively, given by

$$f_I(t) = r(t)\cos[\phi(t)], \qquad (2.6\text{-}11)$$

$$f_Q(t) = r(t)\sin[\phi(t)]. \qquad (2.6\text{-}12)$$

Both of these components are bandlimited to the band $|\omega| \leq W_f/2$. By a few simple steps it should become clear to the reader that the network of Fig. 2.6-2(a) will generate $f_I(t)$ and $f_Q(t)$. The products can be implemented with balanced modulators. The lowpass filters are to remove spectral components centered at $\pm 2\omega_0$, while passing the band $-W_f/2 \leq \omega \leq W_f/2$.

Each of the signals $f_I(t)$ and $f_Q(t)$ may be sampled at a rate of $W_f/2\pi$ samples per second and reconstructed as shown in Fig. 2.6-2(b). By forming the products indicated in (b) we may recover $f(t)$. Thus we may sample an arbitrary bandlimited bandpass signal at a *total* rate of W_f/π samples per second, using preprocessing, regardless of the ratio of $\omega_0 + (W_f/2)$ and $\omega_0 - (W_f/2)$, and recover $f(t)$. Notice that this is different from the sampling discussed before because we now have *two* samples being transmitted, each at a rate $W_f/2\pi$ samples per second, instead of one sample at a rate $2W_f/2\pi$.

Because of the preprocessing of $f(t)$, the components $f_I(t)$ and $f_Q(t)$ are independent and may be independently sampled according to the *lowpass* sampling theorem. The two trains of sampling pulses do not have to have the same timing; one can be staggered relative to the other. They may be interlaced, forming a composite sample train at a rate $\omega_s \geq 2W_f$, which alternately carries samples of $f_I(t)$ and $f_Q(t)$. In quadrature sampling a means must be provided in the receiver to separate the two sample trains.

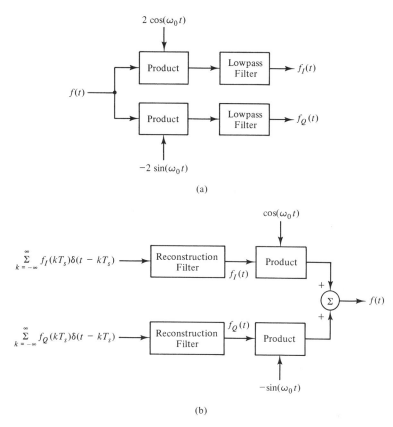

Figure 2.6-2. Quadrature sampling of a bandpass signal $f(t)$. (a) Quadrature signal generation prior to sampling and (b) waveform recovery from samples [6].

Bandpass Sampling Using Hilbert Transforms

Another form of preprocessing prior to sampling uses Hilbert transforms [2, 21]. As usual, let $f(t)$ be a bandpass signal bandlimited to $W_0 \leq |\omega| \leq W_0 + W_f$, where its center frequency is $\omega_0 = W_0 + (W_f/2)$, and let $\hat{f}(t)$ be its Hilbert transform.† The signal $\hat{f}(t)$ can be generated by passing $f(t)$ through a constant phase shift of $-\pi/2$ for $\omega > 0$ and $\pi/2$ for $\omega < 0$. Samples of both $f(t)$ and $\hat{f}(t)$ are adequate to determine $f(t)$ [2] according to

$$f(t) = \sum_{k=-\infty}^{\infty} [f(kT_s)h_1(t - kT_s) + \hat{f}(kT_s)h_2(t - kT_s)], \qquad (2.6\text{-}13)$$

† The Hilbert transform of a waveform $f(t)$ is defined by

$$\hat{f}(t) = \frac{1}{\pi} \int_{-\infty}^{\infty} \frac{f(\tau)}{t - \tau} \, d\tau.$$

where

$$h_1(t) = \text{Sa}\left[\frac{W_f t}{2}\right]\cos(\omega_0 t) \qquad (2.6\text{-}14)$$

$$h_2(t) = -\text{Sa}\left[\frac{W_f t}{2}\right]\sin(\omega_0 t). \qquad (2.6\text{-}15)$$

The sampling period assumed in (2.6-13) is

$$T_s = \frac{2\pi}{W_f}, \qquad (2.6\text{-}16)$$

which corresponds to $\omega_s = W_f$, half the Nyquist rate. However, there are two functions being sampled so the total (average) sampling rate is equal to the Nyquist (minimum) rate.

It can be shown (see Prob. 2-39) that (2.6-13) is exactly equivalent to quadrature sampling when samples of $f_I(t)$ and $f_Q(t)$ occur simultaneously and are taken at the minimum rates.

Sampling of Bandpass Random Signals

Because the quadrature sampling technique is a general one, and since it reduces the sampling representation to one of representing lowpass functions (in-phase and quadrature components), it can be used for random signals as well. We show this fact in the following development.

We shall draw on work in Appendix B. Let $N(t)$ be a zero-mean, wide-sense stationary random process bandlimited to the band $W_0 \leq |\omega| \leq W_0 + W_N$ centered at a carrier frequency $\omega_0 = W_0 + (W_N/2)$. $N(t)$ has the general representation

$$N(t) = N_c(t)\cos(\omega_0 t) - N_s(t)\sin(\omega_0 t) \qquad (2.6\text{-}17)$$

from (B.8-2). Here $N_c(t)$ and $N_s(t)$ are zero-mean, jointly wide-sense stationary processes bandlimited to $|\omega| \leq W_N/2$. By applying (2.2-5) to $N_c(t)$ and $N_s(t)$ we have

$$\begin{aligned}
\tilde{N}(t) &= \tilde{N}_c(t)\cos(\omega_0 t) - \tilde{N}_s(t)\sin(\omega_0 t) \\
&= \frac{W_N T_s}{2\pi}\left\{\sum_{k=-\infty}^{\infty} N_c(kT_s)\text{Sa}\left[\frac{W_N(t - kT_s)}{2}\right]\right\}\cos(\omega_0 t) \\
&\quad - \frac{W_N T_s}{2\pi}\left\{\sum_{k=-\infty}^{\infty} N_s(kT_s + t_0)\text{Sa}\left[\frac{W_N(t - KT_s - t_0)}{2}\right]\right\}\sin(\omega_0 t)
\end{aligned}$$
$$(2.6\text{-}18)$$

as the quadrature sampling representation of $N(t)$. It can be shown (see Prob. 2-41) that $\tilde{N}(t) = N(t)$ with zero mean-squared error. The parameter t_0 is a constant representing the shift in timing of the samples of $N_s(t)$ relative to the samples of $N_c(t)$; it is given by $-T_s \leq t_0 \leq T_s$. The sampling period T_s in (2.6-18) must satisfy $T_s \leq 2\pi/W_N$ because the Nyquist rate *per quadrature component* is W_N (rad/s).

★2.7 OTHER SAMPLING THEOREMS

Although there will be no need to invoke them in the following work, we shall briefly study some other sampling theorems.

It will be helpful to recall and use the network view of the sampling process. For ideal sampling of a lowpass signal we found that the sampling theorem representation for a signal $f(t)$ could be modeled as shown in Fig. 2.1-2. We interpreted the filter as a reconstruction filter in the receiver which used instantaneous (impulsive) samples as its input.

Even in cases where impulses were not transmitted, such as with natural or flat-topped pulses or the generalization to an arbitrary sample pulse $q(t)$, we still found that an ideal lowpass filter was required.† The effect of the shape of $q(t)$ only caused the need for an equalizing filter with transfer function $1/Q(\omega)$ with $q(t) \leftrightarrow Q(\omega)$. These comments are incorporated in the network interpretation of the lowpass sampling theorem shown in Fig. 2.7-1(a). Since $Q(\omega)H_{eq}(\omega) = 1$, there is no overall effect on the

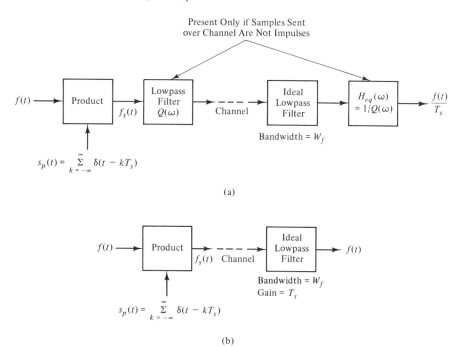

(a)

(b)

Figure 2.7-1. Transmission and reception of sampled signals. (a) System using samples having arbitrary sampling waveform and (b) equivalent system [6].

† This fact is strictly true only if $\omega_s = 2W_f$.

recovered output due to arbitrary pulse shape, and the equivalent representation of (b) applies to any of these sampling methods based on instantaneous samples.†

Higher-Order Sampling Defined

The lowpass sampling theorem may be summarized by writing $f(t)$ in the form

$$f(t) = \sum_{n=-\infty}^{\infty} f(nT_s)h(t - nT_s), \qquad (2.7\text{-}1)$$

where we use the model of Fig. 2.7-1(b). The filter impulse response must be

$$h(t) = \frac{W_f T_s}{\pi} \text{Sa}[W_f t] \qquad (2.7\text{-}2)$$

in order for (2.7-1) to equal (2.1-9).

More generally, we may allow $f(t)$ to be either lowpass or bandpass, but still a bandlimited, function. We now define $g_1(t)$, equal to the right side of (2.7-1), as a *first-order sampling* of $f(t)$ [18]. Thus first-order sampling involves a single train of uniformly separated samples of $f(t)$. By extending the idea we define a *pth-order sampling* of $f(t)$ as

$$f(t) = \sum_{i=1}^{p} g_i(t) = \sum_{i=1}^{p} \sum_{n=-\infty}^{\infty} f(nT_{si} + \tau_i)h_i(t - nT_{si} - \tau_i). \qquad (2.7\text{-}3)$$

Here we have p functions like (2.7-1). Each has uniformly separated samples T_{si} seconds apart. Each train has a time displacement τ_i relative to the chosen origin.

The general problem in higher-order sampling is to find the $h_i(t)$ which make (2.7-3) valid [18]. Only certain special cases are of usual interest. These are: (1) first-order lowpass signal sampling, (2) second-order lowpass signal sampling, (3) first-order bandpass signal sampling, and (4) second-order bandpass signal sampling. We have already discussed (1) and (3) in the preceding work. In the following paragraphs we discuss (2) and (4). In addition we shall also consider a sampling theorem for lowpass sampling in two dimensions.

Second-Order Sampling of Lowpass Signals

Here $p = 2$. Let $\tau_1 = 0$, $\tau_2 = \tau$, $T_{s2} = T_{s1} = 2\pi/W_f$, the maximum sample interval (smallest sample rate allowable for each pulse train). The sample times of the first train are then $n2\pi/W_f$, while those of the second train are $(n2\pi/W_f) + \tau$. Figure 2.7-2(a) and (b) show the two trains of

---

† Note that we have assigned a gain T_s to the filter for convenience.

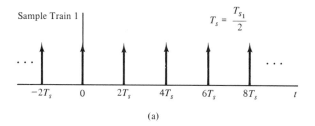

(a)

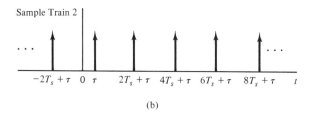

(b)

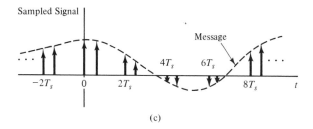

(c)

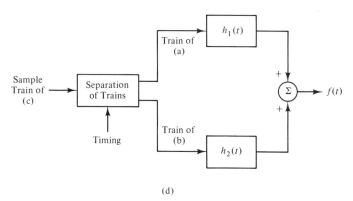

(d)

Figure 2.7-2. (a) and (b) are sampling pulse trains used to produce the second-order sampled signal of (c). The message reconstruction method is shown in (d) [6].

sampling impulses, where we define $T_s = T_{s_1}/2$. The composite sampled signal is

$$f_s(t) = \sum_{n=-\infty}^{\infty} f(nT_{s_1})\delta(t - nT_{s_1}) + \sum_{n=-\infty}^{\infty} f(nT_{s_1} + \tau)\delta(t - nT_{s_1} - \tau), \quad (2.7\text{-}4)$$

which is illustrated in (c).

For a real time function $f(t)$ the necessary reconstruction filter impulse responses are [2]

$$h_1(t) = \frac{\cos\left[W_f\left(t - \dfrac{\tau}{2}\right)\right] - \cos(W_f\tau/2)}{W_f t \sin(W_f\tau/2)}, \quad (2.7\text{-}5)$$

$$h_2(t) = h_1(-t). \quad (2.7\text{-}6)$$

The receiver reconstruction of $f(t)$ is illustrated in Fig. 2.7-2(d).

Probably the main advantage of second-order sampling of real lowpass time functions is that a nonuniform sampling may be used.

Example 2.7-1

Consider the special case of second-order sampling where the time difference between sampling pulse trains is

$$\tau = \frac{T_{s_1}}{2} = T_s = \frac{\pi}{W_f}.$$

In this case the composite train of pulses is uniformly separated by the sampling period T_s, and one might suspect that second-order sampling would reduce to first order. The filter responses (2.7-5) and (2.7-6) reduce to

$$h_1(t) = h_2(t) = \frac{\sin(W_f t)}{W_f t} = \text{Sa}(W_f t).$$

When this expression is used in (2.7-3) the result is found to be the same as (2.1-9) when $T_s = \pi/W_f$.

Second-Order Sampling of Bandpass Signals

The advantage of second-order versus first-order sampling of bandpass waveforms is that the minimum (Nyquist) rate of sampling is allowed for any choices of W_0 and W_f. Furthermore, the samples of $f(t)$ are independent.

Again letting $\tau_1 = 0, \tau_2 = \tau, T_{s_1} = T_{s_2} = 2T_s = 2\pi/W_f$, the reconstructed signal is given by (2.7-3) with [2]

$$h_1(t) = \frac{\cos[mW_f\tau - (W_0 + W_f)t] - \cos\{mW_f\tau - [(2m-1)W_f - W_0]t\}}{W_f t \sin(mW_f\tau)}$$

$$+ \frac{\cos\{[(2m-1)W_f\tau/2] - [(2m-1)W_f - W_0]t\}}{W_f t \sin[(2m-1)W_f\tau/2]} \quad (2.7\text{-}7)$$

$$- \frac{\cos\{[(2m - 1)W_f\tau/2] - W_0 t\}}{W_f t \sin[(2m - 1)W_f\tau/2]},$$

$$h_2(t) = h_1(-t). \tag{2.7-8}$$

Here m is the largest integer for which $(m - 1)W_f < W_0$ and τ is arbitrary, except that it may not be an integral multiple of π/W_f unless $(m - 1)W_f = W_0$ [2]. In the latter case W_0/W_f is an integer, and a development based on first-order sampling can be used to give

$$f(t) = \sum_{n=-\infty}^{\infty} f(nT_s)h(t - nT_s), \tag{2.7-9}$$

$$h(t) = \frac{T_s}{\pi t} \{\sin(mW_f t) - \sin[(m - 1)W_f t]\} \tag{2.7-10}$$

where samples are independent.

We observe that (2.7-10) agrees with (2.6-2) if we allow for the fact that $(m - 1)W_f = W_0$.

Lowpass Sampling Theorem in Two Dimensions

As a final topic in sampling theory we shall consider the uniform sampling theorem in two dimensions for lowpass functions [22]. Although the theorem has little application to ordinary radio communication systems, it is important in optical communication systems, antenna theory, picture processing, image enhancement, pattern recognition, and other areas.

It is difficult to state a completely general theorem owing to factors which we subsequently discuss.† However, we state the most useful and widely applied theorem.

A lowpass function $f(t, u)$, bandlimited such that its Fourier transform $F(\omega, \sigma)$ is nonzero only within, at most, the region bounded by $|\omega| \leq W_{f\omega}$ and $|\sigma| \leq W_{f\sigma}$, may be completely reconstructed from uniform samples separated by an amount $T_{st} \leq \pi/W_{f\omega}$ in t and an amount $T_{su} \leq \pi/W_{f\sigma}$ in u.

In following paragraphs we shall prove this theorem and discuss some additional fine points. In particular, we shall find that

$$f(t, u) = \sum_{k=-\infty}^{\infty} \sum_{n=-\infty}^{\infty} f(kT_{st}, nT_{su})\mathrm{Sa}\left[\pi\left(\frac{t}{T_{st}} - k\right)\right]\mathrm{Sa}\left[\pi\left(\frac{u}{T_{su}} - n\right)\right], \tag{2.7-11}$$

which is the two-dimensional extension of the one-dimensional theorem. Although we only consider two dimensions and sampling on a rectangular lattice, the theorem can be generalized to N-dimensional Euclidean space

---

† Recall that, even in the one-dimensional case, the lowpass theorem was not the most general theorem which could have been stated.

with sampling on an N-dimensional parallelepiped [22] as well as other lattices [23]. It can also be extended to nonuniform samples [24].

The proof of (2.7-11) amounts to postulating the function

$$f(t, u) = \sum_{k=-\infty}^{\infty} \sum_{n=-\infty}^{\infty} f(t_k, u_n) h(t - t_k, u - u_n), \qquad (2.7\text{-}12)$$

where

$$t_k = kT_{st}, \qquad (2.7\text{-}13)$$

$$u_n = nT_{su}, \qquad (2.7\text{-}14)$$

and finding an appropriate function $h(t, u)$ which will make (2.7-12) true for bandlimited $f(t, u)$. The function $h(t, u)$ is called an *interpolating function* and, as will be found, its solution is not unique.

We first extend the definition of the delta function to two dimensions as follows

$$\int_{-\infty}^{\infty} \int_{-\infty}^{\infty} \phi(\xi, \eta) \delta(\xi - x, \eta - y) \, d\xi \, d\eta = \phi(x, y), \qquad (2.7\text{-}15)$$

where $\phi(\xi, \eta)$ is an arbitrary function, continuous at (x, y). Now using the fact that

$$\delta(\xi - x, \eta - y) = \delta(\xi - x)\delta(\eta - y), \qquad (2.7\text{-}16)$$

we may apply (2.7-15) to write (2.7-12) in the form

$$f(t, u) = \sum_{k=-\infty}^{\infty} \sum_{n=-\infty}^{\infty} \int_{-\infty}^{\infty} \int_{-\infty}^{\infty} f(\xi, \eta) h(t - \xi, u - \eta)$$
$$\cdot \, \delta(\xi - t_k, \eta - u_n) \, d\xi \, d\eta$$
$$= \int_{-\infty}^{\infty} \int_{-\infty}^{\infty} f(\xi, \eta) h(t - \xi, u - \eta) \sum_{k=-\infty}^{\infty} \delta(\xi - t_k)$$
$$\cdot \sum_{n=-\infty}^{\infty} \delta(\eta - u_n) \, d\xi \, d\eta, $$

$$(2.7\text{-}17)$$

if it is assumed that the order of integrations and summations may be reversed. It is known that

$$\sum_{k=-\infty}^{\infty} \delta(\xi - t_k) = \sum_{k=-\infty}^{\infty} \delta(\xi - kT_{st}) = \frac{1}{T_{st}} \sum_{k=-\infty}^{\infty} e^{jk\omega_{st}\xi}, \qquad (2.7\text{-}18)$$

where the sample rate in t is

$$\omega_{st} = 2\pi/T_{st}. \qquad (2.7\text{-}19)$$

By defining the sample rate in u as

$$\omega_{su} = 2\pi/T_{su}, \qquad (2.7\text{-}20)$$

an expression similar to (2.7-18) can be written. Substitution of the two expressions into (2.7-17) gives

$$f(t, u) =$$

$$\int_{-\infty}^{\infty} \int_{-\infty}^{\infty} h(t - \xi, u - \eta) \frac{1}{T_{st}T_{su}} \sum_{k=-\infty}^{\infty} \sum_{n=-\infty}^{\infty} f(\xi, \eta) e^{jk\omega_{st}\xi + jn\omega_{su}\eta} \, d\xi \, d\eta.$$

$$(2.7\text{-}21)$$

This expression is recognized as the two-dimensional convolution of the two functions $h(t, u)/T_{st}T_{su}$ and

$$f_s(t, u) = \sum_{k=-\infty}^{\infty} \sum_{n=-\infty}^{\infty} f(t, u) e^{jk\omega_{st}t + jn\omega_{su}u}. \qquad (2.7\text{-}22)$$

If the Fourier transform of $f_s(t, u)$ is defined as $F_s(\omega, \sigma)$, the frequency shifting property of Fourier transforms gives

$$F_s(\omega, \sigma) = \sum_{k=-\infty}^{\infty} \sum_{n=-\infty}^{\infty} F(\omega - k\omega_{st}, \sigma - n\omega_{su}). \qquad (2.7\text{-}23)$$

Next, we recognize that the Fourier transform of a convolution of two functions in the t, u domain is the product of individual transforms in the ω, σ domain. The transform of (2.7-21) is then

$$F(\omega, \sigma) = \frac{H(\omega, \sigma)}{T_{st}T_{su}} \sum_{k=-\infty}^{\infty} \sum_{n=-\infty}^{\infty} F(\omega - k\omega_{st}, \sigma - n\omega_{su}), \qquad (2.7\text{-}24)$$

where $H(\omega, \sigma)$ is defined as the transform of $h(t, u)$.

This expression allows us to determine the required properties of the interpolating function $h(t, u)$. Figure 2.7-3 will aid in its interpretation. The bandlimited spectrum $F(\omega, \sigma)$ is illustrated in (a). The double sum of terms in (2.7-24) is illustrated in (b). Clearly, if the right side of (2.7-24) must equal $F(\omega, \sigma)$, two requirements must be satisfied. First, $\omega_{st} \geqslant 2W_{f\omega}$ and $\omega_{su} \geqslant 2W_{f\sigma}$ are necessary so that the replicas in (b) do not overlap, and second, the function $H(\omega, \sigma)/T_{st}T_{su}$ must equal unity over the *aperture* region in the ω, σ plane occupied by $F(\omega, \sigma)$ and must be zero in all regions occupied by the replicas. In the space between these regions $H(\omega, \sigma)/T_{st}T_{su}$ may be arbitrary. Hence there is no unique interpolating function in general. The first requirement establishes the sampling intervals stated in the original theorem.

Regarding the second condition, we may ask what interpolating function should be used. There is no one-correct answer. However, suppose we select

$$H(\omega, \sigma) = T_{st}T_{su} \, \text{rect}\left(\frac{\omega}{\omega_{st}}\right) \text{rect}\left(\frac{\sigma}{\omega_{su}}\right). \qquad (2.7\text{-}25)$$

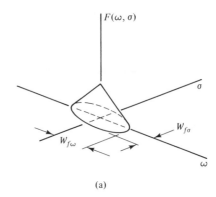

(a)

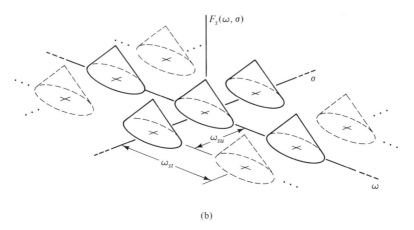

(b)

Figure 2.7-3. (a) A two-dimensional, bandlimited Fourier transform and (b) its
periodic version representing the spectrum of the sampled signal [6].

This choice has the advantage of admitting *all* aperture shapes within the
prescribed rectangular boundary. By inverse transformation

$$h(t, u) = \frac{1}{(2\pi)^2} \int_{-\infty}^{\infty} \int_{-\infty}^{\infty} H(\omega, \sigma) e^{j\omega t + j\sigma u} \, d\omega \, d\sigma$$

$$= \frac{T_{st}}{2\pi} \int_{-\omega_{st}/2}^{\omega_{st}/2} e^{j\omega t} \, d\omega \, \frac{T_{su}}{2\pi} \int_{-\omega_{su}/2}^{\omega_{su}/2} e^{j\sigma u} \, d\sigma \qquad (2.7\text{-}26)$$

$$= \mathrm{Sa}\!\left(\frac{\pi t}{T_{st}}\right) \mathrm{Sa}\!\left(\frac{\pi u}{T_{su}}\right).$$

By substituting this expression back into (2.7-12), we have finally proved
the original theorem embodied in (2.7-11).

The interpolating function defined by either (2.7-25) or (2.7-26) is called the *canonical interpolating function* [22]. It is *orthogonal* in the t, u space, which means that samples of $f(t, u)$ are linearly independent.

For the rectangular sampling plan the canonical interpolating function may give 100% *sampling efficiency*. Let sampling efficiency η be defined as the ratio of the area A_a in the ω, σ plane over which $F(\omega, \sigma)$ is nonzero to the area A_s of the rectangle defining the repetitive "cell" due to sampling [22]. Since $A_s = \omega_{st}\omega_{su}$, we have

$$\eta = \frac{A_a}{A_s} = \frac{A_a}{\omega_{st}\omega_{su}}. \tag{2.7-27}$$

Efficiency is maximized by sampling at the minimum rates. By assuming this to be the case, A_a is maximum for $F(\omega, \sigma)$ existing over a rectangular aperture. The corresponding efficiency is $\eta = 1.0$, or 100%. For comparison, it can be shown that maximum efficiency is 78.5% for either a circular aperture, or an elliptical aperture with major axis in either ω or σ directions.

Example 2.7-2

We note that the filter transfer function of (2.7-25) is arbitrary in the sense that the ideal filter can be more narrowband if $W_{f\omega} < \omega_{st}/2$ and $W_{f\sigma} < \omega_{su}/2$. We shall assume this to be the case and choose

$$H(\omega, \sigma) = T_{st}T_{su} \operatorname{rect}(\omega/2W_{f\omega})\operatorname{rect}(\sigma/2W_{f\sigma})$$

so that a form of the sampling theorem that is more directly in the form of (2.1-9) can be stated. By inverse transformation

$$h(t, u) = \frac{W_{f\omega}T_{st}}{\pi}\operatorname{Sa}(W_{f\omega}t)\, \frac{W_{f\sigma}T_{su}}{\pi}\operatorname{Sa}(W_{f\sigma}u).$$

From (2.7-12) we have

$$f(t, u) = \frac{W_{f\omega}T_{st}}{\pi}\sum_{k=-\infty}^{\infty}\frac{W_{f\sigma}T_{su}}{\pi}\sum_{n=-\infty}^{\infty} f(kT_{st}, nT_{su})\operatorname{Sa}[W_{f\omega}(t - kT_{st})]\operatorname{Sa}[W_{f\sigma}(u - nT_{su})]$$

which is the two-dimensional form of (2.1-9).

2.8 TIME DIVISION MULTIPLEXING

As mentioned at the start of this chapter, one of the greatest benefits to be derived from sampling is that *time division multiplexing* (TDM) is possible. TDM amounts to using sampling to simultaneously transmit many messages over a single communication link by interlacing trains of sampling pulses.

In this section we briefly describe time multiplexing of flat-top samples of N similar messages (such as telephone signals). The concepts are applicable to other waveforms to be developed in later work.

The Basic Concept

The conceptual implementation of the time multiplexing of N similar messages $f_n(t)$, $n = 1, 2, \ldots, N$, is illustrated in Fig. 2.8-1. Sampled signals

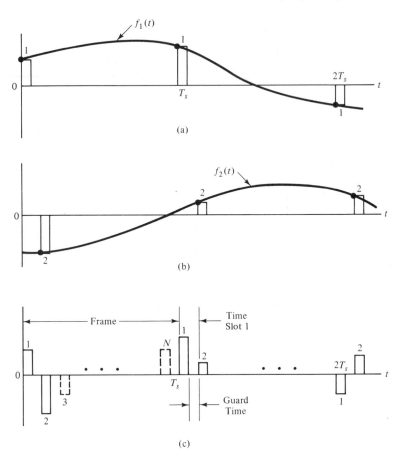

Figure 2.8-1. Time division multiplexing of flat-top sampled messages. Pulse
trains of: (a) message 1, (b) message 2, and (c) the multiplexed train.

(pulse trains) for messages one and two are shown in (a) and (b). The pulse
train of (b) is delayed slightly from the train of (a) to prevent overlap.
Other messages are treated similarly. When the N total trains are combined
(multiplexed), the waveform of (c) is obtained. The time allocated to one
sample of one message is called a *time slot*. The time interval over which
all messages are sampled at least once is called a *frame*. The portion of
the time slot not used by the sample pulse is called the *guard time*. In
Fig. 2.8-1 all time slots are occupied by message samples. In a practical
system some time slots may be allocated to other functions (for example,
signaling, monitoring, and synchronization).

Example 2.8-1
Suppose we want to time-multiplex $N = 50$ similar telephone messages. Assume
each message is bandlimited to 3.3 kHz. Thus $W_f = 2\pi(3.3)10^3$, and we must

sample each message at a rate of at least $6.6(10^3)$ samples per second. From practical considerations let us be limited to a sampling rate of 8 kHz and use a guard time equal to the sample pulse duration τ. We find the required value of τ.

By equating 2τ, the slot time per sample, to T_s/N, the allowed slot time, we get $\tau = T_s/(2N)$. But $T_s = 1/8$ ms, so $\tau = 10^{-3}/(8 \cdot 100) = 1.25$ μs.

Synchronization

To maintain proper positions of sample pulses in the multiplexer, it is necessary to synchronize the sampling process. Because the sampling operations are usually electronic, there is typically a *clock* pulse train that serves as the reference for all samplers. At the receiving station there is a similar clock that must be synchronized with that of the transmitter. Clock synchronization can be derived from the received waveform by observing the pulse sequence over many pulses and averaging the pulses (in a closed loop with the clock as the voltage-controlled oscillator).

Clock synchronization does not guarantee that the proper *sequence* of samples is synchronized. Proper alignment of the time slot sequence requires *frame synchronization*. A simple technique is to use one or more time slots per frame for synchronization. By placing a special pulse that is larger than the largest expected message amplitude in time slot 1, for example, the start of a frame can readily be identified using a suitable threshold circuit.

2.9 SUMMARY AND DISCUSSION

In this chapter we have demonstrated one most important fact, that a continuous waveform representing an information source can be completely reconstructed in a receiver using only periodic *samples* of the waveform. Thus the original waveform can be reformed, even at times between the samples, from just its samples. It is necessary only that the waveform be bandlimited (bandwidth W_f) and that instantaneous samples be taken at a high enough rate (the minimum rate $2W_f$ rad/s is the Nyquist rate). The sampling theorem was developed to prove this rather remarkable fact.

In practice, waveforms are never bandlimited to have zero spectral components outside the band W_f. However, there is always some frequency above which spectral components are negligible and can be approximated as zero; this practical frequency becomes W_f, the signal's spectral extent.

For baseband waveforms the sampling theorem is developed for both deterministic (Sec. 2.1) and random, or noiselike, waveforms (Sec. 2.2). Even if samples are not instantaneous, sampling theorems are shown to apply when using practical sampling techniques (Secs. 2.3 and 2.4) and when using practical reconstruction methods in the receiver (Sec. 2.5). Sampling theorems for bandpass signals are also given (Sec. 2.6) and generalized theorems are stated for both baseband and bandpass waveforms (Sec. 2.7).

By interleaving samples of several source waveforms in time, it is possible to transmit enough information to a receiver via only *one channel* to recover all message waveforms. The technique, called time division multiplexing, is briefly discussed (Sec. 2.8). Time multiplexing is one of the principal applications of sampling.

In many modern-day communication problems, the information waveform is analog, whereas the communication system is digital. The sampling theorem forms the basic theory that allows the waveform to be converted to a form suitable for use in such systems. We continue to develop these ideas in the next chapter.

REFERENCES

[1] Jerri, A. J., The Shannon Sampling Theorem—Its Various Extensions and Applications: A Tutorial Review, *Proceedings of the IEEE,* Vol. 65, No. 11, November, 1977, pp. 1565–1596.

[2] Linden, D. A., A Discussion of Sampling Theorems, *Proceedings of the IRE,* July 1959, pp. 1219–1226.

[3] Black, H. S., *Modulation Theory,* McGraw-Hill Book Co., New York, 1953.

[4] Whittaker, E. T., On the Functions which are Represented by the Expansion of Interpolating Theory, *Proceedings of the Royal Soc. of Edinburgh,* Vol. 35, 1915, pp. 181–194.

[5] Shannon, C. E., Communications in the Presence of Noise, *Proceedings of the IRE,* Vol. 37, No. 1, January 1949, pp. 10–21.

[6] Peebles, Jr., P. Z., *Communication System Principles,* Addison-Wesley Publishing Co., Inc., Reading, Massachusetts, 1976. (Figures 2.1-1, 2.1-2, 2.1-3, 2.3-1, 2.3-2, 2.3-3, 2.4-1, 2.5-1, 2.5-2, 2.5-3, 2.6-1, 2.6-2, 2.7-1, 2.7-2, and 2.7-3 have been adapted.)

[7] Reza, F. M., *An Introduction to Information Theory,* McGraw-Hill Book Co., New York, 1961.

[8] Gagliardi, R., *Introduction to Communications Engineering,* John Wiley & Sons, New York, 1978.

[9] Stremler, F. G., *Introduction to Communication Systems,* 2nd ed., Addison-Wesley Publishing Co., Reading, Massachusetts, 1982.

[10] Balakrishnan, A. V., A Note on the Sampling Principle for Continuous Signals, *IRE Transactions on Information Theory,* Vol. IT-3, No. 2, June 1957, pp. 143–146.

[11] Lloyd, S. P., A Sampling Theory for Stationary (Wide Sense) Stochastic Processes, *Transactions American Mathematical Society,* Vol. 92, 1959, pp. 1–12.

[12] Balakrishnan, A. V., Essentially Band-Limited Stochastic Processes, *IEEE Transactions on Information Theory,* Vol. IT-11, 1965, pp. 145–156.

[13] Rowe, H. E., *Signals and Noise in Communication Systems,* Van Nostrand Reinhold Co., New York, 1965.

[14] Sakrison, D. J., *Communication Theory: Transmission of Waveforms and Digital Information,* John Wiley & Sons, New York, 1968.

[15] Shanmugam, K. S., *Digital and Analog Communication Systems,* John Wiley & Sons, New York, 1979.

[16] Haykin, S., *Communication Systems,* 2nd ed., John Wiley & Sons, New York, 1983.

[17] Ragazzini, J. R., and Franklin, G. F., *Sampled-Data Control Systems,* McGraw-Hill Book Co., New York, 1958.

[18] Kohlenberg, A., Exact Interpolation of Band-Limited Functions, *Journal of Applied Physics,* December, 1953, pp. 1432–1436.

[19] Brown, Jr., J. L., First-Order Sampling of Bandpass Signals, *IEEE Transactions on Information Theory,* Vol. IT-26, No. 5, September 1980, pp. 613–615.

[20] Berkowitz, R. S. (Editor), *Modern Radar, Analysis, Evaluation, and System Design,* John Wiley & Sons, New York, 1965. See Part II, Chapter 5 by R. S. Berkowitz.

[21] Goldman, S., *Information Theory,* Prentice-Hall, Inc., New York, 1953.

[22] Petersen, D. P., and Middleton, D., Sampling and Reconstruction of Wave-Number-Limited Functions in *N*-Dimensional Euclidean Spaces, *Information and Control,* Vol. 5, 1962, pp. 279–323.

[23] Mersereau, R. M., The Processing of Hexagonally Sampled Two-Dimensional Signals, *Proceedings of the IEEE,* Vol. 67, No. 6, June 1979, pp. 930–949.

[24] Gaarder, N. T., A Note on Multi-Dimensional Sampling Theorem, *Proceedings of the IEEE,* Vol. 60, No. 2, February 1972, pp. 247–248.

[25] Abramowitz, M., and Stegun, I. A. (Editors), *Handbook of Mathematical Functions with Formulas, Graphs, and Mathematical Tables,* National Bureau of Standards Applied Mathematics Series 55, U.S. Government Printing Office, Washington, D.C., 1964.

PROBLEMS

2-1. A signal $f(t) = A\,\mathrm{Sa}(\omega_f t)$ is bandlimited to $\omega_f/2\pi$ Hz. From the sampling theorem justify that only *one* sample is adequate to reconstruct $f(t)$ for all time.

2-2. The spectrum of the signal of Prob. 2-1 is $F(\omega) = (A\pi/\omega_f)\mathrm{rect}(\omega/2\omega_f)$. $f(t)$ is applied to a filter with transfer function

$$H(\omega) = \mathrm{rect}\!\left(\frac{\omega}{2\omega_f}\right)\!\left[1 + \sum_{n=1}^{N} a_n\cos\!\left(\frac{n\pi\omega}{\omega_f}\right)\right].$$

(a) Inverse transform the filter's output spectrum to show the effect of distorting the signal's spectrum. (b) How many samples are required if the filter's output signal is to be reconstructed exactly from its samples? (c) What is the effect of replacing the cosines by sines?

2-3. Begin with (2.1-14) and show that the spectrum of the ideally sampled signal is given by (2.1-19):

$$F_s(\omega) = \frac{1}{T_s} \sum_{k=-\infty}^{\infty} F(\omega - k\omega_s), \qquad \omega_s = \frac{2\pi}{T_s}.$$

2-4. A nonrandom signal, bandlimited to 17.5 kHz, is to be reconstructed exactly from its samples. It is sampled by ideal sampling and it is recovered in the receiver by an ideal lowpass filter. (a) What is the filter's minimum bandwidth allowed? (b) At what minimum rate must the signal be sampled?

2-5. A system transmits ideal samples of its input message. The sampling rate is 25 kHz and the receiver uses an ideal lowpass filter with 10 kHz bandwidth. The system is to be used with a nonrandom message bandlimited to 13 kHz. Can this message be reconstructed without distortion? If not, state why and discuss how you might send the message over the system (with changes).

2-6. Show by application of the lowpass sampling theorem that the bandlimited signal

$$f(t) = \frac{\pi\cos(W_f t)}{(\pi/2)^2 - (W_f t)^2}$$

is the sum of two sampling functions separated in time by the sampling time $T_s = \pi/W_f$. (*Hint:* Choose $T_s/2$ and $-T_s/2$ as the sample times.)

2-7. Prove that (2.1-14) is true.

2-8. A message has a bandlimited spectrum $F(\omega) = 10 \, \text{tri}(\omega/W_f)$. It is ideally sampled at a rate ω_s of three times its Nyquist rate. (a) Sketch the spectrum of the sampled signal. (b) In the receiver a nonideal filter is used that has a transfer function $H(\omega) = [1 + (\omega/2W_f)^4]^{-1}$. Neglect the distortion caused by roll off due to $H(\omega)$ in the band $-W_f \leq \omega \leq W_f$ but find how far down the filter attenuates the edge of the first replica components at $|\omega| = \omega_s - W_f$.

2-9. Let $f(t)$ be bandlimited to $-W_f \leq \omega \leq W_f$ and have the sampling representation of (2.1-9). (a) Find the bandwidth of the response $g(t)$ of a square-law device where $g(t) = f^2(t)$. (b) Write an equation for the sampling representation of $f^2(t)$ in terms of samples of $f(t)$.

2-10. A signal $f(t)$ is bandlimited to 5 kHz; it is sampled at a very high rate of 100 kHz using ideal sampling (impulses). At another location the sampled signal is filtered by an ideal *bandpass* filter that passes a band of 15 kHz width centered on 100 kHz. (a) Write an equation for the exact filter response signal, $f_0(t)$. (b) Discuss the form of the signal $f_0(t)$.

2-11. A signal $f_1(t) = 6u(t)\exp(-3t)$ has a Fourier transform $F_1(\omega) = 6[3 + j\omega]^{-1}$. Clearly, $f_1(t)$ is not bandlimited. (a) Find the bandwidth of an *ideal* filter that will allow the *filtered* signal, denoted by $f(t)$, to contain 99% of the total energy in $f_1(t)$. (b) What is the minimum allowable rate that $f(t)$ can be sampled without aliasing?

2-12. Work Prob. 2-11 for the signal

$$f_1(t) = 3tu(t)e^{-4t} \leftrightarrow F_1(\omega) = 3[4 + j\omega]^{-2}.$$

2-13. Work Prob. 2-11 for the signal

$$f_1(t) = 2 \, e^{-5|t|} \leftrightarrow F_1(\omega) = 20[25 + \omega^2]^{-1}.$$

2-14. A signal is given by $f(t) = 3\cos^3[2\pi(10^4)t]$. (a) Find the spectrum $F(\omega)$ of $f(t)$. (b) What is the Nyquist rate for ideal sampling of $f(t)$?

2-15. (a) Determine the minimum (Nyquist) sampling rate for the signal $f(t) = f_1(t) + f_2(t)$ where $f_1(t)$ is bandlimited to $W_1(\text{rad/s})$ and $f_2(t)$ is bandlimited to $3W_1$ with $W_1 > 0$ a constant. (b) Discuss how the minimum rate of part (a) compares to the respective minimum rates of sampling $f_1(t)$ and $f_2(t)$ individually.

2-16. How would the conclusions of Prob. 2-15 change if $f_1(t)$ and $f_2(t)$ were replaced by their respective squares, $f_1^2(t)$ and $f_2^2(t)$? Discuss.

2-17. A lowpass bandlimited signal $f(t)$ has energy E_f. Find E_f in terms of the ideal samples of $f(t)$ taken at the Nyquist rate. [*Hint:* Use (2.1-11).]

2-18. The signal $f(t)$ of Prob. 2-14 is ideally sampled at a 60 kHz rate and the sampled waveform is applied to an ideal lowpass filter with bandwidth 30.1 kHz. Find an equation for the filter's response. Is there aliasing? Explain.

★2-19. A rectangular time function $f(t) = A\,\text{rect}(t/T)$ that is *not* bandlimited is ideally sampled at a rate $\omega_s = 2W_f$, where W_f is the bandwidth of an ideal filter to which the sampled signal is applied. (a) Find W_f so that the fractional aliasing error is $1/12$. (b) How many samples occur over the duration of $f(t)$ for the sample rate $\omega_s = 2W_f$? (*Hint:* Use $\int_0^x \text{Sa}^2(\xi)\,d\xi = \text{Si}(2x) - x^{-1}\sin^2(x)$ where the *sine integral* $\int_0^\alpha \text{Sa}(\xi)d\xi \triangleq \text{Si}(\alpha)$ is a tabulated quantity [25].)

2-20. A nonrandom signal $f(t)$ has an energy density spectrum $\mathscr{E}_f(\omega) = 24\omega^2[1 + (\omega/W)^2]^{-2}$, where $W > 0$ is a constant. If $f(t)$ is ideally sampled at a rate $\omega_s = 5W$, what fractional aliasing error occurs?

2-21. A nonrandom signal $f(t)$ has an energy density spectrum $\mathscr{E}_f(\omega) = A[1 + (\omega/W)^2]^{-1}$, where $W > 0$ is a constant. At what rate must $f(t)$ be ideally sampled if the fractional aliasing error is to be 0.01 (or 1%)?

2-22. A waveform exists only over a total time interval of 80 μs. If its spectrum is to be represented exactly by its samples taken every $W_s(\text{rad/s})$ apart, what is the largest that W_s can be?

2-23. A baseband noise waveform is reconstructed exactly from its samples taken at a 60-kHz rate. If samples occur at three times the Nyquist rate, what is the spectral extent of the noise?

2-24. A message $f(t) = 6\cos(10^3 t)$ is naturally sampled at a rate $\omega_s = 2.5(10^3)$ rad/s using pulses of duration 60 μs. The amplitude of the sampling pulse train is $K = 2$ V. (a) Write an expression for the response of a receiver's ideal filter (lowpass) of bandwidth $W_f = 1010$ rad/s to the train of samples. (b) If W_f is changed to 990 rad/s what is the response?

2-25. Find and plot the spectrum of the sampled signal of Prob. 2-24.

2-26. Show that if a message is a sinusoid of frequency ω_f that the ideal filter in the receiver requires no equalization for distortion-free signal recovery so long as the sampling rate exceeds $2\omega_f$ and the filter's bandwidth is slightly larger than ω_f. What then is the effect of leaving out equalization? Assume flat-top sampling.

★2-27. Assume flat-top sampling and find and plot the magnitude of the equalizing filter's transfer function to follow a lowpass filter for message recovery when the sample pulse shapes are as given.

(a) $p(t) = \begin{cases} A\left[1 - \left(\dfrac{2t}{\tau_p}\right)^2\right], & |t| \leq \dfrac{T_p}{2} \\ 0, & \text{elsewhere} \end{cases}$

(b) $p(t) = \begin{cases} A\cos^2\left(\dfrac{\pi t}{\tau_p}\right), & |t| < \dfrac{T_p}{2} \\ 0, & \text{elsewhere} \end{cases}$

(c) $p(t) = A\,\mathrm{tri}\left(\dfrac{2t}{\tau_p}\right).$

2-28. Start with (2.4-1) and show that the spectrum of $s_p(t)$ may be written as

$$S_p(\omega) = \omega_s \sum_{n=-\infty}^{\infty} P(n\omega_s)\delta(\omega - n\omega_s),$$

where $P(\omega)$ is the Fourier transform of $p(t)$.

2-29. Show that C_0 of (2.4-4) equals $K\tau/T_s$ with τ given by (2.4-5).

2-30. Pulses of the form given in Prob. 2-27(b) are used to construct a periodic sampling pulse train for natural sampling of a signal bandlimited to W_f. What smallest pulse duration τ_p is required if τ_p is adjusted so that the spectral replica at $3\omega_s$ is to disapppear (be made zero) in the sampled message when sampling is at the Nyquist rate? Sketch the spectrum of the sampled signal.

2-31. Find the durations of the equivalent rectangular pulses of the waveforms of Prob. 2-27(a) and (b).

2-32. Show that (2.5-4) follows from Fourier transformation of (2.5-2).

2-33. Plot $|H_{eq}(\omega)|$ of (2.5-5) for $0 \leq \omega \leq W_f$ when $H(\omega)$ is given by (2.5-4) with $\omega_s = 6\,W_f$. How much does $|H_{eq}(\omega)|$ vary over the band? Would you consider the equalizer necessary in practice for this problem?

2-34. The spectrum $F(\omega)$ of a signal $f(t)$ is given by

$$F(\omega) = 8\,\mathrm{tri}\left(\frac{\omega - 25}{5}\right) + 8\,\mathrm{tri}\left(\frac{\omega + 25}{5}\right).$$

(a) Sketch the spectrum of the ideally sampled version of $f(t)$ for sampling rates $\omega_s = 20, 25,$ and 50 rad/s. (b) Can $f(t)$ be recovered exactly from any of the three sampled signals in (a) by a bandpass filter that passes the band $20 \leq |\omega| \leq 30$? If so, state which ones.

⋆2-35. A bandpass nonrandom signal is bandlimited such that $W_0/W_f = 2.8$ and $W_f/2\pi = 10^4$ Hz. It is directly sampled. What is the smallest allowable sampling rate if perfect recovery from its samples is to be achieved?

⋆2-36. An engineer has a bandpass nonrandom message $f(t)$ bandlimited to a bandwidth of 6 MHz. He wants to transmit ideal direct samples of $f(t)$ at the Nyquist rate. (a) If he has the freedom of choosing only the center frequency of $f(t)$, what is the smallest and the next three higher frequencies that he may select? (b) If he chooses the lowest allowable center frequency, is the waveform being sampled still bandpass?

★2-37. A signal $f(t) = 3 \cos(100t)\cos(350t)$ is to be sampled at the smallest allowable rate. (a) What is the sample rate? (b) Sketch the spectrum of the (ideally) sampled signal and verify that a bandpass filter passing the bands $250 \leqslant |\omega| \leqslant 450$ rad/s will recover $f(t)$ without distortion.

★2-38. A bandpass signal, bandlimited to a bandwidth W_f, is directly sampled at a minimum rate 30% higher than the Nyquist (minimum possible) rate. Find the largest value that is possible for W_0/W_f. How many values of W_0/W_f are possible? [*Hint:* Use (2.6-6).]

★2-39. Show that (2.6-13) is equivalent to quadrature sampling of $f(t)$ if samples of $f_I(t)$ and $f_Q(t)$ occur at the same times and sampling is at the minimum possible rate. [*Hint:* Use the product property of Hilbert transforms. It states that the Hilbert transform of a product $f(t)c(t)$ is $f(t)\hat{c}(t)$ if $f(t)$ is lowpass, bandlimited to $|\omega| \leqslant W_f$, and $c(t)$ has a nonzero spectrum only for $|\omega| > W_f$.]

★2-40. Let $N(t)$ be a zero-mean, wide-sense stationary noise bandlimited to $W_0 \leqslant |\omega| \leqslant W_0 + W_N$ and have the representation of (2.6-17) with $\omega_0 = W_0 + (W_N/2)$. (a) Show that the cross-correlation functions $R_{N_cN_s}(\tau)$ and $R_{N_sN_c}(\tau)$ are both bandlimited to $|\omega| \leqslant W_N/2$. [*Hint:* Assume (B.8-3e) applies and prove (B.8-3g) is true.] (b) Show that the cross-correlation functions are both zero when $\tau = 0$.

★2-41. Prove that $\tilde{N}(t)$, given by (2.6-18), equals $N(t)$ of (2.6-17) with zero mean-squared error; that is, prove that $\overline{\varepsilon^2} = E\{[N(t) - \tilde{N}(t)]^2\} = 0$. (*Hint:* Use results of Prob. 2-40.)

★2-42. Use (2.7-5) and (2.7-6) in (2.7-3) with $p = 2$, $\tau_1 = 0$, $\tau_2 = \tau$ and $T_{s_2} = T_{s_1}$ and allow $\tau \to 0$. Show that [2]

$$f(t) = \sum_{n=-\infty}^{\infty} \{f(nT_{s_1}) + (t - nT_{s_1})\dot{f}(nT_{s_1})\}\mathrm{Sa}^2\left[\frac{W_f(t - nT_{s_1})}{2}\right]$$

in the limit, where $\dot{f}(t) = df(t)/dt$. This result shows that $f(t)$ can be recovered from a sequence of its samples and samples of its time derivative, each taken at a rate $\omega_{s_1} = W_f$, or half the Nyquist rate. The average number of samples per second still equals the Nyquist rate, however, because there are two samples being taken in each sampling interval.

★2-43. A lowpass function $f(t, u)$ has a Fourier transform $F(\omega, \sigma)$ that is nonzero only over a diamond shaped region having two of its points on the ω axis at $\pm W_{f\omega}$ and the other two points on the σ axis at $\pm W_{f\sigma}$. (a) If sampling rates are $\omega_{su} = 3W_{f\sigma}$ and $\omega_{st} = 2.3W_{f\omega}$, find the sampling efficiency. (b) To what value will sampling efficiency increase if *both* sampling rates are reduced to the smallest allowed values?

★2-44. Work Prob. 2-43 except let $F(\omega, \sigma)$ be nonzero over an elliptically shaped region in the ω, σ plane.

★2-45. Work Prob. 2-43 except assume $F(\omega, \sigma) \neq 0$ only over a circular region in the ω, σ plane with radius $W_{f\omega} = W_{f\sigma} = W_f$.

2-46. Devise a method of interlacing samples at one summing junction so that Nyquist rate sampling of five signals occurs when four signals are bandlimited to 5 kHz and one is bandlimited to 20 kHz.

2-47. Work Prob. 2-46 using two summing junctions, one that sums samples of only the four 5-kHz signals and a second that multiplexes only samples of the 20-kHz signal and the output line of the first multiplexer. Discuss synchronization of the two multiplex operations.

2-48. A TDM system uses flat-top sampling pulses 0.7 μs wide. If a guard time of 0.34 μs is allowed and $N = 120$ telephone messages are multiplexed, what is the largest allowable message spectral extent?

Chapter 3

Baseband Digital Waveforms

3.0 INTRODUCTION

After the somewhat extensive discussions in the preceding chapter on sampling theorems, it seems appropriate to pause and place the direction of future developments in perspective. Again we refer to Fig. 1.3-1.†

The overall purpose of this chapter is to describe the various functions of Fig. 1.3-1 as they relate to a *baseband* digicom system. We shall be concerned primarily with the analog-to-digital (A/D) converter, source and channel encoders, and the generation of the transmitted waveform (in the modulator). Discussions of the receiving subsystem functions are brief because they are basically just the inverse operations to those in the transmitter subsystem. When the transmitter functions are developed, the use and implementations of the inverse operations should be more or less obvious to the reader.

Our discussions ultimately lead to the description of a number of important digital waveforms for transmission over the channel. However, because one of the significant advantages of a digicom system is its ability to interlace in time, or *time-multiplex,* digital waveforms of many messages, we also introduce the basic elements of this technique.

---

† Other functions are sometimes included in the digicom system. Later we mention time-multiplexing of many waveforms. Other functions, such as data encryption and frequency spreading (*spread spectrum*), are not covered here. Sklar [1] has given a good summary of these functions.

Because the digital conversion of analog signals is the only operation in the transmitter subsystem that involves analog waveforms and since the preceding sampling theory forms the initial part of this operation, we begin with this topic. All succeeding discussions can then deal entirely with digital concepts.

3.1 DIGITAL CONVERSION OF ANALOG MESSAGES

The A/D conversion of an analog message involves first sampling the message and then quantizing the samples. We assume the message $f(t)$ to be bandlimited to W_f (rad/s) so that it can be reconstructed without error from its samples, according to the sampling theorem (Chap. 2), if samples are taken at a rate W_f/π (samples/s) or faster. Thus sampling produces no error in the reconstruction of the message, in principle. However, the process of quantization, the rounding of sample values to give a finite number of discrete values, discards some information present in the continuous samples, and the reconstructed signal can be only as good as the quantized samples allow. In other words, there remains some error, the *quantization error,* that is related to the number of levels used in the quantizer.

Quantization of Signals

Let the analog message $f(t)$ be modeled as a random waveform. Define the probability density function (Appendix B) of $f(t)$ at any given time t as $p_f(f)$. A possible function is illustrated in Fig. 3.1-1. In (a) we have a message that possesses definite extreme values, whereas the message of (b) has no well-defined extremes, such as a Gaussian signal.

The process of quantization subdivides the range of values of f into discrete intervals. If a particular sample value of $f(t)$ falls anywhere in a given interval it is assigned a single discrete value corresponding to that interval. In Fig. 3.1-1 the intervals fall between boundaries denoted by $f_1, f_2, \ldots, f_{L+1}$, where we assume L intervals. The quantized (assigned) values are denoted by $l_1, l_2, \ldots, l_L$ and are called *quantum levels.* The width of a typical interval is $f_{i+1} - f_i$ and is called the interval's *step size.* If all step sizes are equal and, in addition, the quantum level separations $l_{i+1} - l_i$ are all the same (constant), we have a *uniform quantizer;* otherwise we have a *nonuniform* quantizer.

If all values of a message do not fall in the range of the quantizer's intervals, which in Fig. 3.1-1(b) is from f_1 to f_{L+1}, these values *saturate* or *amplitude overload* the quantizer.† The message of (a) does not overload

---

† A practical quantizer for this type of message must be designed purposely to allow a small, controlled amount of overload.

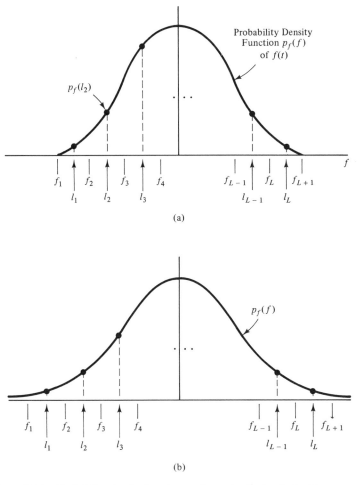

Figure 3.1-1. Probability density functions and quantization levels for (a) messages with well-defined extreme values and (b) those without such extremes.

the quantizer but could if the quantizer levels were established on the basis of a certain message power level and then the incoming message's power suddenly increased. An increase in message power corresponds to an attendant spread in the density function $p_f(f)$. Thus a quantizer must be designed not only for a *form* of message density but for a specific density that results when the message has a design power level. We call this power level the signal's *reference power level*, denoted by $\overline{f_r^2(t)}$.

Even when $f(t)$ does not overload the quantizer, there is still error—the rounding, or quantization, error—associated with every interval of the quantizer. Consider an exact sample value f that falls in interval i, that is,

$f_i \leq f \leq f_{i+1}$. The quantizer output is level l_i, which differs from the exact sample value by the *error*

$$\varepsilon_i = f - l_i, \qquad i = 1, 2,..., L. \qquad (3.1\text{-}1)$$

Once quantized levels l_i are generated and transmitted by whatever system is used, the best that any receiver can ever hope to do is recover these levels exactly. Actual message values can never be recovered. When a receiver reconstructs the message from its quantized levels, it will contain errors related to the various errors ε_i that occur during quantization. These errors place a limit on the performance of the overall system.

Quantization Error and Its Performance Limitation

We are interested in finding an overall *mean-squared quantization error*, denoted by $\overline{\varepsilon_q^2}$, which results from quantization. Let $p_f(f)$ represent the probability density function of $f(t)$ and consider a quantizer defined by intervals and levels shown in Fig. 3.1-1. The mean value and power in the analog signal can be written as†

$$\overline{f(t)} = \bar{f} = \int_{-\infty}^{\infty} f p_f(f)\, df = \sum_{i=1}^{L} \int_{f_i}^{f_{i+1}} f p_f(f)\, df \qquad (3.1\text{-}2)$$

$$\overline{f^2(t)} = \overline{f^2} = \int_{-\infty}^{\infty} f^2 p_f(f)\, df = \sum_{i=1}^{L} \int_{f_i}^{f_{i+1}} f^2 p_f(f)\, df. \qquad (3.1\text{-}3)$$

The quantizer output for any one sample can be treated as a discrete random variable, denoted by f_q, that has possible levels $l_1, l_2, \ldots, l_L$. The probability that f_q will have a typical value, say l_i, is just the probability, denoted by P_i, that f falls in interval i. The mean value and power present in the quantizer output are obtained by averaging the discrete random variable f_q over all its quantum levels:

$$\bar{f_q} = \sum_{i=1}^{L} l_i P_i = \sum_{i=1}^{L} l_i \int_{f_i}^{f_{i+1}} p_f(f)\, df \qquad (3.1\text{-}4)$$

$$\overline{f_q^2} = \sum_{i=1}^{L} l_i^2 P_i = \sum_{i=1}^{L} l_i^2 \int_{f_i}^{f_{i+1}} p_f(f)\, df \qquad (3.1\text{-}5)$$

where

$$P_i = \int_{f_i}^{f_{i+1}} p_f(f)\, df. \qquad (3.1\text{-}6)$$

Of course, it would be desirable for the mean value of the quantizer

---

† Interval boundaries f_1 and f_{L+1} are shown finite for ease of writing equations. All practical quantizers assign levels l_1 and l_L when $f \leq f_2$ and $f_L < f$, respectively; this result is equivalent to setting $f_1 = -\infty$ and $f_{L+1} = \infty$, which we assume in some of the following work.

output to equal the mean value of the analog input signal. If we require this to be true, (3.1-4) must equal (3.1-2). By solving the equality we have

$$l_i \int_{f_i}^{f_{i+1}} p_f(f)\, df = \int_{f_i}^{f_{i+1}} f p_f(f)\, df, \qquad i = 1, 2,\ldots, L, \qquad (3.1\text{-}7)$$

or

$$l_i = \frac{\displaystyle\int_{f_i}^{f_{i+1}} f p_f(f)\, df}{\displaystyle\int_{f_i}^{f_{i+1}} p_f(f)\, df}, \qquad i = 1, 2, \ldots, L. \qquad (3.1\text{-}8)$$

In other words, the quantizer should choose its quantum levels equal to the conditional mean values of each interval.

We are now able to find the mean-squared quantization error. From (3.1-1) the mean-squared error when the sample value falls in interval i is

$$\overline{\varepsilon_i^2} = \int_{f_i}^{f_{i+1}} (f - l_i)^2 p_f(f|i)\, df \qquad (3.1\text{-}9)$$

where $p_f(f|i)$ is the probability density of f given that f falls in interval i. It is given by

$$p_f(f|i) = \frac{p_f(f)}{P_i}. \qquad (3.1\text{-}10)$$

By averaging these mean-squared errors over all intervals we obtain the overall mean-squared error

$$\overline{\varepsilon_q^2} = \sum_{i=1}^{L} \overline{\varepsilon_i^2}\, P_i = \sum_{i=1}^{L} \int_{f_i}^{f_{i+1}} (f - l_i)^2 p_f(f)\, df. \qquad (3.1\text{-}11)$$

By expanding the right side of (3.1-11) and substituting (3.1-7) and (3.1-5), we have

$$\overline{\varepsilon_q^2} = \overline{f^2} - \overline{f_q^2}. \qquad (3.1\text{-}12)$$

In other words, the overall quantization mean-squared error equals the difference between the power in the analog signal input to the quantizer and the average power in the quantizer output.

To gain insight into quantization error and to interpret (3.1-12), we note that sampling and quantizing samples is equivalent to quantizing first and then sampling. By using the latter viewpoint we construct a possible quantized waveform prior to sampling, as illustrated in Fig. 3.1-2. For illustrative purposes we assume a uniform quantizer with step size δv. Clearly, sampling the quantized message $f_q(t)$ is the same as sampling the waveform

$$f_q(t) = f(t) + \varepsilon_q(t) \qquad (3.1\text{-}13)$$

where $-\varepsilon_q(t)$ is an equivalent quantization error waveform having sample

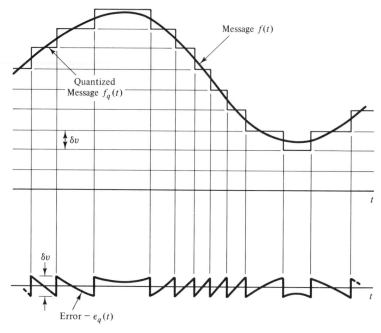

Figure 3.1-2. A message, its quantized version, and the quantization error [2].

values given by (3.1-1). A receiver can reconstruct only $f_q(t)$ from its samples. The power in $f_q(t)$ is

$$S_q = \overline{f_q^2(t)} = \overline{f_q^2}.$$ (3.1-14)

We may define a signal-to-quantization noise ratio for the receiver-recovered message according to

$$\frac{S_q}{N_q} = \frac{\overline{f_q^2}}{\overline{\varepsilon_q^2}} = \frac{\overline{f^2}}{\overline{\varepsilon_q^2}} - 1,$$ (3.1-15)

where (3.1-12) has been used and

$$N_q = \overline{\varepsilon_q^2}$$ (3.1-16)

is treated as a noise power present in the output.

Typically, $\overline{f^2}/\overline{\varepsilon_q^2} \gg 1$ in any practical system, so (3.1-15) can be written as

$$\left(\frac{S_q}{N_q}\right) \approx \left(\frac{S_o}{N_q}\right) = \frac{\overline{f^2}}{\overline{\varepsilon_q^2}}$$ (3.1-17)

where

$$S_o = \overline{f^2} = \overline{f^2(t)}$$ (3.1-18)

is the power in an undistorted message. Overall system performance is limited by the mean-squared quantization noise power $N_q = \overline{\varepsilon_q^2}$.

3.2 DIRECT QUANTIZERS FOR A/D CONVERSION

Uniform Quantizers—Nonoptimum

We next examine a quantizer with a *large* number of quantum levels having constant step size $\delta v = f_{i+1} - f_i$, $i = 1, 2, \ldots, L$. In the next subsection we shall return to this *uniform quantizer* to discuss its performance with a smaller number of levels and see how it can be optimized by minimizing its mean-squared error. For the present discussion there will be L levels centered in equally spaced intervals between two extreme values $f_{\min}$ and $f_{\max}$. Thus

$$\delta v = \frac{f_{\max} - f_{\min}}{L} \tag{3.2-1}$$

$$l_i = f_{\min} + \left(i - \frac{1}{2}\right)\delta v, \qquad i = 1, 2, \ldots, L. \tag{3.2-2}$$

Mean-squared quantization error, from (3.1-11), becomes

$$\overline{\varepsilon_q^2} = \int_{-\infty}^{f_{\min}} (f - l_1)^2 p_f(f)\, df + \int_{f_{\min}}^{f_2} (f - l_1)^2 p_f(f)\, df + \cdots$$
$$+ \int_{f_L}^{f_{\max}} (f - l_L)^2 p_f(f)\, df + \int_{f_{\max}}^{\infty} (f - l_L)^2 p_f(f)\, df. \tag{3.2-3}$$

The first and last terms represent overload noise,† which we denote by $N_{q,ol}$. If δv is small enough (large L) so that $p_f(f) \approx$ constant $= p_f(l_i)$ for each interval and $N_{q,ol}$ is relatively small, it is easy to show that the middle terms in (3.2-3) are approximately equal to $(\delta v)^2/12$ (see Prob. 3-5). Hence

$$\overline{\varepsilon_q^2} = \frac{(\delta v)^2}{12} + N_{q,ol}, \tag{3.2-4}$$

where

$$N_{q,ol} = \int_{-\infty}^{f_{\min}} (f - l_1)^2 p_f(f)\, df + \int_{f_{\max}}^{\infty} (f - l_L)^2 p_f(f)\, df. \tag{3.2-5}$$

Signal-to-quantization noise power ratio becomes

$$\left(\frac{S_o}{N_q}\right) = \frac{12\,\overline{f^2(t)}}{(\delta v)^2} \cdot \frac{1}{1 + [12\,N_{q,ol}/(\delta v)^2]}. \tag{3.2-6}$$

For a given message power $\overline{f^2(t)}$, (3.2-6) clearly shows that performance without overload is better for smaller step size δv. At the same time the

---

† Note that errors relative to level 1 corresponding to $f_{\min} < f \leq f_2$ are not treated as overload errors. Only errors relative to l_1 corresponding to $f \leq f_{\min}$ are overload. Thus interval 1 is broken into two parts for convenience in defining overload. A similar division of interval L has been adopted.

relative effect of a given overload noise power $N_{q,ol}$ is enhanced by smaller step size.

A special, but very useful, case of (3.2-6) can be given for messages having a definite maximum amplitude. If the message has zero mean with a symmetric probability density function and crest factor K_{cr}, we let $|f_r(t)|_{max}$, the maximum magnitude of $f(t)$ at its reference power level,† establish the extreme range of the quantizer. In this case we have $-f_{min} = f_{max} = |f_r(t)|_{max}$ and

$$\delta v = \frac{2|f_r(t)|_{max}}{L} = \left(\frac{2K_{cr}}{L}\right)\sqrt{\overline{f_r^2(t)}} \qquad (3.2\text{-}7)$$

so

$$\left(\frac{S_o}{N_q}\right) = \frac{3L^2}{K_{cr}^2}\left[\frac{\overline{f^2(t)}}{\overline{f_r^2(t)}}\right]\frac{1}{1 + [3L^2 N_{q,ol}/K_{cr}^2 \overline{f_r^2(t)}]}. \qquad (3.2\text{-}8)$$

Here $N_{q,ol} = 0$ if $\overline{f^2(t)} \leqslant \overline{f_r^2(t)}$.

Example 3.2-1

We let $f(t)$ be uniformly distributed on $(-|f(t)|_{max}, |f(t)|_{max})$ and solve (3.2-8). Message power is

$$\overline{f^2(t)} = \int_{-|f|_{max}}^{|f|_{max}} f^2\frac{df}{2|f|_{max}} = \frac{|f|_{max}^2}{3}.$$

Reference level power is then $\overline{f_r^2(t)} = |f_r(t)|_{max}^2/3$. Hence $K_{cr}^2 = 3$. As long as $\overline{f_r^2(t)} \geqslant \overline{f^2(t)}$, we have $N_{q,ol} = 0$ so

$$\left(\frac{S_o}{N_q}\right) = L^2\left[\frac{\overline{f^2(t)}}{\overline{f_r^2(t)}}\right], \qquad \frac{\overline{f^2(t)}}{\overline{f_r^2(t)}} \leqslant 1.$$

From (3.2-5) we have

$$N_{q,ol} = \int_{-|f|_{max}}^{-|f_r|_{max}}\left[f + |f_r|_{max} - \frac{\delta v}{2}\right]^2\frac{df}{2|f|_{max}}$$
$$+ \int_{|f_r|_{max}}^{|f|_{max}}\left[f - |f_r|_{max} + \frac{\delta v}{2}\right]^2\frac{df}{2|f|_{max}}, \qquad |f|_{max} > |f_r|_{max}.$$

Solutions of these integrals are straightforward. By using the fact that $2|f_r(t)|_{max}/\delta v = L$, (3.2-8) becomes

$$\left(\frac{S_o}{N_q}\right) = \frac{L^2 R^2}{1 + \frac{3L^2}{R}\left[\frac{R^3}{3} - R^2\left(\frac{L-1}{L}\right) + R\left(\frac{L-1}{L}\right)^2 - \left(\frac{L^2 - 3L + 3}{3L^2}\right)\right]},$$

---

† Reference power level is the maximum power level for which the quantizer does not saturate, that is, $N_{q,ol} = 0$ for $\overline{f^2(t)} \leqslant \overline{f_r^2(t)}$. Crest factor for a signal $f(t)$ is defined by $K_{cr}^2 = |f(t)|_{max}^2/\overline{f^2(t)}$.

for $R > 1$, where we define

$$R^2 = \frac{\overline{f^2(t)}}{\overline{f_r^2(t)}}.$$

A plot of (S_o/N_q) is shown in Fig. 3.2-1 for $L = 16$ and 128. Note that (S_o/N_q) decreases abruptly as message power $\overline{f^2(t)}$ exceeds the reference (design) level $\overline{f_r^2(t)}$, which corresponds to the onset of amplitude overload.

The previous example served to illustrate the use of (3.2-8), which applies to messages having a well-defined maximum magnitude. For messages with no such maximum (Gaussian, for example), one can still use (3.2-6) when L is large. However, there now exists the problem of where to set the extreme quantum levels (l_1 and l_L) in relation to the probability density function of the message when it is at some reference power level. For a given number of levels (either large or finite) too small a quantization range results in excessive overload noise power. A large range gives less saturation but gives increased in-range quantization error. These extremes suggest that an optimum design may exist somewhere in between that will minimize overall quantization error power. We next consider this problem.

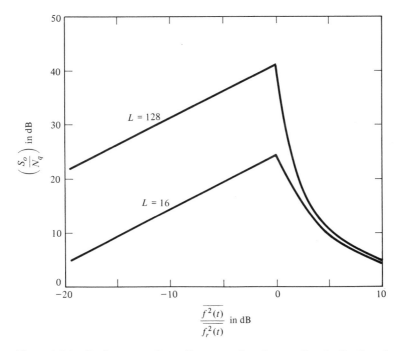

Figure 3.2-1. Performance of a uniform quantizer for a uniformly distributed input message.

Optimum Quantizers

It is well known [3, 4, 5] that quantizers having nonuniformly separated quantum levels can be found that give smaller mean-squared quantization error, $\overline{\varepsilon_q^2}$, than those having uniform levels. Equation (3.1-11) gives $\overline{\varepsilon_q^2}$ for any quantizer and can be written as

$$\overline{\varepsilon_q^2} = \int_{-\infty}^{f_2} (f - l_1)^2 p_f(f)\, df + \sum_{i=2}^{L-1} \int_{f_i}^{f_{i+1}} (f - l_i)^2 p_f(f)\, df$$

$$+ \int_{f_L}^{\infty} (f - l_L)^2 p_f(f)\, df. \qquad (3.2\text{-}9)$$

The optimum quantizer is one that chooses levels l_i and interval boundaries f_i so that $\overline{\varepsilon_q^2}$ is minimized.

By applying Leibniz's rule, (3.2-9) can be differentiated to obtain a set of necessary conditions that must hold for the optimum quantizer [5]:

$$\frac{\partial \overline{\varepsilon_q^2}}{\partial l_i} = -2 \int_{f_i}^{f_{i+1}} (f - l_i) p_f(f)\, df = 0, \qquad i = 1, 2, \ldots, L, \qquad (3.2\text{-}10)$$

$$\frac{\partial \overline{\varepsilon_q^2}}{\partial f_i} = [(f_i - l_{i-1})^2 - (f_i - l_i)^2] p_f(f_i) = 0, \qquad i = 2, 3, \ldots, L. \quad (3.2\text{-}11)$$

where $f_1 = -\infty$ and $f_{L+1} = \infty$. Equation (3.2-11) is equivalent to

$$f_i = \frac{(l_i + l_{i-1})}{2}, \qquad i = 2, 3, \ldots, L, \qquad (3.2\text{-}12)$$

which says that interval boundaries f_i should fall midway between the adjacent quantum levels. An alternative form for (3.2-12) is

$$l_i = 2f_i - l_{i-1}, \qquad i = 2, 3, \ldots, L. \qquad (3.2\text{-}13)$$

Equation (3.2-10) is readily solved for l_i

$$l_i = \frac{\displaystyle\int_{f_i}^{f_{i+1}} f p_f(f)\, df}{\displaystyle\int_{f_i}^{f_{i+1}} p_f(f)\, df}, \qquad i = 1, 2, \ldots, L. \qquad (3.2\text{-}14)$$

This result was previously found in (3.1-8) to be necessary for the mean value of the quantizer's output to equal the mean value of the input message, a fact noted by Bucklew, et al. [6] and others [4, 7].

The solution of (3.2-13) and (3.2-14) for the general nonuniform quantizer is difficult. However, a procedure to obtain a solution by computer iteration has been given [4, 5]. For a specified message probability density $p_f(f)$ and fixed value of L, l_1 is first selected arbitrarily (it should be a value in the tail of $p_f(f)$ that corresponds to small f). With $f_1 = -\infty$, we solve

(3.2-14) for f_2. Next, f_2 and l_1 are used in (3.2-13) to obtain l_2. The cycle is then repeated to obtain f_3 from (3.2-14) and l_3 from (3.2-13). Continued iteration finally stops when l_L is obtained from (3.2-13). If l_1 has been correctly guessed, then l_L will satisfy (3.2-14) with $f_{L+1} = \infty$. If it does not, l_1 is corrected to a new value and the process repeated until l_L satisfies (3.2-14). This procedure satisfies the conditions (3.2-13) and (3.2-14), which are necessary for optimality. Sufficient conditions have also been given for the equations to be the optimal set [8]. Recently, more general work has proved the existence of optimal quantizers [9–11].

Max [5] used the above procedure to find the quantum levels for a zero-mean Gaussian message and all values of L from 1 to 36. Paez and Glisson [12] used the procedure to find optimum levels for signals having either a gamma density,† given by

$$p_f(f) = \frac{\sqrt{k}\, e^{-k|f|}}{2\sqrt{\pi|f|}},$$ (3.2-15)

where

$$k = \left[\frac{0.75}{\overline{f_r^2(t)}} \right]^{1/2},$$ (3.2-16)

or a Laplace density given by

$$p_f(f) = \frac{\alpha}{2}\, e^{-\alpha|f|},$$ (3.2-17)

where

$$\alpha = \left[\frac{2}{\overline{f_r^2(t)}} \right]^{1/2}.$$ (3.2-18)

Levels in [12] are given for $L = 2, 4, 8, 16,$ and 32. For these optimum nonuniform quantizers, Fig. 3.2-2 illustrates where the highest quantum level l_L is set in relation to the rms value $[\overline{f_r^2(t)}]^{1/2}$ of the reference signal for which the quantizer is optimized.‡ The resulting signal-to-quantization noise power ratio is shown in Fig. 3.2-3. We note that the curves are not more than 3.2 dB apart for any value of L shown. Thus performance does not change radically for the densities illustrated. However, the required ratio of l_L to $[\overline{f_r^2(t)}]^{1/2}$, called the *loading factor*, changes greatly with both L and message type.

The procedure of Max [5] can also be applied under a constraint that steps $l_i - l_{i-1}$ are constant for $i = 2, 3, \ldots, L$. The result is an *optimum uniform quantizer*. Figures 3.2-4 and 3.2-5 illustrate loading factor and (S_o/N_q) for Gaussian, Laplace, and gamma densities. Performance now

---

† The gamma density is a reasonable approximation to real speech messages [13, 12].
‡ Because of symmetry in $p_f(f)$, $l_1 = -l_L$.

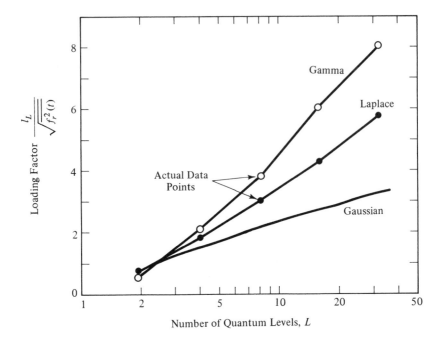

Figure 3.2-2. Loading factor for optimum nonuniform quantizer for several types of message (after [5, 12]).

depends more heavily on the type of message than in the nonuniform quantizers, but loading factor is not as sensitive to message type.

Other comparisons of uniform and nonuniform optimum quantizers are given in [14]. For large values of L, an approximation of Mauersberger [15] for the minimum value of $\bar{\varepsilon}_q^2$ applicable to a nonuniform quantizer can be used to extend the curves of Fig. 3.2-3. We have

$$\left(\frac{S_o}{N_q}\right) \approx \frac{L^2}{1 + [a(\nu) - 1][1 - (N_b + 1)e^{-N_b}]}, \qquad (3.2\text{-}19)$$

where

$$N_b = \log_2(L) \qquad (3.2\text{-}20)$$

$$a(\nu) = \frac{27^{1/\nu} [\Gamma(1 + 1/\nu)]^3}{\Gamma(1 + 3/\nu)} \qquad (3.2\text{-}21)$$

and ν is a real number that determines the form of $p_f(f)$ through the *generalized Gaussian distribution* [15]

$$p_f(f) = \frac{\nu\,\eta\,\exp(-\,\eta|f|^\nu)}{2\,\Gamma(1/\nu)}. \qquad (3.2\text{-}22)$$

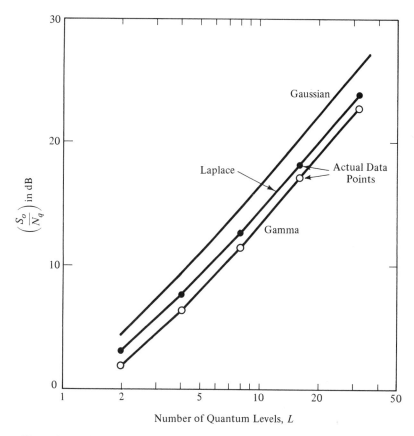

Figure 3.2-3. Performance of optimum nonuniform quantizer for several types of message (after [5, 12]).

Here

$$\eta = \left[\frac{\Gamma(3/\nu)}{\Gamma(1/\nu)\,\overline{f_r^2(t)}} \right]^{1/2}, \qquad (3.2\text{-}23)$$

and $\Gamma(x)$ is the gamma function [16]. When $\nu = 1$ and 2, (3.2-22) gives the Laplace and Gaussian densities, respectively. As $\nu \to \infty$, $p_f(f)$ tends to the uniform distribution [15]. Equation (3.2-19) is claimed to be accurate to about 2.5% (or 0.11 dB) for $\nu > 0.6$ with accuracy degrading to about 15% (or 0.61 dB) for $\nu = 0.3$.

Example 3.2-2
We shall use (3.2-19) to verify the performance of a nonuniform quantizer for a message having a Laplace density and $L = 32$ levels. Here $\nu = 1$, so (3.2-21)

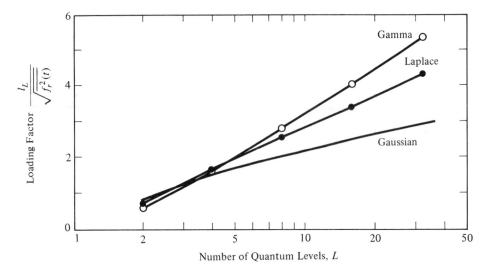

Figure 3.2-4. Loading factors for optimum uniform quantizer for several types of message (after [5, 12]).

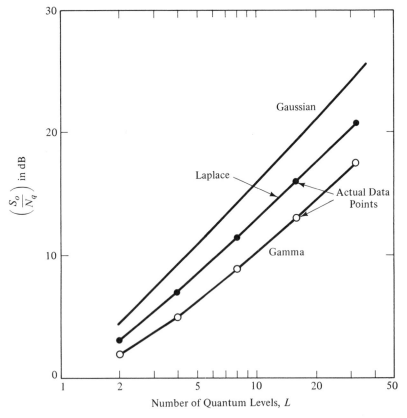

Figure 3.2-5. Performance of optimum uniform quantizer for several types of message (after [5, 12]).

gives

$$a(v) = \frac{27\Gamma(2)}{\Gamma(4)} = \frac{27}{3!} = \frac{27}{6}.$$

From (3.2-20) $N_b = \log_2(32) = 5$. Thus

$$\left(\frac{S_o}{N_q}\right) \approx \frac{(32)^2}{1 + \left[\frac{27}{6} - 1\right][1 - 6e^{-5}]} = 234.943 \quad \text{(or 23.71 dB)}.$$

The actual value in Fig. 3.2-3 is 23.83 dB, from [12]. The approximate result is then accurate to -0.12 dB if the result in [12] is assumed to be the true value.

When the message has a uniform density it can be shown by considering (3.2-14) that the optimal quantizer is uniform.

3.3 COMPANDED QUANTIZERS

Compandors and Their Optimization

Let us reconsider (3.1-11), the basic equation describing the mean-squared quantization error $\overline{\varepsilon_q^2}$. Again, (3.2-13) and (3.2-14) become conditions that must be satisfied by any quantizer giving minimum mean-squared error. We now assume that L is large enough that $p_f(f)$ is nearly constant in every interval and that the message at its reference power level has definite extreme values, so that $f_1 \leq f_r(t) \leq f_{L+1}$. This last assumption can be relaxed if $p_f(f)$ exceeds the extremes f_1 and f_{L+1} with negligible probability.

If Δ_i is defined as the width of the quantization intervals—that is, if $\Delta_i \triangleq f_{i+1} - f_i$—we find from (3.2-14) that l_i falls in the center of the interval. Thus

$$f_{i+1} - l_i = \frac{\Delta_i}{2} \tag{3.3-1}$$

$$f_i - l_i = -\frac{\Delta_i}{2} \tag{3.3-2}$$

for $i = 1, 2, \ldots, L$. With these results and our assumptions, (3.1-11) becomes

$$\overline{\varepsilon_q^2} = \sum_{i=1}^{L} \int_{f_i}^{f_{i+1}} (f - l_i)^2 p_f(f) \, df \approx \sum_{i=1}^{L} p_f(l_i) \int_{f_i}^{f_{i+1}} (f - l_i)^2 \, df \tag{3.3-3}$$

$$= \frac{1}{12} \sum_{i=1}^{L} \Delta_i^2 \, p_f(l_i)\Delta_i.$$

At this point we observe that a nonuniform quantizer can be constructed by first passing the message through a suitable nonlinear network and then using a *uniform* quantizer [3, 7]. Figure 3.3-1 illustrates the network's characteristic, denoted by $v(f)$, and the way in which input and output

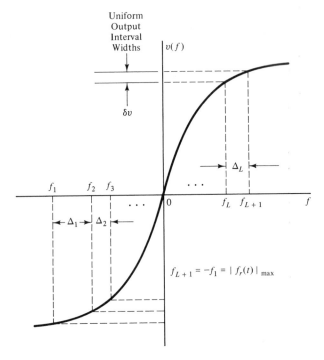

Figure 3.3-1. Compressor characteristic for a bounded message with symmetric probability density function.

quantization intervals are related. A bounded message with symmetrical density, $p_f(-f) = p_f(f)$, is assumed in the sketch. The network is called a *compressor* because its action is to compress message peaks and enhance small amplitudes. In effect it decreases the crest factor of the message. After reconstruction in the receiver of the quantized samples, it is necessary to use an *expandor* having the inverse characteristic $v^{-1}(\cdot)$. The combination of the *compressor* and *expandor* is called a *compandor*, and the system is said to use companding.

If the slope of $v(f)$ is defined as $w(f)$, then

$$w(f) = \frac{dv(f)}{df} \approx \frac{\delta v}{\Delta_i} \tag{3.3-4}$$

where δv denotes the constant step size of the uniformly quantized output. Since L is large, all interval widths Δ_i are small, so (3.3-3) can be written as

$$\overline{\varepsilon_q^2} \approx \frac{(\delta v)^2}{12} \int_{f_1}^{f_{L+1}} w^{-2}(f) p_f(f) \, df \tag{3.3-5}$$

which we seek to minimize by finding the optimum $v(f)$. The necessary procedures are beyond our scope,† so we shall give only the results. For a symmetrical message where $p_f(-f) = p_f(f)$, the compressor characteristic has odd symmetry, $v(-f) = -v(f)$, and the minimum $\overline{\varepsilon_q^2}$ and optimum $v(f)$ are given by [3, 4, 7, 18]

$$\overline{\varepsilon_q^2} \approx \frac{2}{3L^2} \left[\int_0^V p_f^{1/3}(f) \, df \right]^3 \tag{3.3-6}$$

$$v(f) = \frac{V \int_0^f p_f^{1/3}(\xi) \, d\xi}{\int_0^V p_f^{1/3}(f) \, df}, \qquad 0 \leqslant f \leqslant V. \tag{3.3-7}$$

Here V is a constant defining the extreme input-output levels according to

$$-V = f_1 \leqslant f \leqslant f_{L+1} = V, \tag{3.3-8a}$$

$$-V = v(f_1) \leqslant v(f) \leqslant v(f_{L+1}) = V \tag{3.3-8b}$$

and

$$\delta v = \frac{2V}{L}. \tag{3.3-9}$$

To summarize, (3.3-7) is the compressor characteristic that gives the smallest possible mean-squared quantization error $\overline{\varepsilon_q^2}$, given by (3.3-6), when a large number of fixed levels is used. The message is assumed to have zero mean, a symmetric probability density, and to be bounded such that $|f(t)| < V$. The quantizer's boundaries are set at $\pm V$ according to (3.3-8). If the message can have amplitudes of magnitude larger than V, the levels $-V$, V need to be set so that the probability of these amplitudes giving overload is negligible.

Of course, the minimum value of $\overline{\varepsilon_q^2}$ occurs only for the message's reference (design) power level $\overline{f_r^2(t)}$. If message power $\overline{f^2(t)}$ falls below or above the reference value the quantizer is said to be *variance (power) mismatched* [19] and $\overline{\varepsilon_q^2}$ increases. The increase for $\overline{f^2(t)} \leqslant \overline{f_r^2(t)}$ is not nearly so severe, however, as when $\overline{f^2(t)} > \overline{f_r^2(t)}$ occurs. *Shape mismatch* can also occur if $p_f(f)$ does not have the same shape (form) as that assumed in quantizer design. Shape mismatch is most serious when $p_f(f)$ has tails that are larger than those for which the quantizer is designed (for equal message powers).

We shall consider an example to demonstrate some of the above points and to illustrate one of the disadvantages of the optimum quantizer.

---

† The procedure uses the method of Lagrange multipliers, which leads in our case to the application of the Euler-Lagrange condition [17, pp. 627–633].

Example 3.3-1

The Laplace density

$$p_f(f) = \frac{1}{\sqrt{2\overline{f^2}}} \exp\left(\frac{-\sqrt{2}\,|f|}{\sqrt{\overline{f^2}}}\right),$$

where $\overline{f^2} = \overline{f^2(t)}$, is a fairly representative model of the statistical characteristics of a speech message. We find the optimum compressor and mean-squared quantization error for this compressor when message power can vary from its reference level $\overline{f_r^2}$ for which the the quantizer is designed.

For $v(f)$ we solve (3.3-7) when the density $p_f(f)$ is defined at the reference power level $\overline{f_r^2}$. Easily solved integrals are involved and the result is [18]

$$v(f) = \frac{V\,[1 - \exp(-\sqrt{2}f/3\sqrt{\overline{f_r^2}})]}{1 - \exp(-\sqrt{2}\,V/3\sqrt{\overline{f_r^2}})}.$$

The slope of $v(f)$, by differentiation, becomes

$$w(f) = \frac{\sqrt{2V^2/9\overline{f_r^2}}\,\exp[-\sqrt{2f^2/9\overline{f_r^2}}]}{1 - \exp[-\sqrt{2V^2/9\overline{f_r^2}}]}.$$

On substitution of $p_f(f)$ *for arbitrary message power* $\overline{f^2}$ and $w(f)$ into (3.3-5) and using (3.3-9), we obtain $\overline{\varepsilon_q^2}$. Signal-to-quantization noise power then is found to be

$$\left(\frac{S_o}{N_q}\right) = \frac{\overline{f^2}}{\overline{\varepsilon_q^2}}$$

$$= \frac{2L^2[3 - 2\sqrt{\overline{f^2}/\overline{f_r^2}}](\overline{f^2}/\overline{f_r^2})}{9\,[1 - \exp(-\sqrt{2V^2/9\overline{f_r^2}})]^2[1 - \exp\{-[\sqrt{\overline{f_r^2}/\overline{f^2}} - (2/3)]\sqrt{2V^2/\overline{f_r^2}}\}]}.$$

This function is plotted in Fig. 3.3-2 for $V/\sqrt{\overline{f_r^2}} = 10, 20,$ and 40. By selecting $L = 128$, for example, the best performance occurs for $\overline{f^2} = \overline{f_r^2}$ and is

$$\left(\frac{S_o}{N_q}\right) \text{ in dB} = 10\log_{10}(128^2) - 6.53 = 35.61 \text{ dB}.$$

It is clear from the curves of Fig. 3.3-2 that power mismatch limits the dynamic range over which $\overline{f^2}$ can vary while maintaining a minimum prescribed value of (S_o/N_q). Mismatch is most severe for $\overline{f^2(t)} > \overline{f_r^2(t)}$. Limited dynamic range is a major disadvantage of the optimum compandor.

Logarithmic Companding

Compressor characteristics can be found that differ from (3.3-7) and give less sensitivity to signal power mismatch than the optimum. Naturally there will be some loss in performance [smaller (S_o/N_q)], but the added dynamic range may make the loss acceptable if it is not too large. The μ-

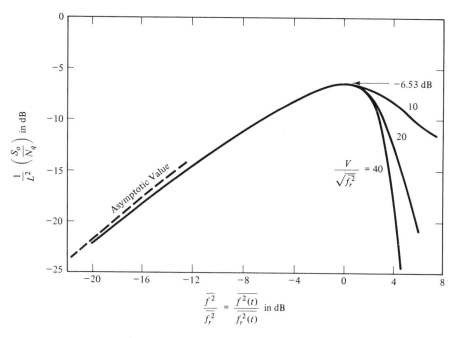

Figure 3.3-2. Performance of optimum quantizer for a message having Laplace probability density function.

law compressor has such desirable properties; its characteristic is

$$v(f) = V \, \text{sgn}(f) \frac{\ln[1 + (\mu|f|/V)]}{\ln(1 + \mu)} \qquad (3.3\text{-}10)$$

where sgn($\cdot$) is the signum function,[†] ln($\cdot$) represents the natural logarithm, $\mu \geq 0$ is a constant, and V is the largest input signal magnitude that can be quantized without overload. Figure 3.3-3 shows the behavior of (3.3-10). In practice the value $\mu = 255$ is used in telephone systems in the United States [20].

If we again assume that the number of levels L is large and the message's probability density is symmetric, we may find quantization error $\overline{\varepsilon_q^2}$ by use of (3.3-5). It can be written as

$$\overline{\varepsilon_q^2} = \frac{V^2}{3L^2} \int_{-V}^{V} w^{-2}(f) p_f(f) \, df, \qquad (3.3\text{-}11)$$

where

$$w(f) = \frac{dv(f)}{df} = \frac{\mu[1 + (\mu|f|/V)]^{-1}}{\ln(1 + \mu)} \qquad (3.3\text{-}12)$$

[†] The signum function is defined by sgn(f) = 1 for $f > 0$ and -1 for $f < 0$.

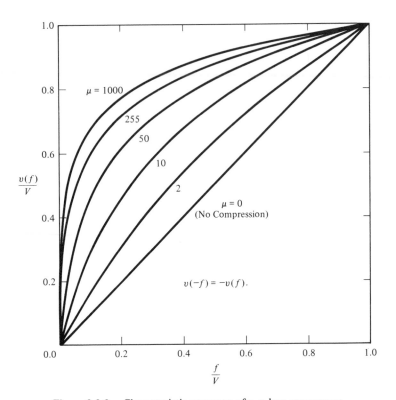

Figure 3.3-3. Characteristic response of a μ-law compressor.

from (3.3-10). By substitution of (3.3-12) in (3.3-11), we obtain

$$\overline{\varepsilon_q^2} = \frac{1}{3}\left[\frac{V\ln(1 + \mu)}{\mu L}\right]^2\left(1 + \frac{2\mu}{V}\,\overline{|f|} + \frac{\mu^2}{V^2}\,\overline{f^2}\right), \tag{3.3-13}$$

where

$$\overline{|f|} = \int_{-V}^{V} |f|p_f(f)\,df. \tag{3.3-14}$$

Hence

$$\left(\frac{S_o}{N_q}\right) = \frac{\overline{f^2}}{\overline{\varepsilon_q^2}} = \frac{3L^2\,[\mu/\ln(1 + \mu)]^2\,\overline{f^2}/V^2}{1 + 2\mu(\overline{|f|}/V) + \mu^2(\overline{f^2}/V^2)}. \tag{3.3-15}$$

To illustrate the behavior of (3.3-15), let $p_f(f)$ be the Laplace density of Example 3.3-1. We find

$$\overline{|f|} = \sqrt{\frac{\overline{f^2}}{2}}\left[1 - \left(1 + \sqrt{\frac{2V^2}{\overline{f^2}}}\right)e^{-\sqrt{2V^2/\overline{f^2}}}\right]. \tag{3.3-16}$$

If $\overline{f^2}/V^2 \to 0$, we have $\overline{|f|} \to [\overline{f^2}/2]^{1/2}$. Figure 3.3-4 illustrates a plot of

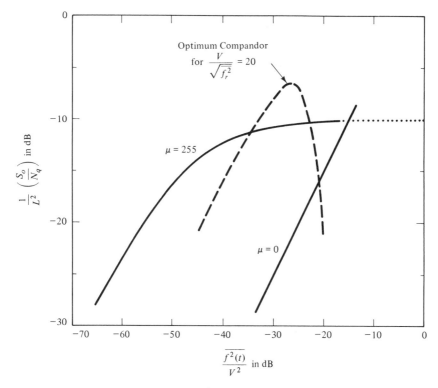

Figure 3.3-4. Performance of μ-law compandor for $\mu = 255$ and $\mu = 0$ when the message has a Laplace probability density. The optimum compandor's curve is shown for comparison.

(3.3-15), when (3.3-16) is used. The case $\mu = 255$ is shown, since this value and the Laplace density both apply to voice transmission. Also shown for comparison is the response of the optimum compandor of Example 3.3-1.† The μ-law compressor results in about a 4-dB loss in performance but clearly has an improved dynamic range‡ compared to either the optimum compandor or a system with no companding ($\mu = 0$ curve).

Another common logarithmic compressor used in Europe has a char-

---

† The abscissa in Fig. 3.3-4 is obtained by subtracting $10 \log_{10}(V^2/\overline{f_r^2})$ from the abscissa of Fig. 3.3-2.

‡ Not shown are the effects of amplitude overload, which become significant in the μ-law device for $\overline{f^2(t)}/V^2$ larger than about -18 dB (for $\mu = 255$ and L up to 256).

acteristic called *A-law* [21] given by

$$
v(f) = \begin{cases} V(1 + \ln A)^{-1}\left(\dfrac{Af}{V}\right), & 0 \le f \le \dfrac{V}{A} \\[3mm] V(1 + \ln A)^{-1}\left[1 + \ln\left(\dfrac{Af}{V}\right)\right], & \dfrac{V}{A} \le f \le V \end{cases} \tag{3.3-17}
$$

$$
v(-f) = -v(f), \tag{3.3-18}
$$

where $A \ge 1$ is a constant. $A = 1$ corresponds to no companding. Curves of (3.3-17) appear similar in form to those of Fig. 3.3-3. *A*-law companding gives a slightly higher signal-to-quantization noise ratio (by tenths of a decibel) and a slightly flatter response within its dynamic range. However, its dynamic range is smaller than that of the μ-law system [22]. These comparisons assume $A = 100$ and $\mu = 255$.

3.4 SOURCE ENCODING OF DIGITAL MESSAGES

The result of the sampling and quantization discussed earlier is a digital signal or message. It is digital in the sense that, for any given time, only a discrete number of symbols (output levels of the quantizer) are possible. Almost all the remaining parts of this chapter will deal with digital signal processing methods designed to work with these digital messages. Of course, these processors do not care if the digital message came from a sampler-quantizer or from a digital source directly. However, because of the way we have chosen to define a digital signal (waveform with a finite number of discrete levels) the first function of the processor is to convert each discrete level into a suitable digital representation. This function is called *source encoding*.

Natural Binary Encoding

The decimal (base 10) number system uses the *digits* 0, 1, . . . , 9. A *binary* (base 2) number system uses only two digits, **0** and **1**, called *binary digits*. The decimal number 13 (a two-digit number) has the equivalent binary representation **1101**, which we call a *codeword*. This codeword has four *bits*† (after *binary digit*), and all possible 4-bit codewords as a group form a *binary code*. An N_b-bit code, or sequence of digits, $b_{N_b} \ldots b_3 b_2 b_1$ can represent any decimal integer number N from 0 to $2^{N_b} - 1$ according to

$$
N = b_{N_b}(2^{N_b - 1}) + \cdots + b_2(2^1) + b_1(2^0). \tag{3.4-1}
$$

---

† We adopt this name due to common usage. Do not confuse our use of bit with the unit of information. Some texts [23] use *binit* as the name of the message *binary digit*. We shall rely on context to resolve any conflicts that may arise in definitions.

The b_ns are either **0** or **1**. Here b_1 is the *least significant bit* (LSB) while b_{N_b} is the *most significant bit* (MSB).

A simple form of source encoding is to assign a binary code to represent the levels available from the source. When codeword digits are arranged in the sequence shown in (3.4-1), the code is a *natural binary code*. Figure 3.4-1 illustrates the assignment of natural codewords to a digital source having not more than eight quantum levels. Actual levels may change every T_s seconds, which is the sampling interval for an analog source or the symbol duration for a digital source. Clearly, an N_b-digit code has $L_b = 2^{N_b}$ distinct codewords. When used with a source with L possible symbols (or levels), we require

$$L \leqslant L_b = 2^{N_b} \tag{3.4-2}$$

for unique coding.

The source levels obviously can represent negative as well as positive sample values. We simply label the levels decimally without regard to the polarity of the message sample. The first column of Table 3.4-1 illustrates a decimal assignment for a 16-level quantization in which half the levels are for positive and half for negative samples. Column 2 shows the corresponding natural binary code. We take an example to illustrate the choice of L, L_b, and N_b.

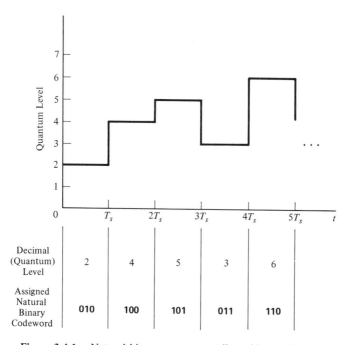

Figure 3.4-1.   Natural binary source encoding with a 3-bit code.

TABLE 3.4-1. Various Binary and Gray Codewords for a 4-Bit Encoder.

Decimal Level Number	Natural Binary Code $b_4b_3b_2b_1$	Folded Binary Code	Inverted Folded Binary	Gray Code $g_4g_3g_2g_1$	Comments
15	1111	1111	1000	1000	↑
14	1110	1110	1001	1001	Levels
13	1101	1101	1010	1011	assigned
12	1100	1100	1011	1010	to
11	1011	1011	1100	1110	positive
10	1010	1010	1101	1111	message
9	1001	1001	1110	1101	samples
8	1000	1000	1111	1100	↓
7	0111	0000	0111	0100	↑
6	0110	0001	0110	0101	Levels
5	0101	0010	0101	0111	assigned
4	0100	0011	0100	0110	to
3	0011	0100	0011	0010	negative
2	0010	0101	0010	0011	message
1	0001	0110	0001	0001	samples
0	0000	0111	0000	0000	↓

↑   ↑   ↑

Sign bits for messages having
positive and negative values

Example 3.4-1

A message has zero mean value and a maximum magnitude $|f(t)|_{max}$ = 10 V. It is to be quantized using a quantum step size of 0.1 V with one level equal to 0 V. We find the necessary values of L, L_b, and N_b.

Since this signal always exists between −10 and +10 V, there are L = (20/0.1) + 1 = 201 levels required. The closest L_b value is 256 from (3.4-2), requiring an N_b = 8-bit code. The actual range of signal that could be handled is (256 − 1)0.1 = 25.5 V, corresponding to −12.5 to 13.0 V, for example. There is then an amplitude margin of safety of 25%.

Other codes are used [20] that are derived from the natural binary code. The *folded binary code* assigns the first (left) digit to sign and the remaining digits to code magnitude, as shown in the third column of Table 3.4-1. This code is superior to the natural code in masking transmission errors when encoding speech. If only the amplitude digits of a folded binary code are complemented (1s changed to 0s and 0s to 1s), an *inverted folded binary code* results (column 4); this code has the advantage of a higher density of ones for small-amplitude signals, which are most probable for voice messages. The higher density of ones relieves some system timing errors but does lead to some increase in crosstalk in multiplexed systems. Advantages outweigh disadvantages, however, and the inverted folded code has been used in some telephone systems [20, p. 1684].

Gray Encoding

With natural binary encoding, a number of codeword digits can change even when a change of only one source level occurs. In Table 3.4-1, for example, a change from level 7 to level 8 entails *every* digit changing in the 4-bit illustrated code. In some applications this behavior is undesirable and a code is desired for which only one digit changes when any transition occurs between adjacent levels. The *Gray code* shown in Table 3.4-1 has this property if we consider extreme levels adjacent.

Digits of the Gray code, denoted by g_k, can be derived from those of the natural binary code by

$$g_k = \begin{cases} b_{N_b}, & k = N_b \\ b_{k+1} \oplus b_k & k < N_b, \end{cases} \qquad (3.4\text{-}3)$$

where $\oplus$ represents modulo-2 addition of binary digits ($0 \oplus 0 = 0$, $0 \oplus 1 = 1$, $1 \oplus 0 = 1$, $1 \oplus 1 = 0$).

The reverse behavior of the Gray code does not hold. That is, a change in any one code digit does *not* necessarily result in a change of only one code level. For example, a change in digit g_4 from **0** when the code is **0001** (level 1) to a **1** now corresponds to level 14 codeword **1001**, a change spanning almost the full quantizer's range.

M-ary Encoding

In some systems it is desirable to encode the L source levels to other than a binary base. For a code with N_M digits with each digit having M possible levels, we require

$$L \leq L_M = M^{N_M} \qquad (3.4\text{-}4)$$

for unique encoding. We refer to this as an *M-ary code*. Table 3.4-2 illustrates a 3-ary (ternary) code that has $N_M = 2$ digits. In the special case where $N_M = 1$, the code base equals the number of message levels.

Source encoding amounts to mapping a source described by one code base to another. In all the digital sources discussed so far, the original source was equivalent to a one-digit, base-L codeword (decimal number representation of the L levels was used) and its codewords were mapped into another base (binary, ternary, *M*-ary). In every case *each* resulting codeword corresponded to a *single* source codeword (sample). There is another way in which *M*-ary encoding finds very practical use; it involves encoding a *sequence* of several source symbols (codewords) into one *M*-ary codeword. Consider a binary source ($L = 2$) that generates one binary

TABLE 3.4-2. Three-ary Encoding of a
Source with L = 9 Levels.

Source Level Number	3-ary Codeword
8	22
7	21
6	20
5	12
4	11
3	10
2	02
1	01
0	00

digit each symbol interval. A sequence of N_b such 1-bit binary words represents $2^{N_b} = L_b$ distinct "words" that are possible during the N_b symbol intervals. This sequence can be encoded into a *single* M-ary code having N_M digits if

$$M^{N_M} = L_M \geq L_b = 2^{N_b}. \tag{3.4-5}$$

Note that with $M = 2$, $N_M = N_b$ corresponds to mapping one binary code to another, such as converting a natural to a folded binary code. Of course, the above ideas are not restricted to binary sources.

Source Information and Entropy

Everyone has some intuitive idea of what information is. It will be necessary, however, to state a quantitative measure or definition of information for our future uses. A weather forecast for 12 days of cool, soaking rain for the Sahara desert would obviously carry more information than the usual forecast of clear, hot, and dry. Thus we would expect information to be larger for messages with *small* probabilities of occurrence. Furthermore, from intuition we would expect that the information contained in two separate and *independent* events should be the sum of the information in both events. Finally, we would expect information to be a positive quantity and be zero when an event is certain to occur. There is only one function satisfying all four of these conditions [23, p. 343]. It is the logarithm, and the chosen base determines the unit of information. Base 2 is commonly used in communication systems because it represents the smallest measure of choice.

Thus we define the information I_k contained in a source symbol having probability of occurrence P_k as

$$I_k = -\log_2(P_k) = \log_2\left(\frac{1}{P_k}\right), \tag{3.4-6}$$

where the unit is the *bit*. The *average* information of a source, called *source entropy,* is obtained by averaging the information in all possible symbols. For L symbols this entropy, denoted by H, is†

$$H = -\sum_{k=1}^{L} P_k \log_2(P_k) = -\frac{1}{\ln(2)} \sum_{k=1}^{L} P_k \ln(P_k) \qquad \text{(bits/symbol).} \qquad (3.4\text{-}7)$$

Example 3.4-2

A quantizer output has four levels that occur with probabilities 0.1, 0.3, 0.4, and 0.2. We find the source entropy to be

$$H = -[0.1 \log_2(0.1) + 0.3 \log_2(0.3) + 0.4 \log_2(0.4) + 0.2 \log_2(0.2)]$$

$$= 1.8464 \text{ bits/symbol.}$$

If the quantizer symbols had equal probabilities, entropy would be

$$H = -4(\tfrac{1}{4})\log_2(\tfrac{1}{4}) = 2.0 \text{ bits/symbol.}$$

Example 3.4-2 shows that the source with equal-probability output symbols has the largest entropy. This will be true in general [24].

Source entropy of (3.4-7) can also be applied to a *sequence* of symbols emitted over time. If the source is stationary‡ and its sequential symbols are statistically independent, the average information in a long string of symbols is again given by (3.4-7). Such a source is called a *zero-memory source.*

Optimum Binary Source Encoding

The entropy of a source, given by (3.4-7), is the smallest average number of binary digits that can be used per symbol to encode the source. Practical encoders typically require a larger average number of digits per symbol, which we denote by $\overline{N}$. We define *code efficiency* by

$$\eta_{\text{code}} = \frac{H}{\overline{N}}. \qquad (3.4\text{-}8)$$

The binary codes discussed earlier all used a fixed number of bits (N_b) for each source symbol; for these codes $\overline{N} = N_b$ and $\eta_{\text{code}} \leq 1$, with the equality holding only if source levels are equally probable and the number of source levels is a power of 2 ($L = L_b = 2^{N_b}$). The efficiency for the source of Example 3.4-2 is $1.8464/2.0 = 0.9232$, or 92.32%, for example.

If we abandon the requirement that codewords be of fixed length, it is possible to obtain more optimum codes that increase code efficiency. Huffman [25] has devised an encoding procedure that is optimum in the sense that it gives the smallest possible average number of binary digits

---

† The second form of (3.4-7) is more convenient in computations than the first. Here $\ln(\cdot)$ stands for the natural logarithm; $\ln(x) = \log_e(x)$.

‡ A stationary source has symbol probabilities that do not change with time.

per source symbol subject to the constraint that the code be uniquely decodable.

The Huffman procedure uses the following steps for binary encoding:

1. Arrange the L source symbols in decreasing order of their probabilities.
2. Combine the two symbols with smallest probabilities using a T junction to form a new symbol with probability equal to the sum of the two probabilities. Label the two input branches of the T with a **0** (lower branch) and a **1**.†
3. Rearrange the original unused symbols and the new symbol in decreasing order of probabilities.
4. Repeat Steps 2 and 3 until finally only one symbol remains with probability 1.0.
5. The codeword for each source symbol is obtained by backtracing through the path generated in Steps 1–4 that leads to the symbol of interest. Each time a T is encountered, a new digit is added to the symbol's codeword.

The Huffman procedure is illustrated in Fig. 3.4-2 for an eight-symbol source. The number of source symbols does not have to be a power of 2, however. Observe that symbols with higher probability are assigned shorter codewords. The average word length becomes $\overline{N}$ = 2(0.35) + 2(0.23) + 3(0.15) + 3(0.10) + 3(0.08) + 4(0.05) + 5(0.03) + 5(0.01) = 2.55 digits per source symbol. A fixed length code would require 3.0 digits per source symbol. From (3.4-7) this source has entropy 2.4863 bits per source symbol, so the Huffman code efficiency is 2.4863/2.55 = 0.975, whereas the 3-bit code gives only 2.4863/3.0 = 0.8288.

The Huffman procedure can also be applied to M-ary encoding [17]. Basically the preceding procedure is used except with T junctions having M input branches labeled **0, 1, . . ., M − 1**. For arbitrary L the last T will have M input branches only if $L = M + n(M − 1)$, where n is an integer. If this relationship is not satisfied, *dummy* source symbols, all with zero probability, are added so that the last T has M inputs. The encoding procedure is then followed but the dummy symbols are discarded at the end.

Source Extension

When the number of source symbols is small and the symbol probabilities are significantly different, the optimum Huffman encoding procedure of the preceding subsection may still yield low code efficiencies. By using a technique called *source extension,* code efficiency can be increased. We

---

† This labeling is arbitrary. We could also have assigned **0** to the upper branch.

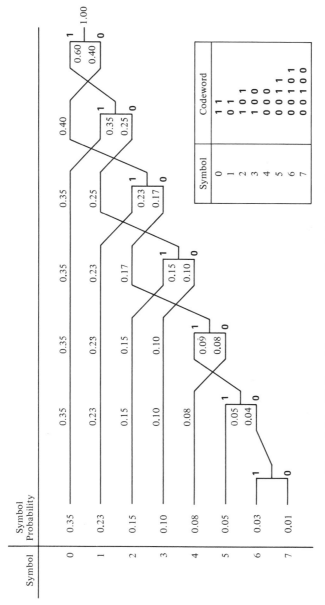

Figure 3.4-2. Huffman binary source coding of eight-symbol source.

discuss the technique under the assumption that source symbols are statistically independent over time.

By the preceding developments, an L-symbol source is encoded into a variable-length code using Huffman's procedure, which we now refer to as a first-order extension of the source. Clearly, nothing has really changed in first-order extension. However, in a second-order extension we consider *two* source symbols at a time for encoding. For L possible symbols per symbol interval and two intervals, there are L^2 possible two-element symbols. If all such two-symbol sequences are treated as a new source, this source has L^2 statistically independent symbols and its entropy is twice that of the original source. Huffman's procedure can be applied to the new source to produce a binary encoding with efficiency typically larger than that of the original source. We consider an example of second-order extension.

Example 3.4-3

Consider a source of two symbols A and B having probabilities of occurrence of $P(A) = 0.8$ and $P(B) = 0.2$. Clearly, the Huffman procedure simply assigns a code **1** to symbol A and code **0** to B. Entropy is readily found to be $H = 0.7219$ bits per symbol, so code efficiency is $\eta_{\text{code}} = 0.7219$.

The second-order extension of the source is defined in Table 3.4-3. Pairs of source symbols are now encoded by the Huffman procedure as though they were single symbols from a new source. Source entropy of the new source is readily found to be $H = 1.4439$ bits per new symbol, so efficiency becomes $\eta_{\text{code}} = H/\overline{N} = 1.4439/1.56 = 0.9255$. This efficiency is 28.2% larger than that realized by encoding of the binary source on a per-symbol basis.

To generalize the preceding procedure, we define the p-order extended source as the new source having L^p symbols, which result from consideration of groups of p symbols from the L-symbol original source. Thus we encode sequences of p symbols from the L-symbol source. For large p, efficiency converges to unity. We illustrate these comments by developing the third-order extension of the binary source of Example 3.4-3.

TABLE 3.4-3. Second-order Extension of a Binary Source with Symbol Probabilities $P(A) = 0.8$ and $P(B) = 0.2$.

Second-order Extended Source Symbols	Symbol Probability	Huffman Codeword	Probability · [Number of Code Symbols]
AA	0.64	**1**	0.64
AB	0.16	**0 0**	0.32
BA	0.16	**0 1 1**	0.48
BB	0.04	**0 1 0**	0.12
			$\overline{N} = 1.56$ digits per symbol

Example 3.4-4

The third-order extension of the binary source of Example 3.4-3 is summarized in Table 3.4-4. Here three source symbols at a time are Huffman-encoded to form an eight-symbol source of entropy $H = 2.1658$ bits per new symbol. Average codeword length is $\overline{N} = 2.184$ binary digits per new codeword. Efficiency becomes $\eta_{code} = 2.1658/2.184 = 0.9917$, which is near the theoretical limit of 1.0.

Other Encoding Methods

All source encoding methods described earlier relate to zero-memory sources. When a source has memory so that output symbols in a sequence are not independent, source encoding methods can still be devised, usually based on *Markov models* (where any given symbol depends on at most a finite number of past symbols). These methods are considered to be beyond our scope. We shall, however, describe some special forms of source encoders for sources with memory. These descriptions are deferred to Chap. 4 because they are best handled as part of the discussion of an overall system.

TABLE 3.4-4.   Third-order Extension of a Binary Source with Symbol Probabilities $P(A) = 0.8$ and $P(B) = 0.2$.

Third-order Extended Source Symbols	Symbol Probability	Huffman Codeword	Probability · [Number of Code Symbols]
AAA	0.512	1	0.512
AAB	0.128	0 1 1	0.384
ABA	0.128	0 1 0	0.384
BAA	0.128	0 0 1	0.384
ABB	0.032	0 0 0 1 1	0.160
BAB	0.032	0 0 0 1 0	0.160
BBA	0.032	0 0 0 0 1	0.160
BBB	0.008	0 0 0 0 0	0.040
			$\overline{N} = 2.184$ digits per symbol

3.5 CHANNEL ENCODING FUNDAMENTALS

The purpose of the channel encoder of Fig. 1.3-1 is to convert the source code to a form that will allow the receiver to reduce the number of errors that occur in its output due to channel noise. The theory surrounding channel encoding is vast and we shall make no effort to survey the topic. We only present a few of the more important and fundamental concepts and techniques that are within our scope and that give at least a flavor for what channel encoding involves.

The channel encoder adds redundancy to the source code by inserting extra code digits in a controlled manner so that the receiver can possibly detect and correct channel-caused errors. The resulting channel codes are usually binary. This is the only case we shall consider, although much theory has also been developed for nonbinary codes. The encoding process usually falls into two classes: *block codes* (often called *group codes*) and *convolutional codes* (also known as *tree codes*).

Block codes correspond to subdividing the sequence of source (binary) digits into sequential blocks of k digits. Each k-digit block is mapped into an n-digit block of output digits, where $n > k$. The ratio k/n is called the *code rate,* or *code efficiency.* The difference $1 - (k/n)$ is called *redundancy.* The encoder is said to produce an (n, k) code. Thus a $(7, 4)$ block code involves seven output digits per codeword for every four-digit input data codeword; the $7 - 4 = 3$ extra digits are called *check digits.* Block codes are *memoryless codes* because each output codeword depends on only one source k-bit block and not on any preceding blocks of digits. Typical parameter values range from three to several hundred for k and from $\frac{1}{4}$ to $\frac{7}{8}$ for k/n [26, p. 9].

In contrast, convolutional codes involve memory implemented in the form of a binary shift register having K cascaded registers, each with k stages. The sequence of source digits is shifted into and along the overall register, k bits at a time. Appropriate taps from the various register stages are connected to n modulo-2 adders, as illustrated in Fig. 3.5-1. The output code becomes the sequence of n digits at the output of these adders generated once for every input shift of k source digits. The ratio k/n is still called the *code rate,* and K is called the *constraint length* [27]. Clearly, each n-bit output codeword depends on the most recent k source bits stored in the first (left) k-stage shift register as well as $K - 1$ earlier blocks of k source bits that are stored in the other registers. A listing [27] of some of the better convolutional codes shows typical values of $2 \le K \le 14$, $1 \le k \le 4$, and $2 \le n \le 8$ and rates from $\frac{1}{8}$ to $\frac{4}{5}$.

The main advantage of channel encoding is that system performance is improved relative to no coding. For two systems (coded and uncoded) that operate on the same channel (same noise level) and give the same rate of errors, performances are compared through the required transmitter energies required per bit. The ratio of required energies is called *coding gain.* Thus if the coded system requires half as much energy as the uncoded system, its coding gain is 2 (or 3 dB). For practical error rates channel encoding gives positive coding gains that range from less than 1 dB [26, p. 36] to well over 4 dB [26, pp. 248–249], depending on type of encoding and specific values of K, k, and n.

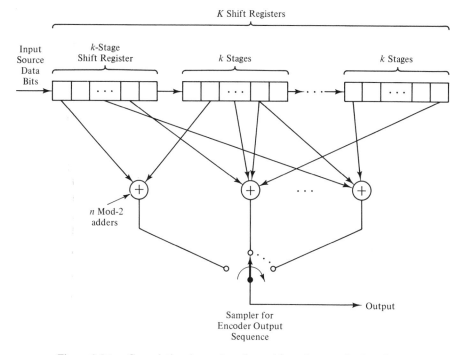

Figure 3.5-1. Convolutional encoder of rate k/n and constraint length K.

Block Codes

The simplest block code adds a single bit to the k data bits. The added bit, called a *parity check* bit, is chosen to make the total number of 1s in the codeword always either even or odd. This simple coding procedure is capable of detecting only errors involving an odd number of bits. Errors in an even number of bits go undetected. In many systems it is unlikely that more than one error per word will occur; in these systems the simple parity check has value.

The single parity check bit code has no error *correction* capability. By adding more check digits, the ability to correct for errors becomes possible. For correction of all patterns of t_e or fewer errors and no others, the total number n of code digits is related to the number of data codeword digits k by [26, p. 6]

$$2^k \le \frac{2^n}{\displaystyle\sum_{i=0}^{t_e} \binom{n}{i}}, \tag{3.5-1}$$

where $\binom{n}{i} = n!/[i!(n-i)!]$ is the binomial coefficient. The equality is achieved only for *perfect codes* (*Hamming codes* discussed later are perfect codes). Most known codes are not perfect. For a $k = 3$-digit data code (8 source symbols), for example, (3.5-1) indicates at least 3 check digits ($n = 6$) are necessary to correct only one error ($t_e = 1$), which is a (6, 3) code.

Block Codes for Single Error Correction

Suppose we examine channel codes where the first k digits are made identical to the source (data) code and the $r = n - k$ parity check digits are appended at the end. This digit arrangement is called a *systematic code*. All other linear (n, k) codes are equivalent to, and can be converted to, a systematic code [27, p. 245] by elementary operations, so our developments apply to any linear (n, k) code with suitable conversions. If the input binary codeword is written in the form $b_1 b_2 \ldots b_k$, where $b_j = 0$ or 1 for all $1 \leqslant j \leqslant k$, the output codeword has the form

$$b_1 b_2 \ldots b_k c_1 c_2 \ldots c_r \tag{3.5-2}$$

where c_i is check digit i.

In order for the check digits to be able to correct an error, they must be functionally related to digits b_j. Let the c_i be made linear functions of the b_j as follows:

$$\begin{aligned}
c_1 &= h_{11}b_1 \oplus h_{12}b_2 \oplus \cdots \oplus h_{1k}b_k \\
c_2 &= h_{21}b_1 \oplus h_{22}b_2 \oplus \cdots \oplus h_{2k}b_k \\
&\vdots \qquad\qquad\qquad \vdots \\
c_r &= h_{r1}b_1 \oplus h_{r2}b_2 \oplus \cdots \oplus h_{rk}b_k.
\end{aligned} \tag{3.5-3}$$

Here the h_{ij} are constants (0 or 1) and products $h_{ij}b_j$ are defined as follows: $0 \cdot 0 = 0, 1 \cdot 1 = 1, 0 \cdot 1 = 0, 1 \cdot 0 = 0$. By adding the c_is to both sides of (3.5-3),† it can be written in the form

$$[H][T] = [0], \tag{3.5-4}$$

where $[0]$ is an $r \times 1$ column matrix of all binary zeros,

$$[H] = \begin{bmatrix} h_{11} & h_{12} \ldots h_{1k} & 1 & 0 \ldots 0 \\ h_{21} & h_{22} \ldots h_{2k} & 0 & 1 \ldots 0 \\ \vdots & \vdots \qquad \vdots & \vdots & \vdots \quad \vdots \\ h_{r1} & h_{r2} \ldots h_{rk} & 0 & 0 \ldots 1 \end{bmatrix} \tag{3.5-5}$$

is called the *parity check matrix,* and $[T]$ is the output codeword (vector)

---

† Clearly $c_i \oplus c_i = 0$ because $0 \oplus 0 = 0, 0 \oplus 1 = 1, 1 \oplus 0 = 1$, and $1 \oplus 1 = 0$ in modulo-2 addition.

given by

$$[T] = \begin{bmatrix} b_1 \\ b_2 \\ \vdots \\ b_k \\ c_1 \\ c_2 \\ \vdots \\ c_r \end{bmatrix}. \qquad (3.5\text{-}6)$$

If $[T]$ is the transmitted codeword,† the received codeword, denoted by $[R]$, can be different than $[T]$ in only one digit. Then we can write

$$[R] = [T] \oplus [E], \qquad (3.5\text{-}7)$$

where $[E]$ is an error codeword having all digits **0** except the error digit, which is a **1**. If there were no error, the product $[H][R] = [H][T]$ would be $[0]$ from (3.5-4). With errors, however, we have

$$[S] \triangleq [H][R] = [H][T] \oplus [H][E] = [H][E], \qquad (3.5\text{-}8)$$

where $[S]$ is called the *syndrome*. Thus if $[E]$ represents an error in the ith digit, the syndrome is just the ith column of $[H]$. This means that the receiver simply computes $[S] = [H][R]$ and identifies which column of $[H]$ has been reproduced in $[S]$; that column is the digit in error in $[R]$. The erroneous digit is then corrected to form the final receiver output codeword.

Example 3.5-1

A (7, 4) block code has the parity check matrix

$$[H] = \begin{bmatrix} 1 & 1 & 1 & 0 & \vdots & 1 & 0 & 0 \\ 1 & 1 & 0 & 1 & \vdots & 0 & 1 & 0 \\ 1 & 0 & 1 & 1 & \vdots & 0 & 0 & 1 \end{bmatrix}.$$

Let the input (data) codeword be **1110**. We find the output (channel) codeword and then show that the syndrome identifies an error in position 5 of the channel codeword.

From (3.5-4) we require

$$[H][T] = \begin{bmatrix} 1 & 1 & 1 & 0 & \vdots & 1 & 0 & 0 \\ 1 & 1 & 0 & 1 & \vdots & 0 & 1 & 0 \\ 1 & 0 & 1 & 1 & \vdots & 0 & 0 & 1 \end{bmatrix} \begin{bmatrix} 1 \\ 1 \\ 1 \\ 0 \\ c_1 \\ c_2 \\ c_3 \end{bmatrix} = [0]$$

---

† Many texts use row matrices as codeword representations. Either row or column definitions are valid as long as all equations are consistent.

or

$$
\begin{array}{ccccccccccc}
1 & \oplus & 1 & \oplus & 1 & \oplus & 0 & \oplus & c_1 & = & 0 \\
1 & \oplus & 1 & \oplus & 0 & \oplus & 0 & \oplus & c_2 & = & 0 \\
1 & \oplus & 0 & \oplus & 1 & \oplus & 0 & \oplus & c_3 & = & 0,
\end{array}
$$

which gives $c_1 = 1$, $c_2 = 0$, $c_3 = 0$. Hence $[T]' = [\mathbf{1110100}]$, where $[\cdot]'$ represents matrix transpose.

For an error in digit 5, we have $[R]' = [\mathbf{1110000}]$ and

$$
[S] = [H][R] = \begin{bmatrix} 1 & 1 & 1 & 0 & \vdots & 1 & 0 & 0 \\ 1 & 1 & 0 & 1 & \vdots & 0 & 1 & 0 \\ 1 & 0 & 1 & 1 & \vdots & 0 & 0 & 1 \end{bmatrix} \begin{bmatrix} 1 \\ 1 \\ 1 \\ 0 \\ 0 \\ 0 \\ 0 \end{bmatrix} = \begin{bmatrix} 1 \\ 0 \\ 0 \end{bmatrix}.
$$

Because $[S]$ is the fifth column of $[H]$, the error is in the fifth digit. After correction of the fifth digit, the correct word is decoded.

The procedure used in Example 3.5-1 for finding the transmitted code-word $[T]$ corresponding to an input (data) codeword defined by

$$
[B] = \begin{bmatrix} b_1 \\ b_2 \\ \vdots \\ b_k \end{bmatrix} \tag{3.5-9}
$$

can be formalized. By rewriting (3.5-9) and combining with (3.5-3) to get

$$
\begin{aligned}
b_1 &= 1\, b_1 \oplus 0\, b_2 \oplus \cdots \oplus 0\, b_k \\
b_2 &= 0\, b_1 \oplus 1\, b_2 \oplus \cdots \oplus 0\, b_k \\
&\vdots \qquad\qquad\qquad \vdots \\
b_k &= 0\, b_1 \oplus 0\, b_2 \oplus \cdots \oplus 1\, b_k \\
c_1 &= h_{11}\, b_1 \oplus h_{12}\, b_2 \oplus \cdots \oplus h_{1k}\, b_k \\
&\vdots \qquad\qquad\qquad \vdots \\
c_r &= h_{r1}\, b_1 \oplus h_{r2}\, b_2 \oplus \cdots \oplus h_{rk} b_k
\end{aligned} \tag{3.5-10}
$$

we readily have

$$
[T] = [G]'[B], \tag{3.5-11}
$$

where

$$
[G] = \begin{bmatrix} 1 & 0 & \dots & 0 & h_{11} & h_{21} & \dots & h_{r1} \\ 0 & 1 & \dots & 0 & h_{12} & h_{22} & \dots & h_{r2} \\ \vdots & \vdots & & \vdots & \vdots & \vdots & & \vdots \\ 0 & 0 & \dots & 1 & h_{1k} & h_{2k} & \dots & h_{rk} \end{bmatrix} \tag{3.5-12}
$$

is called the *generator matrix*. Our result (3.5-11) shows that the output codeword is given by the product of the transpose of the generator matrix and the input codeword.

Hamming Block Codes

The design of codes consists, to a large extent, in finding suitable parity check matrices. (Clearly, finding $[H]$ also gives $[G]$ since the two are related from (3.5-5) and (3.5-12).) Hamming [28] has given a procedure for finding perfect block binary codes† that have single error-correction capability. It follows from (3.5-1), with $t_e = 1$, that

$$2^k = \frac{2^n}{(1+n)} \tag{3.5-13}$$

for these codes. Equation (3.5-13) will have a solution only when $n + 1$ is a power of 2, or

$$n = 2^m - 1, \quad m = 1, 2, \ldots . \tag{3.5-14}$$

By substituting (3.5-14) back into (3.5-13), we have

$$2^k = \frac{2^{(2^m - 1)}}{2^m}. \tag{3.5-15}$$

After equating the exponents, we obtain

$$k = 2^m - 1 - m. \tag{3.5-16}$$

By combining (3.5-14) and (3.5-16), the allowable *Hamming codes* are defined by

$$(n, k) = (2^m - 1, 2^m - 1 - m), \quad m = 1, 2, \ldots. \tag{3.5-17}$$

The first five nontrivial codes are (7, 4), (15, 11), (31, 26), (63, 57), and (127, 120). The efficiency (code rate) approaches unity for Hamming codes as m becomes large.

The parity check matrix is readily defined for the Hamming codes. The $n = 2^m - 1$ columns of $[H]$ consist of all possible $(r = m)$-digit binary words except the all-zero word. The resulting matrix $[H]$ is not in systematic form but can be rearranged to be systematic by column permutations. We demonstrate these points by an example.

Example 3.5-2
For the (7, 4) Hamming code, the $n = 7$ columns become all $r = 7 - 4 = $ three-digit binary words except the word having all zeros:

$$[H] = \begin{bmatrix} 0 & 0 & 0 & 1 & 1 & 1 & 1 \\ 0 & 1 & 1 & 0 & 0 & 1 & 1 \\ 1 & 0 & 1 & 0 & 1 & 0 & 1 \end{bmatrix}.$$

Column rearrangement puts $[H]$ in systematic form

$$[H] = \begin{bmatrix} 0 & 1 & 1 & 1 & 1 & 0 & 0 \\ 1 & 0 & 1 & 1 & 0 & 1 & 0 \\ 1 & 1 & 0 & 1 & 0 & 0 & 1 \end{bmatrix}.$$

---

† Recall that a perfect code is defined as one where the equality holds in (3.5-1).

Decoding of Block Codes

Recovery of the original data binary sequence from the codeword produced by the receiver demodulator (Fig. 1.3-1) typically involves two techniques. First, the received codeword may be in error because of channel noise; these errors are corrected as well as possible using techniques such as the syndrome discussed earlier. Second, the corrected codeword is associated as well as possible with the corresponding original data word through knowledge of the code employed.

If the error rate is low and corrected received codewords are exact, decoding is especially easy for systematic codes. The parity check digits may be discarded since the first k digits form the original data codeword.

If the error rate is high enough that all errors cannot be corrected, or if the code has no error-correction capability, decoding can be accomplished by codeword comparisons. By comparing the received codeword to each of the possible codewords, the closest match, according to a suitable measure of closeness, is declared the proper codeword. One often-used measure of closeness is the Hamming distance. Two codewords, R_i and R_j, of equal length that differ in d_{ij} digit positions are said to have *Hamming distance* d_{ij}. Clearly, the smaller the Hamming distance, the more nearly two codewords are the same. When word length is large so that many codeword comparisons are required, this technique becomes less attractive.

For additional details on decoding of block codes, as well as other codes not covered here, the reader is referred to some of the recent literature [26, 29–34].

Convolutional Codes—Tree Diagrams

Three methods—tree diagrams, trellis diagrams, and state diagrams—may be used to describe a convolutional code. We briefly consider each of these methods through the description of an example encoder. In each case we assume the encoder shown in Fig. 3.5-2(a). It consists of a single ($k = 1$) shift register with $K = 3$ stages suitably connected to $n = 2$ modulo-2 adders. The contents of the adders are sampled to provide a sequence of two output binary digits for each input binary digit in a sequence of input digits. As shown, shift register stages S_1 and S_3 are modulo-2 added to produce the first digit in the output sequence. Stages S_1, S_2, and S_3 are all modulo-2 added to obtain the second digit.

For example, let us find the encoder's output for an input sequence of **101**. Initially assume the shift register contains all **0**s (it is said to be *cleared* or *reset*†). Output is begun after the first digit, **1**, arrives. The shift register now contains **100** (left to right) and the adder outputs become $1 \oplus 0 = 1$ (adder 1) and $1 \oplus 0 \oplus 0 = 1$ (adder 2). The output sequence

---

† Actually, the rightmost digit is arbitrary because it shifts out of the register and is lost anyway as soon as the first data digit occurs.

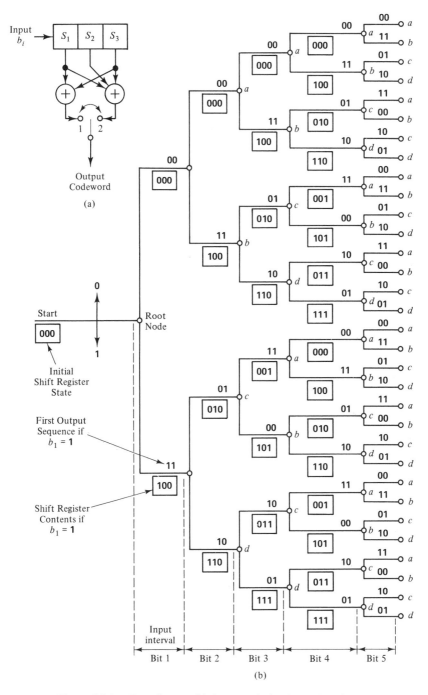

Figure 3.5-2. Tree diagram (b) for convolutional encoder of (a).

11 is generated before the second input digit arrives. After the second digit, **0**, arrives, the register contains **010**; the adders contain $0 \oplus 0 = 0$ (adder 1) and $0 \oplus 1 \oplus 0 = 1$; the output sequence is **01**. The third input digit **1** produces **101** in the shift register, $1 \oplus 1 = 0$ in adder 1, $1 \oplus 0 \oplus 1 = 0$ in adder 2, and an output sequence **00**. Thus the input sequence **101** has generated the sequence **110100**.

A tree diagram is a means of illustrating the preceding sequence of events as well as all other *possible* sequences that may occur in the encoder. Figure 3.5-2(b) shows the tree for the example encoder of (a). Tree operation is as follows: A given sequence of input digits causes a particular path to be followed through the tree from left to right. Each new digit corresponds to branching either upward (for a **0**) or downward (for a **1**) at a node; the chosen branch leads to the next node in the tree path. The shift register's contents (in the boxes) and the output sequence generated are shown for each branch. For example, the first two digits **10** of the sequence of the previous paragraph lead to the first node marked *c*.

Figure 3.5-2(b) shows that the tree diagram is repetitive after the third branching. Only four distinct branch points (nodes) result; they are marked *a*, *b*, *c*, and *d*; these are called the *states* of the encoder. Every time a tree path arrives at node (state) *b*, for example, the upper branch always corresponds to an output **01** and the lower branch generates **10**. Some reflection by the reader will show that the branch chosen at any given state node is determined by the most recent input digit, whereas the type of state node (*a*, *b*, *c*, or *d*) is determined by the two previous digits. Arrival at state a corresponds to previous input digits **00**, *b* to **01**, *c* to **10**, and *d* to **11**.[†]

In our example tree diagram, only four states occur and only two branches emanate from each node. More generally, an encoder of rate k/n and constraint length K will have 2^k branches emanating from each node in the tree and $2^{k(K-1)}$ states [27]. (Why?)

Convolutional Codes—Trellis Diagrams

The number of branches in a code tree doubles each time a new input digit occurs. For a long sequence of input digits to be encoded, the usefulness of the tree diagram is limited. A better approach uses the trellis diagram, shown in Fig. 3.5-3(a) for the encoder of Fig. 3.5-2(a). A trellis carries the same information as a tree but makes use of the fact that the tree is periodic (in the steady state condition) and involves only a finite number of states.

Use of the trellis is similar to that of a tree. Again each node in Fig. 3.5-3(a), for our example encoder ($K = 3$, $k = 1$, $n = 2$), has two branches;

[†] Branches in the tree occurring prior to the first state nodes correspond to a *transient interval* in the encoder. Branching from state nodes to other state nodes corresponds to a *steady-state* condition.

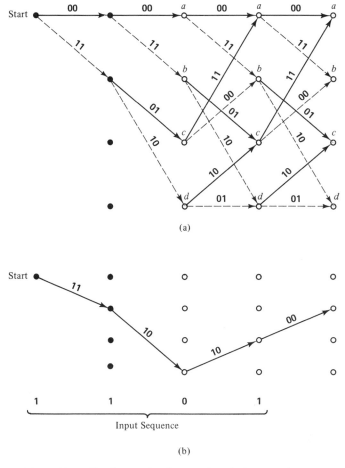

(a)

(b)

Figure 3.5-3.   (a) Trellis diagram for the encoder of Figure 3.5-2(a). (b) Trellis path for input digit sequence **1101.**

upper (shown solid) and lower branches (shown dashed) correspond to input digits of **0** and **1**, respectively. After the first two input digits, the encoder will be in one of its four states *a*, *b*, *c*, or *d*, represented by open dots. Thereafter each new input digit will only cause transitions to occur from one state to another and the trellis becomes repetitive.

Example 3.5-3
The input sequence to the convolutional encoder of Fig. 3.5-2(a) is **1101** when the shift register is initially cleared. We find the output sequence of digits and the trellis path traced by this sequence. From the trellis of Fig. 3.5-3(a), the output sequence is **11101000**. The trellis path is shown in Fig. 3.5-3(b).

The general convolutional encoder of rate k/n and constraint length K will have 2^k branches input to each state node in its trellis and 2^k branches leaving each state node (after the first which terminates the transient interval). The number of possible states is $2^{k(K-1)}$ [27, p. 286].

Convolutional Codes—State Diagrams

From our preceding discussions it is clear that once a steady-state condition is reached in a convolutional encoder, each new input bit causes only a transition from one state to another. These transitions can also be described by a *state diagram,* as shown in Fig. 3.5-4 for our example encoder of Fig. 3.5-2(a). Paths within the diagram correspond to transitions from one state node to another caused by a current bit (solid paths for **0**, dashed paths for **1**). For example, if the encoder is in state a and the currently arriving bit is a **1**, it causes a transition to state b (dashed line) corresponding to a generated output of **11**. (See the tree of Fig. 3.5-2(b).) If the bit were a **0** instead, the only other possible transition (from the tree diagram) is to state a (back to itself), as shown by the solid line; the generated output is now **00**. Transitions from other states are described in a similar way.

Viterbi Decoding of Convolutional Codes

In the receiver, the demodulator will estimate what sequence of binary digits is being received over the channel. Occasionally the demodulator will decide that a **0** was received when a **1** was actually received, and vice versa. In other words, the demodulator's output codewords will occasionally contain erroneous digits. The purpose of the channel decoder is to accept the erroneous sequence of demodulator output digits and produce the most

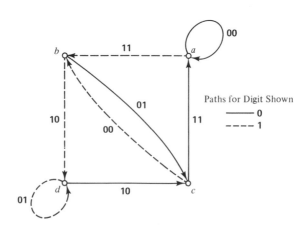

Figure 3.5-4.   State diagram for the encoder of Figure 3.5-2(a).

accurate replica possible of the source sequence that was input to the channel encoder back at the transmitter.

For convolutional codes, the optimum decoding process amounts to finding the single path through the code tree or trellis that most nearly represents the demodulated digit sequence. Obviously the *transmitted* code digits correspond to a *specific* path through the tree or trellis. However, the receiver has no knowledge of the exact path and it can only use the received sequence, which possibly has errors, to find the *most likely* path that corresponds to the received sequence. This most likely path is then used to specify the decoded (data) sequence that would have generated the path. This procedure is called *maximum-likelihood decoding*.†

The Viterbi algorithm [35] is a maximum-likelihood decoding procedure based on finding the tree or trellis path with the smallest distance between its digit sequence and the received sequence. The *Hamming distance* between two codewords of the same length has been defined as the number of digits that differ in the two sequences. For example, the sequence **011010111** differs from the sequence **111001101** in digits 1, 5, 6, and 8, so the Hamming distance is 4.

The Viterbi algorithm is best described through an example using the code's trellis. Again, we shall consider the encoder of Fig. 3.5-2(a) having the trellis of Fig. 3.5-3(a). Now suppose we wish to send the source encoded sequence **01101** to the receiver. For purposes that will be more obvious as we progress, we append two zeros to the end of the sequence, called the *tail*; the source encoded sequence becomes **0110100**. From the code trellis the channel encoded sequence is readily found to be **00111010000111**. These sequences are shown in Fig. 3.5-5(a). Next, assume digits 4 and 10 are erroneously estimated by the receiver so its demodulator output sequence is **00101010010111** as shown in (b). The channel decoder must work with only this sequence and knowledge of the trellis applicable to the channel encoder.

We now describe the Viterbi procedure. Define the trellis' branch levels as 1, 2, 3, . . . , starting with the root node as branch level 1. Paths between any two branch levels in the trellis correspond to one source encoded digit and two channel digits in our example. The first step in our procedure is to draw the trellis paths up to the second level at which state nodes occur (level 4 in our example). There will be eight distinct paths, as shown in Fig. 3.5-6(a). The Hamming distances are computed between the first six received digits (**00 10 10**) and the sequences corresponding to the eight paths. Two paths merge at each of the two state nodes at level 4. Because the succeeding path must be the same for two paths that merge at a state node, we compare the two Hamming distances of the two merging

---

† We describe *hard-decision* decoding. There is also a procedure called *soft-decision* decoding that can further improve performance [31, p. 371].

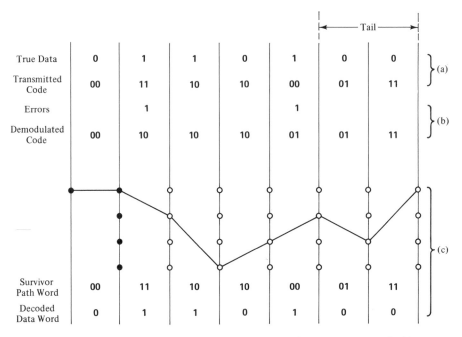

Figure 3.5-5. (a) Transmitter sequences and (b) receiver sequences applicable to Viterbi decoder that gives the survivor path of (c) for the channel encoder of Figure 3.5-2(a).

paths at each node and select the one with smallest distance; it is called the *survivor*. The survivor paths are shown by solid lines in Fig. 3.5-6(a), and discarded paths are shown as dashed lines. Hamming distances are shown for each path with the survivor's distance circled.

The next step is to extend the survivors at branch level 4 to branch level 5, using the eight possible trellis paths and recompute the Hamming distances for the eight paths relative to the received sequence (**00 10 10 10**). Again we determine the four survivors at branch level 5 nodes, as shown in Fig. 3.5-6(b). Here there are *ties* at nodes *a* and *b*. In case of ties we make a random decision and choose the upper of the two paths. This step is repeated until all input regular 2-bit channel sequences are considered (one additional level to branch level 6 in our example). There are always only four survivors at each branch level.

To finally make a choice of one surviving path of the four survivor paths, the two tail digits **00** are used. Since the receiver *knows* that the next two branch intervals correspond to tail **0** digits,† only those paths in the trellis are examined. From the trellis, the branches emanating from

---

† The receiver knows the tail arrival time through synchronization.

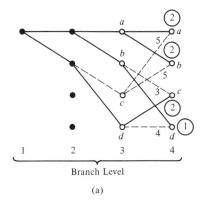

Branch Level

(a)

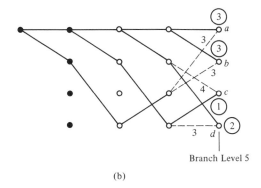

Branch Level 5

(b)

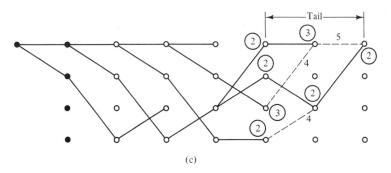

(c)

Figure 3.5-6. Survivor paths applicable to the text's example of the Viterbi decoding algorithm.

nodes a and c will merge at node a for the first tail branch interval, whereas paths from b and d will merge at node c. Thus only two survivors exist (one at node a and one at node c) at the end of the first tail digit's branch interval. Finally, the second tail digit **0** will force the paths emanating from nodes a and c to terminate at node a (at branch level 8).† Figure 3.5-6(c) shows the final paths in the tail intervals and all preceding survivor paths. The final survivor path is shown in Fig. 3.5-5(c) along with the data sequence corresponding to the path. Observe that both channel errors have been corrected by the example code.

The above example used only a 5-digit sequence of data digits. The sequence was terminated by the tail digits **00**, which also reset the decoder shift register for another data sequence. Any length data sequence can be used, but the decoder memory required to store the successive survivor paths becomes large. The delay involved in obtaining the final survivor is also too long for most practical applications. One approach to solving these problems is to determine survivor paths using only a finite (small) number of preceding branch intervals. In effect, the oldest parts of the trellis survivor paths are discarded. It has been found by computer simulation that the loss relative to optimum Viterbi decoding is negligible if the number of retained branch intervals is not less than $5K$ [27, p. 298].

The more general Viterbi decoder for a rate k/n code of constraint length K will have $2^{k(K-1)}$ states in its trellis. Each state will have 2^k paths merging at each state node and 2^k paths emanating from these nodes (in steady state). Thus the number of required computations at each branch interval increases exponentially with k and K. For this reason the Viterbi algorithm is typically used only with small values of k and K.

3.6 WAVEFORM FORMATTING OF DIGITAL SIGNALS

The preceding discussions have centered mainly on the A/D conversion, source encoding, and channel encoding concepts required to produce a suitable sequence of digital symbols for input to the modulator of Fig. 1.3-1. In this section attention is focused on the *modulator,* which converts the binary symbols into a suitable waveform for transmission over the channel. We consider only baseband waveforms. These waveforms are typically sequences of pulses generated according to the input digital code.

We may think of the process of waveform selection as a *formatting* of the digital code. There are many formats (waveforms) that have been developed. We shall discuss only those that are considered the most basic and widely used. In all cases we assume rectangular pulses for clarity of

---

† Recall that two successive data zeros reset the shift register, which is equivalent to forcing all trellis paths to converge at node a.

illustration, even though practical systems often purposely use nonrectangular signals. The more practical waveforms are considered further in Chap. 4.

Unipolar Waveform

A unipolar waveform transmits a pulse for a code **1** and no pulse for a **0**. For a symbol interval of duration T_b and a channel encoded digital sequence as illustrated in Fig. 3.6-1(a), the unipolar waveform is shown in

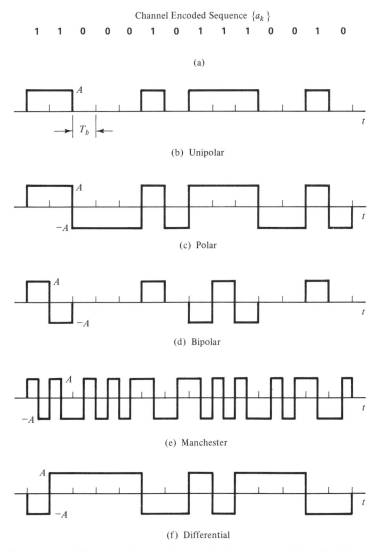

Figure 3.6-1. Waveform formats for the digital sequence of (a): (b) Unipolar, (c) polar, (d) bipolar, (e) Manchester, and (f) differential formats.

(b). The pulse illustrated is said to be a *nonreturn-to-zero* (NRZ) type because it occupies the full symbol interval's duration, T_b. Use of a narrower pulse, typically of duration $T_b/2$, corresponds to a *return-to-zero* (RZ) type. The RZ type waveform requires more bandwidth than the NRZ type. Clearly, the format can be changed to use negative instead of positive pulses if desired.

One of the main advantages of the unipolar format is its simplicity. It also interfaces well with digital logic. Disadvantages are that it does not perform as well with noise as some other formats and is not as easy to synchronize in the receiver.†

For use in Sec. 3.7 to follow, it is desirable to formulate a model for the generation of the unipolar pulse stream. If we let a_k, $k = 0, \pm 1, \pm 2, \ldots$ represent the channel encoded sequence of binary digits ($a_k = \mathbf{0}$ or $a_k = \mathbf{1}$, all k), the transmitted unipolar waveform (NRZ type) can be written as

$$s(t) = \sum_{k=-\infty}^{\infty} \alpha_k \, \text{rect}\left[\frac{t - (T_b/2) - kT_b}{T_b}\right], \tag{3.6-1}$$

where α_k represents the amplitudes of pulses with α_k equal to A or 0 for $a_k = \mathbf{1}$ or $\mathbf{0}$, respectively. We draw on the sampling models of Chap. 2 and observe that (3.6-1) can be generated by the network of Fig. 3.6-2 if the network is defined by

$$p(t) = \text{rect}\left[\frac{t - (T_b/2)}{T_b}\right] \tag{3.6-2}$$

$$P(\omega) = T_b \, \text{Sa}\left(\frac{\omega T_b}{2}\right) e^{-j\omega T_b/2}. \tag{3.6-3}$$

This model is useful in calculating the spectral properties of $s(t)$ in the following section, but it does not necessarily reflect a practical way of generating the channel's waveform. (Think about why this fact is true!)

To demodulate the waveform of (3.6-1), the receiver can sample the received waveform once during each symbol interval. If it senses level A is present, a $\mathbf{1}$ is declared the code symbol received. If level zero (no pulse) is sensed, a $\mathbf{0}$ is declared.

Polar Waveform

Figure 3.6-1(c) illustrates the *polar* signal format. Here a positive pulse is generated for each code $\mathbf{1}$ and a negative pulse is used for a $\mathbf{0}$. Reversal of polarities can also be used. The NRZ type format is shown

† A pulse train, generated by the receiver's *clock*, must be synchronized with the received bit stream so that the receiver will know, in time, when the bit (or symbol) intervals occur.

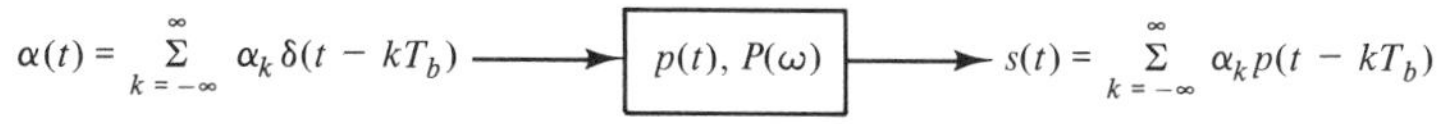

$$\alpha(t) = \sum_{k=-\infty}^{\infty} \alpha_k \, \delta(t - kT_b) \longrightarrow \boxed{p(t), P(\omega)} \longrightarrow s(t) = \sum_{k=-\infty}^{\infty} \alpha_k \, p(t - kT_b)$$

Figure 3.6-2. Model for the generation of a formatted waveform. Amplitudes, α_k, of the impulses are related to the binary sequence of channel encoder output digits.

but an RZ type is also possible for pulses of duration less than the symbol (digit) interval T_b (pulse duration typically $T_b/2$). The polar NRZ signal performs better in noise than the polar RZ waveform (for equal pulse amplitudes and values of T_b). However, the latter can more readily be used in the receiver to provide synchronization because it always has two level transitions per symbol interval.

The model of Fig. 3.6-2 applies to both polar NRZ and polar RZ formats if sequence coefficients α_k now have values $+A$ (for **1**) and $-A$ (for **0**). Equations (3.6-1)–(3.6-3) apply to the NRZ waveform. For the RZ case

$$s(t) = \sum_{k=-\infty}^{\infty} \alpha_k \, \text{rect}\left[\frac{t - (T_b/4) - kT_b}{T_b/2} \right] \tag{3.6-4}$$

$$p(t) = \text{rect}\left[\frac{t - (T_b/4)}{T_b/2} \right] \tag{3.6-5}$$

$$P(\omega) = \frac{T_b}{2} \, \text{Sa}\left(\frac{\omega T_b}{4} \right) e^{-j\omega T_b/4} \tag{3.6-6}$$

if pulses have duration $T_b/2$.

To demodulate the polar waveform, the receiver can sample the signal once each symbol interval. Sample amplitudes of A or $-A$ correspond to binary digits **1** and **0**, respectively. Of course, our comments assume noise is absent. Noise effects are discussed later (Chaps. 4 and 5).

Bipolar Waveform

In the *bipolar* format, code **0**s are transmitted by no pulse and **1**s are sent as pulses of alternating polarity, as shown in Fig. 3.6-1(d) for NRZ type. RZ can also be implemented, usually with pulse durations $T_b/2$. A main advantage of the bipolar format is that it contains no dc component, even if long strings of **0**s or **1**s occur (this is not true for unipolar and polar waveforms). Receiver circuitry can be ac coupled and dc drifts become less of a problem.

The reader can readily verify that the model of Fig. 3.6-2 applies to bipolar formats (NRZ and RZ) if the coefficients α_k now have *three* values $+A$, 0, and $-A$. Because three levels are used, bipolar modulation is sometimes called *pseudoternary* [36]. Another name that has been used is *alternate mark inversion* (AMI) [37].

Although receiver timing can be recovered from the bipolar waveform, a long string of **0**s can cause difficulties [18, p. 476]. Special sequences of pulses can be filled into the no pulse sequence caused by the string of **0**s. The resulting waveforms are called *high-density bipolar*. The added sequences are specially designed to *violate* the bipolar rule so that they can be detected by the receiver. Additional details on these waveforms are given in Spilker [18] and Shanmugam [38].

Receiver demodulation of the bipolar waveform involves examining the amplitudes of samples taken once per symbol interval. For large enough sample magnitude, a **1** is decided. If the sample's magnitude is small enough, a **0** is decided. By examining the sign sequence of samples, the receiver can also gain knowledge of errors since an alternating sign behavior should occur for demodulated **1**s.

Manchester Waveform

The Manchester waveform transmits a **1** as a positive pulse for half the symbol interval, followed by a negative pulse for the remainder of the interval; a **0** is conveyed by the same two-pulse sequence but of opposite polarity. An example waveform is shown in Fig. 3.6-1(e). Other names in use for Manchester coding are *split-phase, twinned-binary, bi-phase,* and *one-out-of-two.* It is also called *bi-phase-level* [39, p. 11] to distinguish it from other waveforms called *bi-phase-mark* and *bi-phase-space.*

The bi-phase-mark signal has a level transition at the start of every symbol interval. A **1** causes a second transition one-half symbol period later, but a **0** causes no second transition. The opposite behavior occurs with bi-phase-space; it also has transitions at the start of each interval, but the second transitions one-half interval later are due to **0**s, whereas **1**s cause no second transitions.

The model of Fig. 3.6-2 applies to Manchester coding if the coefficients α_k are A or $-A$ (as in polar-NRZ), but the network is now defined by

$$p(t) = \text{rect}\left[\frac{t - (T_b/4)}{T_b/2}\right] - \text{rect}\left[\frac{t - (3T_b/4)}{T_b/2}\right] \qquad (3.6\text{-}7)$$

and

$$P(\omega) = j\,T_b\text{Sa}\left(\frac{\omega T_b}{4}\right)\sin\left(\frac{\omega T_b}{4}\right)e^{-j\omega T_b/2}. \qquad (3.6\text{-}8)$$

Practical generation of the Manchester waveform results from first generation of a polar-NRZ signal and then multiplication by a synchronized square-wave clock (amplitudes ± 1) having a period T_b.

Because at least one level transition occurs every symbol interval in the Manchester signal, receiver clock timing can be extracted from the waveform, even in the presence of long strings of **0**s or **1**s. A disadvantage is that it requires twice the bandwidth of a bipolar signal.

At least two ways exist for a receiver to demodulate a Manchester waveform. In the first, two samples can be taken each symbol interval; a **1** is declared if samples in a given interval have sign sequence $+ \ -$, whereas a **0** is decided for signs $- \ +$. A second approach multiplies the input waveform by a synchronized square-wave clock ($+1$, -1 amplitudes) of period T_b. The product will be a polar-NRZ waveform that is then demodulated as previously described.

Differential Waveform

The *differential* waveform is actually the result more of a coding technique than it is a format. The waveform often uses two levels A and $-A$, as shown in Fig. 3.6-1(f) but unipolar versions (levels A and 0) are also possible. Occurrence of a digit **1** causes a level transition at the start of its interval, but a **0** causes no transition. Thus a symbol interval has a level different from that of the previous interval only if it corresponds to a code **1**. This waveform has also been called *NRZ-mark* [39].

A differential waveform, where level transitions occur for **0**s and no transitions for **1**s, can also be constructed. These are sometimes called *NRZ-space* [39].

Waveform generation involves first differentially encoding the given binary sequence $\{b_k\}$ to obtain a new sequence $\{a_k\}$ that then becomes the code for generating a *polar* waveform. Figure 3.6-3(a) shows the necessary encoder. Prior to the start of the input sequence the output is set to an initial value (**0** or **1**) that is arbitrary. Thereafter,

$$a_k = b_k \oplus a_{k-1}. \qquad (3.6\text{-}9)$$

Example sequences are shown in (b) for an initial value **1**.

In the receiver, demodulation is the same as for a polar-NRZ signal except now the demodulated sequence, denoted by $\{\hat{a}_k\}$, must be decoded to obtain the receiver's estimated sequence, denoted by $\{\hat{b}_k\}$, corresponding to the original sequence $\{b_k\}$. The decoder is shown in Fig. 3.6-3(c); its operation is defined by

$$\hat{b}_k = \hat{a}_k \oplus \hat{a}_{k-1}. \qquad (3.6\text{-}10)$$

The reader may wish to use the sequence of (b) to verify that (3.6-10) gives the correct response code when no channel errors occur due to noise.

A principal advantage of differential encoding is that the transmitted waveform can be inverted without affecting decoding accuracy. This property is important when the polar waveform passes through a cascade of networks for which the output signal's polarity may be unknown.

Duobinary Waveform

The *duobinary* waveform is one that again uses a pulse to represent a **1** and a negative pulse to represent a **0**. However, now the pulses differ

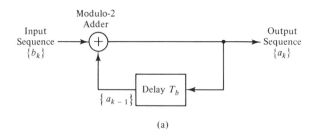

(a)

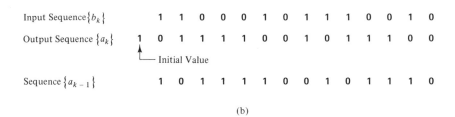

(b)

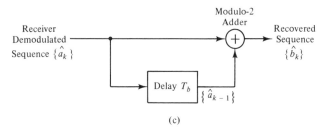

(c)

Figure 3.6-3. (a) Differential encoder, (b) example encoder sequences, and (c) decoder.

from all the waveforms discussed previously in that their durations are *two* symbol intervals rather than one. Thus a code **1** in a given interval corresponds to a pulse occupying the given interval *and* the following interval. Some reflection shows that, after an initial transient interval, the duobinary waveform consists of either positive or negative pulses in some intervals as well as some intervals with no pulses. If the pulses have durations $2T_b$ and amplitudes $A/2$ (for **1**) or $-A/2$ (for **0**), the steady-state waveform pulses have amplitudes $\pm A$. An example channel encoder sequence is shown in Fig. 3.6-4(a), and the corresponding duobinary signal is sketched in (b). The waveforms of (c) and (d) are discussed later and are included only for comparison purposes at this point.

The main advantage of the duobinary waveform is its capability of transmitting a given sequence with only half the bandwidth of unipolar and polar formats (because the basic pulse has duration $2T_b$ instead of T_b).

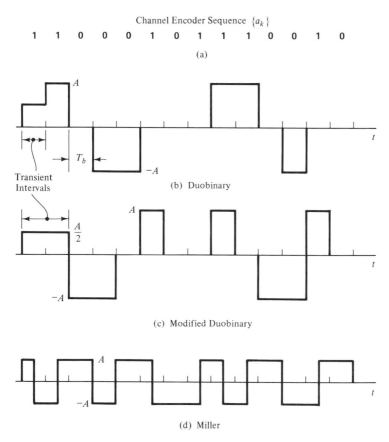

Channel Encoder Sequence $\{a_k\}$

1   1   0   0   0   1   0   1   1   1   0   0   1   0

(a)

(b) Duobinary

(c) Modified Duobinary

(d) Miller

Figure 3.6-4.   Waveforms for the digital sequence of (a): (b) Duobinary, (c) modified duobinary, and (d) Miller.

Alternatively, for the same bandwidth, the rate of symbols transmitted can be doubled with the duobinary system [40–44].

The model of Fig. 3.6-2 can be extended to the duobinary waveform, as shown in Fig. 3.6-5. In the basic method of (a), an impulse is generated in interval k with amplitude $\alpha_k = A/2$ or $-A/2$, according to whether the binary digit in interval k is **1** or **0**, respectively. Each impulse excites the filter with transfer function $P(\omega)$ to generate a pulse of duration T_b. A delayed impulse also generates a second pulse of duration T_b delayed by T_b relative to the first. The net effect is the generation of a pulse of duration $2T_b$ for each binary digit. The reader can readily show that the equivalent system of (b) is valid. For rectangular pulses we require

$$p(t) = \text{rect}\left[\frac{t - (T_b/2)}{T_b}\right] \qquad (3.6\text{-}11)$$

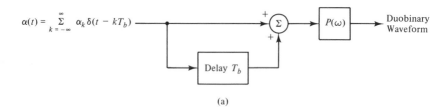

(a)

$$\alpha(t) = \sum_{k=-\infty}^{\infty} \alpha_k \delta(t - kT_b) \longrightarrow \boxed{2P(\omega) \cos\left(\frac{\omega T_b}{2}\right) e^{-j\omega T_b/2}} \longrightarrow \text{Duobinary Waveform}$$

(b)

Figure 3.6-5. Functional block diagrams for generation of duobinary waveforms. (a) Basic method and (b) equivalent system.

$$P(\omega) = T_b \, \text{Sa}\left(\frac{\omega T_b}{2}\right) e^{-j\omega T_b/2}. \qquad (3.6\text{-}12)$$

In the receiver, demodulation follows proper interpretation of samples taken once in each symbol interval. The rule is† as follows: If the sample amplitude is A, the digit is declared a **1** (the preceding digit will also be a **1**); if the amplitude is $-A$, the digit is declared a **0** (the preceding digit is also a **0**); if the sample is zero, the digit is detected as a **1** if the preceding digit decision was **0**, and detected as a **0** if the preceding decision was **1**.

Example 3.6-1

For demodulation of the waveform of Fig. 3.6-4(b) that corresponds to the sequence of (a), we have

Original sequence	**1 1 0 0 0 1 0 1 1 1 0 0 1 0**	
Samples	$\cdot A \; 0 \; -A \; -A \; 0 \; 0 \; 0 \; A \; A \; 0 \; -A \; 0 \; 0$	
Recovered sequence	**1 1 0 0 0 1 0 1 1 1 0 0 1 0**	

Note that digit 1 (a **1**) is recovered since a sample A means the previous digit had to be a **1**. This result was possible because the first two original digits were the same. If the first two are different, the recovered sequence is indeterminate until two original digits are identical. (Why?)

The receiver's demodulation process for a given symbol interval depends on the previous interval, in general. This fact means that errors tend to propagate. However, if the channel encoder's output sequence is first differentially encoded, as described earlier, and the new sequence is used to generate the duobinary waveform, error propagation can be eliminated. The procedure is called *precoding,* and a new detection procedure is required.

† Our simplified discussion presumes no noise.

Now the receiver decides a **0** is being received in any given symbol interval if that interval's sample amplitude is either A or $-A$; it decides a **1** is present if the sample amplitude is zero. No reference to previous intervals is required.

The preceding duobinary waveform generation procedure using precoding can be stated in two other equivalent procedures. In the first procedure, no pulse is used in a symbol interval when the binary sequence symbol is a **1**; pulses are used for **0**s with a given pulse's polarity reversed from that of the previous pulse if an odd number of **1**s (no pulse intervals) falls between. An even number of **1**s results in no sign inversion. In the second procedure the *complement* of the data sequence is precoded. No pulse is used for intervals with **0**s; pulses with sign reversal correspond to **1**s with an odd number of **0**s between and no reversal in sign occurs for an even number of **0**s [40]. The preceding detection procedure for precoding now gives the complement of the original data sequence. The reader is encouraged to verify these procedures using the sequence of Fig. 3.6-4(a).

Modified Duobinary Waveform

If a duobinary waveform is delayed one symbol interval (by T_b) and subtracted from the undelayed waveform, the result is a *modified duobinary waveform*. The signal retains the bandwidth reduction relative to other formats but has the additional advantage that its power spectrum has no dc component [43].

The difference of two duobinary waveforms is equivalent to generating a signal using the basic waveform of Fig. 3.6-6(a) for each symbol **1** in the binary sequence and its negative for each **0**. With this basic waveform, the models of (b) and (c) are readily shown to apply, again with $\alpha_k = A/2$ for **1**s and $\alpha_k = -A/2$ for **0**s. From the behavior of $p_{md}(t)$, it is clear that any given digital symbol will be correlated with symbols *two* intervals later, which means that any receiver symbol is related to the symbol occurring two intervals earlier in time. An example waveform is shown in Fig. 3.6-4(c) for the symbol sequence of (a). The initial transient interval is now two intervals in duration. The filter impulse response $p(t)$ and transfer function $P(\omega)$ are again given by (3.6-11) and (3.6-12).

In receiver decoding (no noise assumed), samples in each symbol interval indicate symbols **0** and **1** if the sample values are $-A$ and $+A$, respectively. If the sample value is zero, the symbol is the same as occurred *two* symbols earlier. If an error occurs when noise is present, it tends to propagate and affect later symbols.

As in duobinary modulation, error propagation can be eliminated by precoding. The coder of Fig. 3.6-3(a) again applies, except the delay is $2T_b$ (*two* symbol intervals). Thus

$$a_k = b_k \oplus a_{k-2}. \qquad (3.6\text{-}13)$$

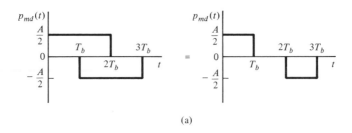

(a)

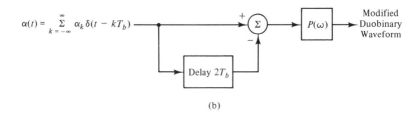

(b)

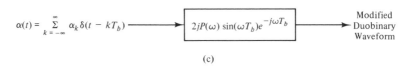

(c)

Figure 3.6-6. (a) Waveforms used in generating a modified duobinary signal with rectangular pulses, (b) block diagram of model system to generate the waveform, and (c) a system equivalent to that of (b).

In the receiver, the sample in any interval is decoded as a binary **0** if its value is zero (no noise case) and as a **1** if its value is either A or $-A$.

Miller Waveform

The *Miller waveform* results when a level transition occurs at the midpoint of each symbol interval corresponding to a **1** in the binary sequence. No transition occurs for a **0** unless followed by a **0**, in which case a transition occurs at the end of the interval. An example waveform is shown in Fig. 3.6-4(d). Miller coding is also called *delay modulation* [39].

A Miller waveform is attractive for magnetic recording and in some phase-shift keyed signals because its power spectrum is small for frequencies near dc and its bandwidth requirements are approximately half those needed by Manchester coding [39].

M-ary Waveform

Our final format example is the *M-ary waveform*. We briefly considered *M*-ary codes in Sec. 3.4. If the input to the modulator is a sequence of

M-ary digits, one digit per signaling interval of duration T_b, a logical extension of the binary formats is to simply assign a pulse of duration T_b to each digit that can have any one of M amplitudes. If all amplitudes are nonnegative, we have a *unipolar* format. In a *polar* format, pulses would be paired so that there is a negative pulse having a magnitude equal to one of the positive pulses.† If M is a power of 2, such as $M = 2^{N_b}$, and each of the M amplitudes is equally probable, the information conveyed in each symbol interval (assumed independent interval to interval) is $I = -\log_2(1/M) = N_b$ bits per symbol. The *rate* at which this waveform can convey information is, therefore, N_b/T_b bits/s.

Another common problem involves the use of an M-ary waveform to represent a sequence of binary digits, one digit per symbol interval of duration T_b. In this case

$$M = 2^{N_b} \qquad (3.6\text{-}14)$$

pulse amplitudes are associated with N_b symbol intervals. For equally probable independent binary digits (symbols) in the sequence, the information

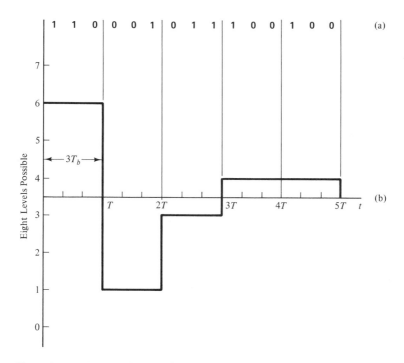

Figure 3.6-7. *M*-ary polar waveform (b) for binary sequence of (a) when $N_b = 3$ and $M = 8$.

† M is assumed to be even here, as would occur in the important case where M is a power of 2.

conveyed is N_b bits per M-ary symbol. However, an M-ary symbol duration
is N_bT_b, so the information *rate* is $1/T_b$ bits/s, the same as in the binary
sequence. Although information rate is no different than if the sequence
had been represented by a polar binary waveform, bandwidth is different.
Because bandwidth is proportional to the reciprocal of symbol waveform
durations (T_b in binary format, N_bT_b in this type of M-ary signal), the binary
system requires larger bandwidth by a factor N_b. Reduced channel bandwidth
requirements is one of the major advantages of M-ary formatting. A major
disadvantage is its greater sensitivity to noise. An example of M-ary formatting
is shown in Fig. 3.6-7 for $M = 8$.

3.7 SPECTRAL CHARACTERISTICS OF DIGITAL FORMATS

In the preceding section we found that several of the digital waveforms
could be generated by the networks of Figures 3.6-2, 3.6-5, and 3.6-6. In
each case a network is excited by the impulse train

$$\alpha(t) = \sum_{k=-\infty}^{\infty} \alpha_k\delta(t - kT_b) \tag{3.7-1}$$

where the coefficients α_k are related to the message's binary digits. These
relationships are summarized in Table 3.7-1. All the networks have the
form shown in Fig. 3.7-1, where $H(\omega)$ is determined by the waveform format
of interest.

Power Spectrum

If $\mathscr{S}(\omega)$ represents the power spectrum of $s(t)$ in Fig. 3.7-1, then it
can be shown that

$$\mathscr{S}(\omega) = \mathscr{S}_a(\omega)|H(\omega)|^2 \tag{3.7-2}$$

TABLE 3.7-1. Coefficients α_k for Various Digital
Formats.

Format	α_k for Binary Digit		Remarks
	0	1	
unipolar	0	A	
polar	$-A$	A	
bipolar	0	$\pm A$	Signs alternate
Manchester	$-A$	A	
Duobinary	$\dfrac{-A}{2}$	$\dfrac{A}{2}$	
Modified Duobinary	$\dfrac{-A}{2}$	$\dfrac{A}{2}$	

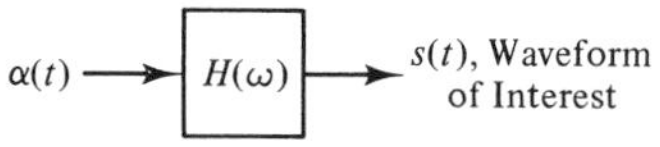

$$\alpha(t) \longrightarrow \boxed{H(\omega)} \longrightarrow \begin{array}{l} s(t), \text{Waveform} \\ \text{of Interest} \end{array}$$

Figure 3.7-1.   Network model for study of spectral properties of digital formats.

where $\mathscr{S}_\alpha(\omega)$ is the power spectrum of $\alpha(t)$ and the network is assumed to have a real impulse response.

Once $\mathscr{S}_\alpha(\omega)$ is found, the spectral properties of a given format are obtained from $\mathscr{S}(\omega)$ when $H(\omega)$ is that which applies to the format. Although somewhat involved, it can be shown (Prob. 3-58) that $\alpha(t)$ has the time-averaged autocorrelation function

$$R_\alpha(\tau) = \frac{1}{T_b} \sum_{r=-\infty}^{\infty} R_r \delta(\tau - rT_b) \tag{3.7-3}$$

and power spectrum

$$\mathscr{S}_\alpha(\omega) = \frac{1}{T_b} \sum_{r=-\infty}^{\infty} R_r e^{-j\omega rT_b}, \tag{3.7-4}$$

where

$$R_r = \lim_{K \to \infty} \frac{1}{2K} \sum_{k=-K}^{K} \overline{\alpha_k \alpha_{k+r}} \tag{3.7-5}$$

and the overbar represents the statistical average.

For the special case where the α_k are independent from interval to interval (over k) and have the same probability distribution in each interval, (3.7-5) reduces to

$$R_r = \begin{cases} \overline{\alpha_k^2}, & r = 0 \\ (\overline{\alpha_k})^2, & r \neq 0. \end{cases} \tag{3.7-6}$$

We next apply (3.7-2) and (3.7-4) to several specific formats. All are assumed to be NRZ type generated from independent binary digits. In some cases (3.7-6) applies. In others we must revert to use of (3.7-5).

Unipolar Format

Here 1s and 0s are assumed to occur with equal probability. Since α_k is a discrete random variable having only the two values 0 and A (Table 3.7-1) we have

$$R_r = \begin{cases} 0\left(\dfrac{1}{2}\right) + A^2\left(\dfrac{1}{2}\right) = \dfrac{A^2}{2}, & r = 0 \\[2mm] \left[0\left(\dfrac{1}{2}\right) + A\left(\dfrac{1}{2}\right)\right]^2 = \dfrac{A^2}{4}, & r \neq 0, \end{cases} \tag{3.7-7}$$

so

$$\mathscr{S}_\alpha(\omega) = \frac{A^2}{4T_b} + \frac{A^2}{4T_b} \sum_{r=-\infty}^{\infty} e^{-j\omega rT_b} = \frac{A^2}{4T_b} + \frac{\pi A^2}{2T_b^2} \sum_{r=-\infty}^{\infty} \delta(\omega - r\omega_b), \tag{3.7-8}$$

where (3.7-4) and results of Prob. A-22 have been used and we define

$$\omega_b = \frac{2\pi}{T_b}. \tag{3.7-9}$$

Finally, $|H(\omega)|^2$ is given by $|P(\omega)|^2$ using (3.6-3), so†

$$\mathcal{S}_u(\omega) = \frac{A^2\pi}{2}\,\delta(\omega) + \frac{A^2T_b}{4}\,\mathrm{Sa}^2\!\left(\frac{\omega T_b}{2}\right). \tag{3.7-10}$$

The continuous part of this function is plotted in Fig. 3.7-2. Bandwidth to the first spectral null is ω_b. The term $(A^2\pi/2)\delta(\omega)$ represents a dc component having half the total power.

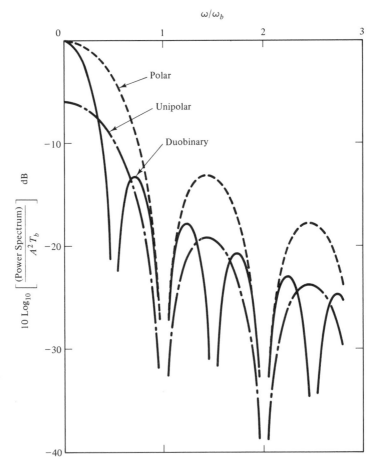

Figure 3.7-2. Normalized power spectrums of unipolar, polar, and duobinary waveforms.

† Subscript u is used to imply unipolar format.

For an RZ format with pulse duration $T_b/2$, the power spectrum $\mathscr{S}_u(\omega)$ has a continuous part with twice the bandwidth of (3.7-10) and has spectral lines at odd multiples of $\pm\omega_b$ (see Prob. 3-59). These lines are useful in deriving clock synchronization.

Polar Format

Here α_k has equally probable values $-A$ and A for equally probable digits **0** and **1**. Thus $R_0 = (-A)^2(\frac{1}{2}) + (A)^2(\frac{1}{2}) = A^2$, $R_r = [(-A)(\frac{1}{2}) + (A)(\frac{1}{2})]^2 = 0$ for $r \neq 0$, so $\mathscr{S}_\alpha(\omega) = A^2/T_b$. Again $|H(\omega)|^2 = |P(\omega)|^2$ of (3.6-3) and †

$$\mathscr{S}_p(\omega) = A^2 T_b \mathrm{Sa}^2\left(\frac{\omega T_b}{2}\right). \tag{3.7-11}$$

This spectrum has the same shape and first-null bandwidth as the continuous part of (3.7-10) for the unipolar format. The power spectrum $\mathscr{S}_p(\omega)/A^2 T_b$ is plotted in Fig. 3.7-2.

Bipolar Format

Even if message sequence binary digits are independent, the amplitude coefficients α_k are not because of the constraint that the pulse polarities assigned to **1**s alternate. Thus more care must be given to the evaluation of $\overline{\alpha_k \alpha_{k+r}}$. From table 3.7-1 α_k has values 0, A, and $-A$; these occur with respective probabilities $\frac{1}{2}$, $\frac{1}{4}$, and $\frac{1}{4}$. Thus when $r = 0$, $\overline{\alpha_k^2} = 0^2(\frac{1}{2}) + A^2(\frac{1}{4}) + (-A)^2(\frac{1}{4}) = A^2/2$ and $R_0 = A^2/2$ from (3.7-5).

When $r = 1$ the message sequence (a_k, a_{k+1}) can have only possible values **(0, 0)**, **(0, 1)**, **(1, 0)**, or **(1, 1)**. The corresponding values of $\alpha_k \alpha_{k+1}$ are 0, 0, 0, and $-A^2$ with probabilities of $\frac{1}{4}$ each. Hence $\overline{\alpha_k \alpha_{k+1}} = 3(0)(\frac{1}{4}) + (-A^2)(\frac{1}{4}) = -A^2/4$, so $R_1 = -A^2/4$ from (3.7-5).

When $r > 1$, half of the possible message sequences $(a_k, a_{k+1}, \ldots, a_{k+r})$ have $a_k = \mathbf{0}$ and lead to $\alpha_k \alpha_{k+r} = 0$. Another one-fourth of the possible sequences have $a_k = \mathbf{1}$ but have $a_{k+r} = \mathbf{0}$, so again $\alpha_k \alpha_{k+r} = 0$. The remaining one-fourth of the possible sequences have both $a_k = \mathbf{1}$ and $a_{k+r} = \mathbf{1}$, so $\alpha_k \alpha_{k+r} = A^2$ or $-A^2$, depending on how many **1**s or **0**s fall between digits k and $k + r$ in the message sequence. There are $r - 1$ digit positions *between* a_k and a_{k+r} so there are 2^{r-1} possible binary "words" in the $r - 1$ digits. Examination shows that half of these words have an odd number of **1**s and half have an even number. An even number results in $\alpha_k \alpha_{k+r} = -A^2$, whereas an odd number gives $\alpha_k \alpha_{k+r} = +A^2$, each half the time. These considerations collectively mean that $\overline{\alpha_k \alpha_{k+r}} = 0$ for $r > 1$.

---

† Subscript p denotes the polar format.

From (3.7-4) we now have

$$\mathcal{S}_\alpha(\omega) = \frac{A^2}{T_b}\left[-\frac{1}{4}e^{j\omega T_b} + \frac{1}{2} - \frac{1}{4}e^{-j\omega T_b}\right]$$

$$= \left(\frac{A^2}{T_b}\right)\sin^2\left(\frac{\omega T_b}{2}\right),$$

(3.7-12)

so the bipolar waveform power spectrum, denoted by $\mathcal{S}_{bi}(\omega)$, is

$$\mathcal{S}_{bi}(\omega) = A^2 T_b \sin^2\left(\frac{\omega T_b}{2}\right) \text{Sa}^2\left(\frac{\omega T_b}{2}\right)$$

(3.7-13)

from (3.7-2) and the fact that $|H(\omega)|^2 = |P(\omega)|^2$ with (3.6-3) applying. A plot of $\mathcal{S}_{bi}(\omega)/A^2 T_b$ is shown in Fig. 3.7-3. Although the first-null bandwidth

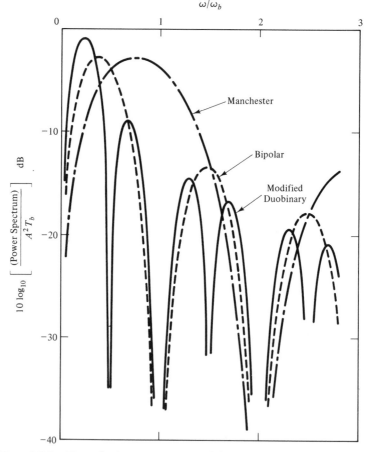

Figure 3.7-3. Normalized power spectrums of bipolar, Manchester, and modified duobinary waveforms.

is ω_b, the same as unipolar and polar formats, there is only small spectral content for frequencies near dc.

Manchester Format

For independent and equally probable message digits, the values A and $-A$ of α_k occur independently with equal probability. We readily find that $R_0 = A^2$ and $R_r = 0$ for $r > 0$. Thus $\mathscr{S}_\alpha(\omega) = A^2/T_b$ and the Manchester waveform's power spectrum, denoted by $\mathscr{S}_{man}(\omega)$, becomes

$$\mathscr{S}_{man}(\omega) = A^2 T_b \sin^2\left(\frac{\omega T_b}{4}\right) \mathrm{Sa}^2\left(\frac{\omega T_b}{4}\right). \tag{3.7-14}$$

In obtaining (3.7-14) we have used (3.6-8) and (3.7-2) with $|H(\omega)|^2 = |P(\omega)|^2$. The first-null bandwidth of $\mathscr{S}_{man}(\omega)$ is $2\omega_b$, twice that required by bipolar signaling. Figure 3.7-3 shows a plot of $\mathscr{S}_{man}(\omega)/A^2 T_b$.

Duobinary Format

Even though duobinary pulses are correlated in adjacent intervals for independent message digits, this correlation is accounted for in the model of Fig. 3.6-5. Thus α_k and α_{k+r} are independent for independent message digits. Since the values of α_k of $A/2$ and $-A/2$ are equally probable for equally probable message digits, we readily find that $R_0 = A^2/4$, $R_r = 0$ for $r > 0$, so $\mathscr{S}_\alpha(\omega) = A^2/4T_b$ and the power spectrum, denoted by $\mathscr{S}_d(\omega)$, becomes

$$\mathscr{S}_d(\omega) = A^2 T_b \cos^2\left(\frac{\omega T_b}{2}\right) \mathrm{Sa}^2\left(\frac{\omega T_b}{2}\right), \tag{3.7-15}$$

where (3.6-12) and the network transfer function of Fig. 3.6-5(b) have been used in (3.7-2).

A plot of $\mathscr{S}_d(\omega)/A^2 T_b$ is shown in Fig. 3.7-2. First-null bandwidth is $\omega_b/2$.

Modified Duobinary Format

Developments for this waveform parallel those for the duobinary format except the waveform model is that of Fig. 3.6-6. The modified duobinary waveform's power spectrum, denoted by $\mathscr{S}_{md}(\omega)$, is now

$$\mathscr{S}_{md}(\omega) = A^2 T_b \sin^2(\omega T_b)\, \mathrm{Sa}^2\left(\frac{\omega T_b}{2}\right). \tag{3.7-16}$$

$\mathscr{S}_{md}(\omega)/A^2 T_b$ is plotted in Fig. 3.7-3 which shows a first-null bandwidth of $\omega_b/2$ and small values for frequencies near dc.

Miller Format

The power spectrum of this format is more difficult to determine than those described earlier. It is given by Lindsey and Simon [39, p. 21] and shows a sharply peaked response near $\omega = \pm 0.38\omega_b$ with low-level values (about 12 dB below peak response) out to about $\omega = \pm 1.7\omega_b$. Other "sidelobe" responses above $\omega = \pm 2\omega_b$ fall off at the same relative rate (as $1/\omega^2$) as the other formats discussed earlier.

3.8 TIME MULTIPLEXING OF BINARY DIGITAL WAVEFORMS

Time division multiplexing was briefly introduced in Sec. 2.8. It was shown how samples from a number of messages could be interlaced in time for combined transmission over a single channel. It was left as more or less obvious to the reader that the same techniques could be applied to other systems. In this section we extend these earlier concepts to show how many messages may be time multiplexed using the pulse formatted waveforms described in the foregoing sections of this chapter. The multiplexer (MUX) involved is a *digital multiplexer*.

Generally, a digital multiplexer can be considered as a *parallel-to-serial converter*. It accepts a set of inputs (or messages, often called *tributaries*) applied in parallel and interlaces the inputs into a single output signal having specific time intervals allocated serially to each message. In a *word-interleaved* multiplexer each of these time intervals contains a complete word for a given message. The serial output would then comprise a word due to message 1, a word for message 2, etc. A *bit-interleaved* multiplexer places a single input bit in each output interval; the output here would consist first of a sequence of the first digits from each of the input words, then a sequence of the second digits from all inputs, and so on. As in Chap. 2 we use the term *frame* to refer to the smallest time interval in the output signal that contains at least one sample (word) from all input messages. A frame can also include some time allocated to other functions, such as synchronization and signaling (dialing and ringing, for example, in telephone systems).

Timing of all operations within a multiplexer is controlled by a highly stable oscillator called the *master clock*. If the digital structure of all the input messages is determined from the same master clock, the multiplexer is called *synchronous*. If different clocks are involved, the multiplexer is said to be *asynchronous*. Of the two, synchronous multiplexers are more straightforward.

In practice, digital multiplexers can be divided into two groups [36]. In the first group are multiplexers designed to combine inputs from a number of low-speed data terminals. Output bit rates of these multiplexers are typically 1200, 2400, 3600, 4800, 7200, or 9600 bits/s when used with *modems* (*mo*dulation-*dem*odulation *s*ystem) designed to interface with commercial voice-grade telephone lines. Lower rates apply to the usual dial-up grade line, and the higher rates apply to specially conditioned privately leased lines.

The second group of multiplexers typically use high bit rates and are of the types used by common carriers such as American Telephone and Telegraph (AT&T) in the United States, Nippon Telegraph and Telephone (NTT) in Japan, and others all over the world. The hierarchies of these systems are subsequently outlined but involve both synchronous and asynchronous multiplexing. The recent trend of AT&T has been toward a nationwide synchronous system with master clock located in Hillsboro, Missouri [37].

AT&T D-Type Synchronous Multiplexers

The basic first-level multiplexer in the AT&T system is a synchronous device. Three models are in use. The oldest, called a D1 *channel bank,* was introduced in the early 1960s [45]; it accepts 24 *analog* voice frequency (VF) messages at its input, samples each at an 8-kHz rate, compresses samples according to a $\mu = 100$ law, and encodes each sample using 8-bit words (7 bits for amplitude, 1 bit for signaling). One additional bit is added per frame for synchronization so $1 + 24(8) = 193$ bits are present per frame of duration $1/8(10^3) = 125\ \mu s$. Output bit rate becomes 193 bits/125 $\mu s = 1.544$ megabits/s. The D1 bank uses word interleaving.

In 1970 the D2 channel bank was introduced [46, 47]; it provided improved performance compared to the D1 bank. The D2 bank accepted 96 analog voice input messages and multiplexed these in groups of 24 onto four 1.544 megabits/s output lines. Thus *each* output line carries 24 VF messages, as in the D1 unit. Multiplex formatting used in the output lines was slightly different than the D1 unit but was the same as that used in a newer channel bank, the D3, introduced in the mid-1970s as a replacement for the D1 bank. Because of similarities, we describe only the D3 bank.

The D3 channel bank [47] multiplexes 24 analog VF input messages into a 1.544 megabits/s output (called a T1 line) that can be combined by additional multiplexing with other similar lines, if desired, or transmitted directly. As in the D1 and D2 systems, message samples are at an 8-kHz rate. Samples in the D3 bank are compressed using a $\mu = 255$ law and coded using 8 bits per sample. One additional bit is added per frame for synchronization for a total of 193 bits per frame. A frame is 125 μs in duration. For 5 frames out of every 6, each message is conveyed by full 8-bit words using a bipolar waveform format. However, in every sixth frame, the eighth digit is stolen for signaling purposes. In some systems *two* signaling paths are required, so a 12-frame sequence called a *multiframe*

[48] (or *superframe* [46]) spans both message and signaling operations. For frame synchronization the framing bit in *odd*-numbered frames has the digital sequence **101010**; thus frames 1, 5, and 9 carry a **1**, whereas frames 3, 7, and 11 carry a **0** in each multiframe. To align the multiframe the *even*-numbered framing bits have the sequence **001110**. This sequence allows frames 6 and 12 to be identified as following transitions **01** and **10**, respectively. Figure 3.8-1 illustrates the format of a typical frame in the D3 channel bank that uses word interleaving.

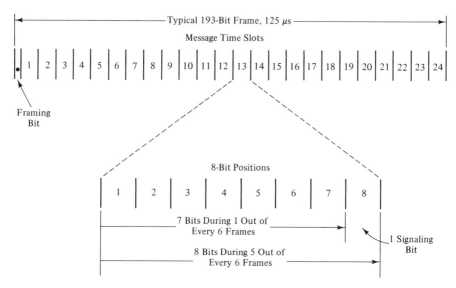

Figure 3.8-1. Format of typical frame for D3 channel bank.

Hierarchies of Digital Multiplexing

Several levels of multiplexing may be combined to form a hierarchy of multiplexers. Three principal hierarchies exist in the world today. In the United States and Canada the system of Fig. 3.8-2 is used. It is known as the T-carrier system of AT&T. At the lowest level the D3 channel bank is commonly used to generate the multiplexed signal that becomes one of four possible inputs (over four T1 lines) to the M12 second level multiplexer. The M12 multiplexer output has a 6.312 megabits/s bit rate and drives a T2 line. Higher multiplexing levels (up to the T5 line) are possible as shown. All the higher multiplexers (M12, M23, and so on) are asynchronous, and the M12 device is discussed in the next subsection as an example of such multiplexers.

Variations exist in the implementations of the hierarchy of Fig. 3.8-2. The M23 multiplexer can be bypassed using an M13 device having

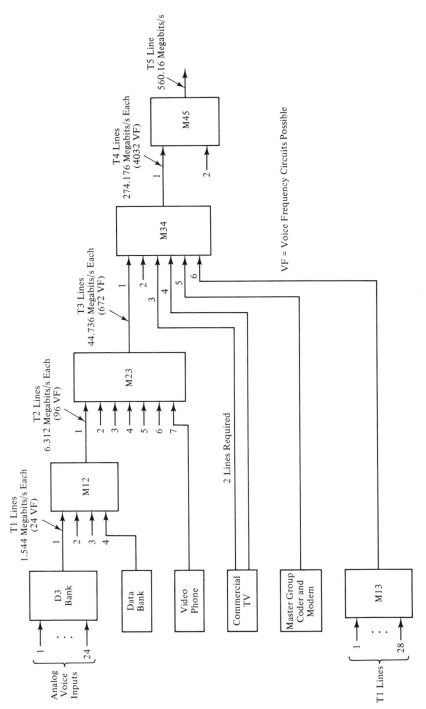

Figure 3.8-2. Digital multiplex hierarchies in use in the United States and Canada.

123

28 T1-line inputs† [49, 50]. Furthermore, the various T-lines do not always have to have analog voice signals as their original source messages. A data bank can be used to multiplex *digital* sources to generate a 1.544 megabits/s output with the correct format to pass over a T1 line. A coder can be used to digitize visual telephone signals for T2 line inputs to M23 multiplexers [51]. A codec (a system for *co*ding and *dec*oding) can be used to make an analog mastergroup signal (a waveform resulting from frequency multiplexing 600 telephone signals) compatible as an input to an M34 multiplexer [45, 51]. Finally, using current-day technology, digital color television signals can be generated with a small enough bandwidth to be compatible with a T3 line [49].‡

The digital hierarchy used in Japan is much the same as that in Fig. 3.8-2 at the lower levels. It differs in the higher levels as shown in Fig. 3.8-3.

Except in North America and Japan, most of the rest of the world uses the hierarchy suggested by the Consultative Committee on International Telegraphy and Telephony (CCITT), as shown in Fig. 3.8-4. As with the AT&T hierarchy, level 3 can be bypassed; in this case an M24 multiplexer generates a level-4 waveform by multiplexing 16 level-2 signals [50].

Asynchronous Multiplexing

In *asynchronous multiplexing,* the input digital bit streams originate from sources with different clock rates. Even if the sources have clocks with the same *average* rates, these rates can change with time due to natural instabilities in the oscillators that generate them. Furthermore, even if all the source clocks are perfectly synchronized, variations in delay of the channels connecting the sources to the multiplexer input can cause variations in the bit rates of the arriving digital signals. A commonly used method for overcoming these problems is to design the multiplexer output bit rate to be at least as large as the sum of the input rates at their largest possible values. With this choice the rate of the input bits (or words) is collectively smaller than the output bit (or word) rate. Times will occur when input data are not available at output clock instants. Such voids are filled by *bit* (or *word*) *stuffing*, the adding of dummy bits or words. *Control bits* in the output data format are used to identify where stuff bits or words are located so the receiver demultiplexer can remove them.

Because the concepts involved in either word or bit stuffing are similar, we discuss only bit stuffing by means of an example system using bit interleaving. Consider a multiplexer such as the AT&T M12 device of Fig. 3.8-2, where all of the input lines (tributaries) have the same average rate,

† The digital signal on a line Ti, $i = 1, 2, \ldots, 5$ is sometimes called a DSi signal; thus a DS3 signal is used on a T3 line.

‡ For details on the M13 and M34 multiplexers, see [52].

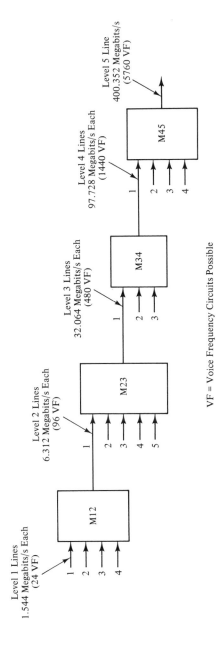

Level 1 Lines
1.544 Megabits/s Each
(24 VF)

Level 2 Lines
6.312 Megabits/s Each
(96 VF)

Level 3 Lines
32.064 Megabits/s Each
(480 VF)

Level 4 Lines
97.728 Megabits/s Each
(1440 VF)

Level 5 Line
400.352 Megabits/s
(5760 VF)

VF = Voice Frequency Circuits Possible

Figure 3.8-3. Digital hierarchy applicable in Japan.

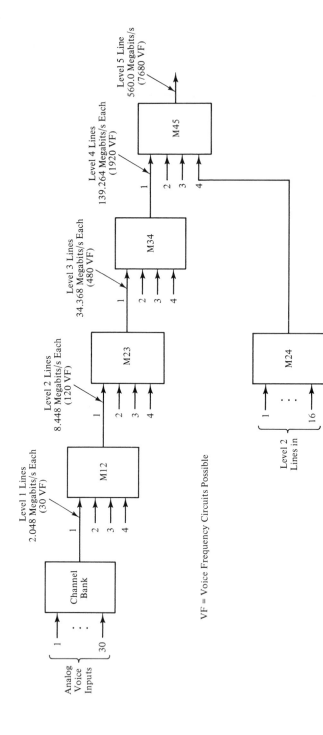

Figure 3.8-4. Digital hierarchy suggested by CCITT applicable to most of the world (United States, Canada, and Japan excluded).

denoted by $\overline{R}_i$. For an output total bit rate R_o and a multiframe time T_{mf}, the total number of bits in a multiframe will be

$$R_o T_{mf} = N(M_d + M_c), \quad (3.8\text{-}1)$$

where N is the number of tributaries, M_d and M_c are the numbers of data and control bits, respectively, *per tributary* per multiframe. Now suppose that the average number of stuff bits added to *each* tributary per multiframe is $\overline{s}$. Then the total number of bits generated from tributaries is $N[\overline{R}_i T_{mf} + \overline{s}]$, which must equal the number of output data bits NM_d (because *all* source bits, even stuff bits, are data to the output line). Hence

$$\overline{R}_i T_{mf} + \overline{s} = M_d. \quad (3.8\text{-}2)$$

By solving (3.8-1) for T_{mf} and substituting into (3.8-2) we have

$$\overline{s} = M_d\left[1 - \frac{N\,\overline{R}_i(M_d + M_c)}{R_o\,M_d}\right]. \quad (3.8\text{-}3)$$

Alternatively,

$$R_o = \frac{N\overline{R}_i(M_d + M_c)}{M_d}\left(\frac{M_d}{M_d - \overline{s}}\right). \quad (3.8\text{-}4)$$

By proper choice of parameters, $\overline{s}$ can be small, whereas a relatively large variation in input bit rate is allowed.

Example 3.8-1

In a given multiplexer suppose $\overline{R}_i = 1.544$ megabits/s, $R_o = 6.312$ megabits/s, $M_d = 288$ data bits per tributary per multiframe, $M_c = 6$ control bits per tributary per multiframe, and $N = 4$ tributaries. From (3.8-3)

$$\overline{s} = 288\left[1 - \frac{4(1.544)294}{6.312(288)}\right] = 0.3346$$

stuff bits occur per tributary per multiframe, on the average.

The values of the above example apply to the AT&T M12 multiplexer, which is designed to allow only one stuff bit, maximum, per tributary per multiframe. By retracing the procedures leading to (3.8-4), we find that the equation applies to instantaneous rates if all tributaries are assumed to have the same rate. That is, we can remove the averaging bars and write

$$R_i = \frac{R_o(M_d - s)}{N(M_d + M_c)}, \quad (3.8\text{-}5)$$

where R_i and s are momentary values. In the M12 multiplexer $0 \leq s \leq 1$, so 1.5404 megabits/s $\leq R_i \leq 1.5458$ megabits/s for a tolerance of 1.544 megabits/s ($+0.116\%$, -0.231%). This allowable range of input bit rates is larger than that expected from current-day stable clocks (easily on the order of ± 50 parts per million or better) but is needed to account for timing jitter and reduce reframe time of the multiplexer [22, p. 613].

The detailed structure of the multiframe format of the M12 multiplexer is shown in Fig. 3.8-5 [22]. Each row will be called a frame and all four rows, transmitted in sequence, are called a multiframe. Positions marked M, F, or C, with subscripts, represent single control bits where the subscripts on M and F denote the bits' digital values (0 or 1). The F-sequence is an alternating 0,1-sequence that allows frame identification for synchronization purposes. After frame lock-on, the receiver's demultiplexer searches the M-sequence to identify its pattern (which is 0111) for multiframe synchronization. After frame and multiframe lock, the demultiplexer identifies data bits that occur in blocks of 48 bits between control bits in Fig. 3.8-5. The C-sequence, which is discussed later, identifies the location of stuff bits.

M_0 (48) C_I (48) F_0 (48) C_I (48) C_I (48) F_1 (48)

M_1 (48) C_{II} (48) F_0 (48) C_{II} (48) C_{II} (48) F_1 (48)

M_1 (48) C_{III} (48) F_0 (48) C_{III} (48) C_{III} (48) F_1 (48)

M_1 (48) C_{IV} (48) F_0 (48) C_{IV} (48) C_{IV} (48) F_1 (48)

Figure 3.8-5. Multiframe format of AT&T M12 multiplexer.

Each 48-bit data block involves interleaving one digit from each of the four inputs and then repeating the pattern until 12 digits are taken from each of the four inputs. Due to design, only one stuff bit per frame can occur; if a bit has been stuffed into input i ($i = 1, 2, 3,$ or 4), it is placed in the first data bit's position (of the 12 for input i) that follows F_1 in frame i. To identify where stuff bits occur, the C-sequence in each frame is used. A C-sequence 111 corresponds to a stuff bit, whereas a normal sequence 000 indicates no stuff bit. Thus control bits C_{III} would have the sequence 111 in frame III if a stuff bit has been inserted; otherwise the C_{III}-sequence is 000. For high reliability, majority logic is used in detecting the C-sequences (two 1s of three is a stuff bit, two 0s of three is no stuff bit) because an error in a stuff-bit decision will result in an error in all bits of the entire frame of the T1 system in question [53, p. 163].

Time Division Multiple Access

Time division multiple access (TDMA)† is a time-multiplexing method important in communication satellite systems. In this system various earth stations may transmit information to a shared satellite during an assigned time interval. All stations transmit on the same frequency (about 6 GHz) and receive down-link transmissions from the satellite on the same frequency

† Although this chapter is mainly on baseband systems whereas TDMA is a carrier (bandpass) system, we include it here because it illustrates the time-multiplexing principle.

(about 4 GHz).† The system is synchronous so that when a station is transmitting, it is the only one and it has access to the satellite's full bandwidth.

To utilize a TDMA system a station must store its input (user) data for the period of time between its transmissions, called *bursts,* and then read the data out for transmission in a burst at a very high bit rate. Because stations occupy different distances from the satellite, which itself can change position with time, careful control over burst timing is essential. These characteristics require stations to have relatively large memory buffers and generally be under computer control.

A typical format for data passing through the TDMA satellite is shown in Fig. 3.8-6. A frame consists of arriving bursts from earth stations being

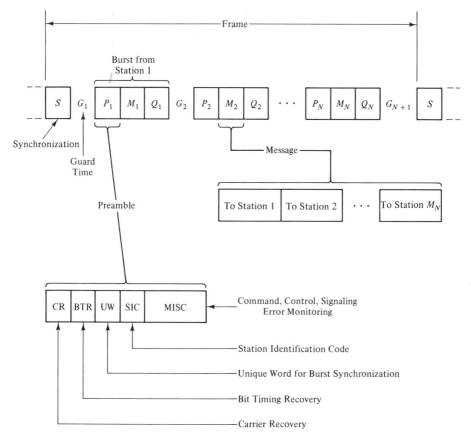

Figure 3.8-6. Frame format for signal processed by a satellite in a TDMA system.

† Recent trends are for newer satellites to operate at higher frequencies (14 GHz up, 12 GHz down) as the lower frequency systems become traffic saturated [54].

served. The bursts do not have to have the same durations and can be longer for heavy-traffic stations. Guard intervals G_1, G_2, . . . , G_{N+1} are present to absorb small timing errors in burst arrivals so that overlap is prevented. An interval S is allocated for frame and system synchronization; this transmission originates from a single station called the *reference station* [31, p. 238]. Bursts consist of a *preamble,* a data portion; and sometimes a *postamble* is present to initialize decoders for the following burst.

The preamble consists of portions used for carrier recovery (CR) in coherent systems, bit timing recovery (BTR), burst synchronization using a *unique word* (UW), station identification coding (SIC), and other miscellaneous items, such as command and control, signaling, and error monitoring. The data or message segment is subdivided into time slots allocated to the stations for which the data are intended, as shown in Fig. 3.8-6.

After amplification and frequency shifting, the satellite retransmits its signal back to the earth stations, which must lock to the bursts received and sort out the data streams unique to their locations.

Other TDM Methods

In the TDMA method just described, the earth stations were required to transmit only during assigned burst time slots. Another TDM method, called ALOHA [55], allows any network station to transmit bursts, called *packets,* at any time. Since the packets are unsynchronized, they may overlap at the satellite, creating what is called a *collision.* Collisions necessitate that the stations retransmit their packets after a random delay period. Repeated collisions require repeated delays and retransmissions. ALOHA makes use of the fact that most stations have significant idle time but it fails when heavy traffic loading of the satellite occurs.

In a modification of the ALOHA system, called *slotted* ALOHA, stations may transmit only during prescribed time slots, but may use any of the allowed slots randomly. In principle, this method eliminates partial collisions but does not prevent full collisions from happening in any given time slot.

3.9 SUMMARY AND DISCUSSION

This chapter is mainly concerned with describing the functions shown in the digicom system block diagram of Fig. 1.3-1 for the special case of baseband transmitted waveforms. The effect of noise on the channel is not considered because this topic is reserved for detailed discussion in the next chapter. Our main efforts concentrate on the ways in which digital waveforms are generated from analog or digital message sources (transmitter problem). Since the functions of the receiver are essentially the inverses of those in the transmitter, these functions receive less attention.

Initially it is shown that an analog message can be sampled and quantized to obtain a digital representation of the analog source. It is found that all quantizers limit the signal-to-noise ratio that can be achieved when message reconstruction is performed. The largest signal-to-noise ratio achievable, denoted by (S_o/N_q), is called the signal-to-quantization noise power ratio. For the very important uniform quantizer, that has a uniform separation δv between adjacent quantum levels, and a message $f(t)$ with average power $\overline{f^2(t)}$, it is shown that

$$\left(\frac{S_o}{N_q}\right) = \frac{12\,\overline{f^2(t)}}{(\delta v)^2} \tag{3.9-1}$$

when the analog message is bounded or when the extreme quantum levels are set so that quantizer overload is negligible. By setting the extreme quantum levels properly (Fig. 3.2-4), the uniform quantizer can be optimized to give the performance shown in Fig. 3.2-5. The curves of Fig. 3.2-5 show clearly that the optimum uniform quantizer is sensitive to shape mismatch, which occurs when the probability density of the message differs in form (shape) from that for which the quantizer is designed.

By generalizing to allow nonuniformly separated quantum levels, a quantizer can be optimized. A procedure is outlined to determine the optimum quantum levels. For the optimum nonuniform quantizer, (S_o/N_q) is given in Fig. 3.2-2. For a large number (L) of quantum levels, (3.2-19) can be used to obtain (S_o/N_q) quite accurately.

The optimum nonuniform quantizer is equivalent to a uniform quantizer operating on a signal that has passed through a properly chosen nonlinear network called a compressor. The companded quantizer of Sec. 3.3, which is defined by the nonlinear function $v(f)$ of (3.3-7), is optimum for large L. Its performance is determined by $(S_o/N_q) = \overline{f^2(t)}/\overline{\varepsilon_q^2}$, where $\overline{\varepsilon_q^2}$ is given by (3.3-6). Often used *nonoptimum* companded quantizers, the μ-law and A-law devices, are also briefly described.

The next major segment of this chapter concerns the conversion (source encoding) of the message levels into a suitable digital representation, usually a sequence of binary digits **0** and **1**. Various encoding methods are described, including the optimum, or Huffman, procedure that minimized the average number of binary digits needed to represent a source.

In some systems additional coding, called channel encoding, is used to improve the system's performance with channel noise present. Both block codes and convolutional codes for this purpose are discussed. By use of these codes, the receiver can correct certain errors it makes in initially determining which binary digits are transmitted.

Next, a number of important waveforms (formats) are discussed that can be assigned to represent the channel-encoded binary digits. The spectrums of these waveforms are also determined. There is no optimum waveform choice for all systems, but some comparisons can be made [56]. In cases where bandwidth conservation is not too important, return-to-zero (RZ) type formats may be attractive. The polar-RZ format, for example, always has two level transitions per symbol interval, which allows extraction of symbol timing. In applications where bandwidth conservation is important the NRZ-type formats are best with duobinary and modified duobinary being the better of the ones discussed. Although symbol synchronization can be derived from these waveforms, problems can arise if long strings of either 0s or 1s occur.

Naturally, bandwidth and synchronization characteristics of a format are not the only important properties of a waveform format. Its performance in noise is also important. We consider this subject in the following chapter.

Finally, this chapter closes on a more practical level by discussing ways of time multiplexing digital waveforms. Examples of synchronous and asynchronous multiplexers are given.

REFERENCES

[1] Sklar, B., A Structured Overview of Digital Communications—A Tutorial Review—Part I, *IEEE Communications Magazine,* Vol. 21, No. 5, August 1983, pp. 4–17. Part II of same title in October 1983 issue, pp. 6–21.

[2] Peebles, Jr., P. Z., *Communication System Principles,* Addison-Wesley Publishing Co., Inc., Reading, Massachusetts, 1976. (Figure 3.1-2 has been adapted.)

[3] Panter, P. F., and Dite, W., Quantization Distortion in Pulse-Count Modulation with Nonuniform Spacing of Levels, *Proceedings of the IRE,* Vol. 39, No. 1, January 1951, pp. 44–48.

[4] Lloyd, S. P., Least Squares Quantization of PCM, *IEEE Transactions on Information Theory,* Vol. IT-28, No. 2, March 1982. This paper is a near-verbatim reproduction of a manuscript prepared prior to July 31, 1957.

[5] Max, J., Quantizing for Minimum Distortion, *IRE Transactions on Information Theory,* Vol. IT-6, No. 1, March 1960, pp. 7–12.

[6] Bucklew, J. A., and Gallagher, Jr., N. C., Some Properties of Uniform Step Size Quantizers, *IEEE Transactions on Information Theory,* Vol. IT-26, No. 5, September 1980, pp. 610–613.

[7] Smith, B., Instantaneous Companding of Quantized Signals, *Bell System Technical Journal,* Vol. 36, No. 3, May 1957, pp. 653–709.

[8] Fleischer, P., Sufficient Conditions for Achieving Minimum Distortion in a Quantizer, *IEEE International Convention Record,* Part I, 1964, pp. 104–111.

[9] Abaya, E., and Wise, G. L., Some Notes on Optimal Quantization, *International Conference on Communications Conference Record,* June 14–18, 1981, pp. 30.7.1–30.7.5.

[10] Bucklew, J. A., and Wise, G. L., Multidimensional Asymptotic Quantization Theory with *r*th Power Distortion Measures, *IEEE Transactions on Information Theory,* Vol. IT-28, No. 2, March 1982, pp. 239–247.

[11] Abaya, E. F., and Wise, G. L., On the Existence of Optimal Quantizers, *IEEE Transactions on Information Theory*, Vol. IT-28, No. 6, November 1982, pp. 937–940.

[12] Paez, M. D., and Glisson, T. H., Minimum Mean-Squared-Error Quantization in Speech PCM and DPCM Systems, *IEEE Transactions on Communications*, Vol. COM-20, No. 2, April 1972, pp. 225–230.

[13] McDonald, R. A., Signal-to-Noise and Idle Channel Performance of Differential Pulse Code Modulation Systems—Particular Applications to Voice Signals, *Bell System Technical Journal*, Vol. 45, No. 7, September 1966, pp. 1123–1151.

[14] Stroh, R. W., and Paez, M. D., A Comparison of Optimum and Logarithmic Quantization for Speech PCM and DPCM Systems, *IEEE Transactions on Communications*, Vol. COM-21, No. 6, June 1973, pp. 752–757.

[15] Mauersberger, W., An Analytic Function Describing the Error Performance of Optimum Quantizers, *IEEE Transactions on Information Theory*, Vol. IT-27, No. 4, July 1981, pp. 519–521.

[16] Abramowitz, M., and Stegun, I. A., *Handbook of Mathematical Functions with Formulas, Graphs, and Mathematical Tables*, Vol. 55, National Bureau of Standards, June 1964.

[17] Thomas, J. B., *An Introduction to Statistical Communication Theory*, John Wiley & Sons, Inc., New York, 1969.

[18] Spilker, Jr., J. J., *Digital Communications by Satellite*, Prentice-Hall, Inc., Englewood Cliffs, New Jersey, 1977.

[19] Mauersberger, W., Experimental Results on the Performance of Mismatched Quantizers, *IEEE Transactions on Information Theory*, Vol. IT-25, No. 4, July 1979, pp. 381–386.

[20] Dammann, C. L., McDaniel, L. D., and Maddox, C. L., D2 Channel Bank: Multiplexing and Coding, *Bell System Technical Journal*, Vol. 51, No. 8, October 1972, pp. 1675–1699.

[21] Freeman, R. L., *Telecommunication Transmission Handbook*, Wiley-Interscience, John Wiley & Sons, Inc., New York, 1975.

[22] Bell Telephone Laboratories Staff, *Transmission Systems for Communications*, Bell Telephone Laboratories, Inc., 4th ed., February 1970.

[23] Carlson, A. B., *Communication Systems, An Introduction to Signals and Noise in Electrical Communication*, 2nd ed., McGraw-Hill Book Co., Inc., New York, 1975. (See also third edition, 1986.)

[24] Gallager, R. G., *Information Theory and Reliable Communication*, John Wiley & Sons, Inc., New York, 1968.

[25] Huffman, D. A., A Method for the Construction of Minimum Redundancy Codes, *Proceedings of the IRE*, Vol. 40, No. 10, September 1952, pp. 1098–1101.

[26] Clark, Jr., G. C., and Cain, J. B., *Error-Correction Coding for Digital Communications*, Plenum Press, New York, 1981.

[27] Proakis, J. G., *Digital Communications*, McGraw-Hill Book Co., Inc., New York, 1983.

[28] Hamming, R. W., Error Detecting and Error Correcting Codes, *Bell System Technical Journal*, Vol. 29, No. 2, April 1950, pp. 147–160.

[29] Blahut, R. E., *Theory and Practice of Error Control Codes*, Addison-Wesley Publishing Co., Reading, Massachusetts, 1983.

[30] Pless, V., *Introduction to the Theory of Error Correcting Codes*, John Wiley & Sons, New York, 1982.

[31] Bhargava, V. K., Haccoun, D., Matyas, R., and Nuspl, P. P., *Digital Communications by Satellite*, John Wiley & Sons, New York, 1981.

[32] Hamming, R. W., *Coding and Information Theory*, Prentice-Hall, Inc., Englewood Cliffs, New Jersey, 1980.

[33] Lin, Shu, and Costello, Jr., D. J., *Error Control Coding Fundamentals and Applications*, Prentice-Hall, Inc., Englewood Cliffs, New Jersey, 1983.

[34] Berlekamp, E. R. (Editor), *Key Papers in the Development of Coding Theory*, IEEE Press, New York, 1974.

[35] Viterbi, A. J., Error Bounds for Convolutional Codes and an Asymptotically Optimum Decoding Algorithm, *IEEE Transactions on Information Theory*, Vol. IT-13, No. 2, April 1967, pp. 260–269.

[36] Couch, II, L. W., *Digital and Analog Communication Systems*, Macmillan Publishing Co., Inc., New York, 1983.

[37] Stremler, F. G., *Introduction to Communication Systems*, Addison-Wesley Publishing Co., Reading, Massachusetts, 1982.

[38] Shanmugam, K. S., *Digital and Analog Communication Systems*, John Wiley & Sons, New York, 1979.

[39] Lindsey, W. C., and Simon, M. K., *Telecommunication Systems Engineering*, Prentice-Hall, Inc., Englewood Cliffs, New Jersey, 1973.

[40] Lender, A., The Duobinary Technique for High-Speed Data Transmission, *IEEE Transactions on Communication and Electronics*, Vol. CE-82, No. 66, May 1963, pp. 214–218.

[41] Kretzmer, E. R., An Efficient Binary Data Transmission System, *IEEE Transactions on Communication Systems*, Vol. CS-12, No. 2, June 1964, pp. 250–251.

[42] Lender, A., Correlative Digital Communication Techniques, *IEEE Transactions on Communication Technology*, Vol. COM-12, No. 4, December 1964, pp. 128–135.

[43] Lender, A., Correlative Level Coding for Binary-Data Transmission, *IEEE Spectrum*, Vol. 3, No. 2, February 1966, pp. 104–115.

[44] Pasupathy, S., Correlative Coding: A Bandwidth-Efficient Signaling Scheme, *IEEE Communications Society Magazine*, Vol. 15, No. 4, July 1977, pp. 4–11.

[45] James, R. T., and Muench, P. E., AT&T Facilities and Services, *Proceedings of the IEEE*, Vol. 60, No. 11, November 1972, pp. 1342–1349.

[46] Henning, H. H., and Pan, J. W., D2 Channel Bank: System Aspects, *Bell System Technical Journal*, Vol. 51, No. 8, October 1972, pp. 1641–1657.

[47] Gaunt, W. B., and Evans, Jr., J. B., The D3 Channel Bank, *Bell Laboratories Record*, August 1972, pp. 229–233.

[48] Owen, F. F. E., *PCM and Digital Transmission Systems,* McGraw-Hill Book Co., New York, 1982.

[49] Feher, K., *Digital Communications: Microwave Applications,* Prentice-Hall, Inc., Englewood Cliffs, New Jersey, 1981.

[50] Inose, H., *An Introduction to Digital Integrated Communications Systems,* University of Tokyo Press, Tokyo, 1979.

[51] Freeman, R. L., *Telecommunication System Engineering Analog and Digital Network Design,* John Wiley & Sons, New York, 1980.

[52] Maunsell, H. I., Robrock, R. B., and von Roesgen, C. A., The M13 and M34 Digital Multiplexes, *ICC 75 Conference Record,* Vol. 3, 1975 International Conference on Communications, June 16–18, San Francisco, California, pp. 48-5 through 48-9.

[53] Schwartz, M., *Information Transmission, Modulation, and Noise,* 3rd ed., McGraw-Hill Book Co., New York, 1980.

[54] Gagliardi, R. M., *Satellite Communications,* Lifetime Learning Publications, Belmont, California, 1984.

[55] Lam, S. S., Satellite Packet Communications, Multiple Access Protocols and Performance, *IEEE Transactions on Communications,* Vol. COM-27, No. 10, Part I, October 1979, pp. 1456–1466.

[56] Stallings, W., Digital Signaling Techniques, *IEEE Communications Magazine,* Vol. 22, No. 12, December 1984, pp. 21–25.

PROBLEMS

3-1. A random analog message has values -8 V $\leq f(t) \leq 8$ V. If samples of $f(t)$ are to be quantized using an 8-bit binary code in a uniform quantizer, what voltage levels form the boundaries of the quantizer's intervals? What is a reasonable set of quantum levels if no further knowledge is available about $f(t)$?

3-2. A message $f(t)$ has a probability density

$$p_f(f) = \begin{cases} \dfrac{\pi}{4V} \cos\left(\dfrac{\pi f}{2V}\right), & -V \leq f \leq V \\ 0, & \text{elsewhere.} \end{cases}$$

It is applied to a uniform quantizer with eight quantum levels ± 1, ± 3, ± 5, and ± 7 V centered in uniform 2-V intervals. (a) What maximum value can V have without amplitude overloading the quantizer? (b) What is the power in $f(t)$ when at its reference level found in (a)?

★**3-3.** A message and quantizer are defined as in Prob. 3-2 except V can have any value. (a) Use (3.1-11) to find an expression for mean-squared quantization error $\overline{\varepsilon_q^2}$ valid for any value of V. (b) Find the power in $f(t)$ for any value of V. (c) Sketch the ratio $\overline{f^2}/\overline{\varepsilon_q^2}$ versus $\overline{f^2}/f_r^2$ where f_r^2 is the signal's reference power level.

★**3-4.** Work Prob. 3-3 except assume a message defined by

$$p_f(f) = \begin{cases} \dfrac{e^{-|f|/V}}{2V(1 - e^{-1})}, & -V \leqslant f \leqslant V \\ 0, & \text{elsewhere.} \end{cases}$$

3-5. If δv is small and L is large show that the middle terms of (3.2-3) will reduce to $(\delta v)^2/12$.

3-6. A message $f(t)$ has a probability density

$$p_f(f) = \begin{cases} \dfrac{\left(1 - \left|\dfrac{f}{4}\right|\right)}{4}, & |f| \leqslant 4 \text{ V} \\ 0, & \text{elsewhere} \end{cases}$$

when at its reference level for a uniform quantizer with a large number of quantum levels. (a) Find the signal's reference power level. (b) What is the signal's crest factor? (c) How many levels are required to provide a maximum value of $(S_o/N_q) = 131{,}072.0$ with no overload?

★**3-7.** A random message with probability density

$$p_f(f) = \frac{1}{\sqrt{2\overline{f^2}}} \exp\left(-\frac{\sqrt{2}|f|}{\sqrt{\overline{f^2}}}\right)$$

is quantized in a quantizer having a large number (L) of levels between extremes $-V$, V. Assume L is so large that extreme quantum levels are $l_1 \approx -V$ and $l_L \approx V$. (a) Solve (3.2-6) for the signal-to-quantization noise power ratio. (b) Plot (S_o/N_q) in decibels versus normalized message power $3\overline{f^2}/V^2$ in decibels for $L = 16$ and 128. (c) Compare your result with Fig. 3.2-1 for a uniformly distributed message by assuming $|f_r(t)|_{max}$ in the figure (and Example 3.2-1) equals V.

3-8. Use Leibniz's rule to derive (3.2-10) and (3.2-11) from (3.2-9).

3-9. Show that (3.2-12) derives from (3.2-11).

3-10. A quantizer has 32 nonuniform quantum levels that are optimally set to work with a Gaussian random message at (reference) power level $\overline{f_r^2(t)} = 0.25$. (a) At what voltages should the extreme quantum levels l_1 and l_{32} be set? (b) What value of (S_o/N_q) can be expected when message recovery occurs in the receiver?

3-11. Work Prob. 3-10 for a message with a Laplace probability density.

3-12. Work Prob. 3-10 for a message with a gamma probability density.

3-13. A quantizer must be designed to operate with a Gaussian message and have optimum uniformly separated quantum levels. (a) If the reconstructed message must have a signal-to-quantization power ratio of 20.3 dB, how many quantum levels are required? (b) What loading factor should the quantizer have? (c) If the extreme quantum levels are to be optimally set to ± 2.35 V what should the signal's reference power be?

3-14. Work Prob. 3-13 for a message with a Laplace probability density and a signal-to-quantization power ratio of 20.6 dB.

3-15. Work Prob. 3-13 for a message with a gamma probability density and a signal-to-quantization power ratio of 17.5 dB.

3-16. For a Gaussian message for which $\nu = 2$ in (3.2-22) show that (3.2-19) will reduce to

$$\left(\frac{S_o}{N_q}\right) \approx \frac{L^2}{1 + \left(\dfrac{\pi\sqrt{27}}{6} - 1\right)[1 - (1 + \log_2 L)e^{-\log_2 L}]}.$$

3-17. Assume $L = 32$ and use the result of Prob. 3-16 to compute the approximate value of (S_o/N_q). Compare your result to the value taken from Fig. 3.2-3.

3-18. The optimum nonuniform quantizer reduces to (becomes) a uniform quantizer when the message has a uniform distribution. Assume such a message and show that (3.2-19) does give the performance of the uniform quantizer (which is given in Example 3.2-1 for $R = 1$). That is, show that

$$\left(\frac{S_o}{N_q}\right) \approx L^2.$$

3-19. Use (3.2-14) to prove that the optimum quantizer for a uniformly distributed message is uniform.

3-20. Sketch the optimum compressor characteristic $v(f)/V$, as given in Example 3.3-1, versus f/V for values of $V/\sqrt{\overline{f_r^2}} = 5$, 10, and 20. Which value of $V/\sqrt{\overline{f_r^2}}$ corresponds to the greatest message compression?

3-21. For a signal having a Laplace probability density show that (3.3-16) is true if V is large enough relative to $\overline{f^2}$ so that $p_f(f)$ is negligible for $|f| > V$.

★3-22. A message $f(t)$, to be quantized after using a μ-law compressor, has a uniform probability density $p_f(f)$ on $-V_f < f < V_f$ where $V_f \leq V$ of the compressor. (a) Find $\overline{f^2}$ and $\overline{|f|}$ for this message. (b) Use (3.3-15) to compute and plot (S_o/N_q) for $\mu = 255$ with $L = 16$ and 128. Compare the results with Fig. 3.3-4 for the message having a Laplace probability density.

3-23. Sketch the A-law compressor characteristic $v(f)/V$ of (3.3-17) versus $0 \leq f/V \leq 1$ for $A = 10$, 100, 1000. Compare your curves with those of Fig. 3.3-3 for μ-law.

3-24. (a) Assume a message has the Laplace density of Example 3.3-1 and is to be quantized by the A-law characteristic of (3.3-17). If $\overline{f^2}/V^2$ is small enough (say 0.01 or smaller) so that $p_f(f)$ is negligible for $|f| > V$, use (3.3-5) with $f_1 = -V$ and $f_{L+1} = V$ to find $(S_o/N_q) = (S_o/\varepsilon_q^2)$. (b) Plot $(S_o/N_q)/L^2$ in dB versus $\overline{f^2}/V^2$ in dB for $\overline{f^2}/V^2 \leq 0.01$ when $A = 100$. Compare your results with the $\mu = 255$ curve of Fig. 3.3-4.

3-25. An eight-digit natural binary code is used to represent the integers $0, 1, \ldots, 255$ where **00000000** represents the number 0. Write the binary codewords for integers (a) 14, (b) 27, (c) 145, (d) 201, and (e) 237.

3-26. (a) What is the minimum number of code digits required to work Prob. 3-25 if a ternary code is used? Find the codewords for the integers (b) 19, (c) 35, (d) 166, and (e) 199.

3-27. Work Prob. 3-26 for a quaternary code.

3-28. A binary code of length 8 is used to represent the levels of a discrete message source. How many levels can the source generate for a unique representation?

3-29. Assume a digital message generates the decimal levels of Table 3.4-1 with the following probabilities:

Levels numbered	0, 15	1, 14	2, 13	3, 12	4, 11	5, 10	6, 9	7, 8
Probability	0.001	0.002	0.007	0.020	0.05	0.08	0.14	0.2

Find the average number of binary 1s transmitted per codeword when coding uses (a) a natural binary code, (b) a folded binary code, (c) an inverted folded binary code, and (d) a Gray code.

3-30. Write the natural binary codewords that correspond to the set of integers 0, 1, 2, . . . , 31. (a) How many digits are required in a 4-ary code to represent these same integers? (b) Write an orderly (increasing numbers) set of 4-ary codewords corresponding to the integers. Note that you have performed a simple binary–to–M-ary code conversion with $M = 4$.

3-31. A digital source can generate any one of four symbols l_1, l_2, l_3, and l_4 that occur with probabilities $P_1 = 1/8$, $P_2 = 1/8$, $P_3 = 1/4$, and $P_4 = 1/2$, respectively. (a) Find the information contained in each source symbol. (b) Find the entropy of this source.

3-32. For the source of Prob. 3-31 what is the code efficiency if source symbols are coded natural binary?

3-33. If binary codewords **0**, **10**, **110**, and **111**, respectively, represent source symbols l_4, l_3, l_2, and l_1, in Prob. 3-31, what is the code's efficiency?

3-34. A digitized source can generate any one of eight symbols at any one time. The symbols l_1, l_2, . . . , l_8 occur with probabilities $P_1 = 0.01$, $P_2 = 0.06$, $P_3 = 0.12$, $P_4 = 0.30$, $P_5 = 0.22$, $P_6 = 0.16$, $P_7 = 0.11$, and $P_8 = 0.02$, respectively. (a) Find the Huffman code for this source. (b) Find the code's efficiency. (c) What would the efficiency be if a Gray code had been used instead?

3-35. A digital source can generate eight symbols l_1, l_2, . . . , l_8 with respective probabilities $P_1 = 0.33$, $P_2 = 0.22$, $P_3 = 0.16$, $P_4 = 0.1$, $P_5 = 0.09$, $P_6 = 0.04$, $P_7 = 0.035$, and $P_8 = 0.025$. Find the M-ary Huffman code for $M = 3$ (ternary coding).

3-36. Symbols A and B of a binary source have probabilities $P(A) = 0.7$ and $P(B) = 0.3$. Symbols over time are statistically independent. Use Huffman source encoding and find the entropy and code efficiency of (a) the source, (b) the second-order extended source, (c) the third-order extended source, and (d) the fourth-order extended source.

3-37. Work Prob. 3-36 except assume $P(A) = 0.9$ and $P(B) = 0.1$.

3-38. A ternary source can generate three symbols, A, B and C with probabilities $P(A) = 0.6$, $P(B) = 0.3$, and $P(C) = 0.1$. Use Huffman binary source encoding and find the entropy, codewords, and code efficiency of (a) the source, and (b) the second-order extension of the source.

3-39. An (n, k) block code with single-error correction capability is desired for a source. Consider values of k of $4 \leq k \leq 12$ and find the smallest number of check digits that can be used for each value. Find each code's efficiency. Which of your codes can be perfect?

3-40. Work Prob. 3-39, except assume a two-error correction code is desired.

3-41. A nonsystematic (n, k) code has the parity check matrix

$$[H] = \begin{bmatrix} 1 1 1 0 0 0 \\ 0 1 0 1 1 0 \\ 1 0 0 1 0 1 \end{bmatrix}.$$

(a) What can you do to put this matrix in systematic form? (b) What are n, k, and r? (c) How many errors, if any, can this code correct? (d) Use the systematic form of $[H]$ and find the code's generator matrix.

3-42. The generator matrix for a (7, 4) code is

$$[G] = \begin{bmatrix} 1 0 0 0 & 1 1 0 \\ 0 1 0 0 & 0 1 1 \\ 0 0 1 0 & 1 1 1 \\ 0 0 0 1 & 1 0 1 \end{bmatrix}.$$

The input data words to the encoder are derived from a 4-bit Gray source encoder. Make a table of the input codewords and their corresponding channel codewords.

3-43. (a) Find the parity check matrix for the encoder of Prob. 3-42. (b) At the receiver the codeword **1011001** is received over the channel. If any errors are present it is unlikely that more than one exists. With this assumption, is the received codeword in error? If so, in which digit?

3-44. Find code efficiency for the first five nontrivial Hamming codes. How many trivial Hamming codes are there? Why would these be called trivial?

3-45. Find both the parity check and generator matrices for a (15, 11) Hamming code.

3-46. Find the Hamming distances between the codeword **0100101111** and each of the codewords **1010101010**, **1110100000**, **1111000000**, and **0101010101**.

3-47. A (6, 3) code has a generator matrix

$$[G] = \begin{bmatrix} 1 0 0 & 1 0 1 \\ 0 1 0 & 0 0 1 \\ 0 0 1 & 1 1 0 \end{bmatrix}.$$

(a) Find all possible channel codewords if the source code is 3-bit natural binary. (b) Find the smallest Hamming distance (called the *minimum distance* d_{min}) of all channel codeword pairs. If d_{min} is at least 3, this code can correct at least one error per codeword.

3-48. A digit sequence **10100** is input to the convolutional encoder of Fig. 3.5-2. Show the path through the code tree and the output sequence generated for this input.

3-49. The encoder of Fig. 3.5-2(a) is modified so that the left and right adder outputs are $S_1 \oplus S_2$ and $S_2 \oplus S_3$, respectively. (a) Sketch the new code tree. (b) Indicate the tree path corresponding to the input sequence **100110** and find the corresponding output sequence.

3-50. Draw the trellis path applicable to the encoder of Prob. 3-49.

3-51. A receiver demodulator for the convolutional encoder of Fig. 3.5-2(a) determines that the received channel sequence is **110100001000011100110010** where the last four are known to be tail digits. Use the Viterbi algorithm with the applicable trellis of Fig. 3.5-3 to find the original data word. Based on the recovered data word being correct, were any channel errors introduced into the received sequence? If so, which digits?

3-52. Find the output sequence of a differential encoder to the input sequence **1011100100011** when the output sequence is initially **0**.

3-53. Let $p(t)$ of (3.6-11) for the duobinary waveform be replaced by

$$p(t) = \text{Sa}\left[\frac{\pi(t - T_b/2)}{T_b}\right].$$

(a) Sketch duobinary waveforms using $p(t)$ and (3.6-11) for the data sequence **01101110001001**. (b) At what times should the two waveforms be sampled in a receiver to produce the same sample values? (c) What advantage does the given waveform have compared to (3.6-11)?

3-54. Precode the data sequence **001011101001** and construct a duobinary waveform. Use waveform samples as indicated in the text (assuming no noise) and verify that the decoding logic ($\pm A$ decodes as **0**, zero decodes as **1**) reproduces the data sequence.

3-55. Work Prob. 3-54 for a modified duobinary waveform where the precoder uses a 2-bit delay and the decoder produces a **0** for sample values of zero and **1** for samples $\pm A$.

★3-56. Replace T_b in Fig. 3.6-5 by $2T_b$ and show that the output waveform is still duobinary. Assume precoding of the input $\{b_k\}$ sequence as in Fig. 3.6-3, except with delay $2T_b$. The proof can consist of showing that the sequence $\{b_k\}$ given by . . . **1001011101100101111000100** . . . is correctly recovered. Samples $\pm A$ decode as **0** and zero decodes as **1**.

3-57. An impulse $\delta(t)$ can be considered as a limit as follows:

$$\delta(t) = \lim_{\varepsilon \to 0} \frac{1}{\varepsilon} \text{rect}\left(\frac{t}{\varepsilon}\right)$$

$$\delta(t) = \lim_{\varepsilon \to 0} \frac{1}{\varepsilon} \text{tri}\left(\frac{t}{\varepsilon}\right)$$

where rect($\cdot$) and tri($\cdot$) are defined by (A.2-1) and (A.2-2), respectively. Use these results to prove that

$$\int_{-\infty}^{\infty} \delta(t - t_1)\delta(t - t_2 - t_0)\, dt = \delta(t_0 - t_1 + t_2)$$

where t_1, t_2, and t_0 are arbitrary real constants.

★3-58. By using the integral of Prob. 3-57 show that $\alpha(t)$ given by (3.7-1) has the time-averaged autocorrelation function of (3.7-3).

3-59. Assume that pulses in a unipolar RZ format have durations $T_b/2$. Find the power spectrum of the pulse stream when the coefficients α_k are 0 or A with equal probabilities. Compare your result with (3.7-10).

3-60. Find the power spectrum of a polar NRZ signal having pulses

$$p(t) = \text{Sa}\left[\frac{\pi}{T_b}\left(t - \frac{T_b}{2}\right)\right]$$

instead of the rectangular pulses of (3.6-2). Plot your result in the form of Fig. 3.7-2 and compare to the rectangular-pulse curve.

3-61. Work Prob. 3-60 for the duobinary waveform.

3-62. A modified duobinary waveform uses the pulse of Prob. 3-60. Find and plot the waveform's power spectrum and compare to that in Fig. 3.7-3 for rectangular pulses.

3-63. A bit interleaved multiplexer is to be designed to operate on a T1 line. Its output bit rate, therefore, is 1.544 megabits/s. The multiplexer's inputs are all at the same average rate of 32 kilobits/s. (a) If the design is required to output 40 data bits for every control bit, applicable to each input channel (so $M_d = 40 \, M_c$), find the number of input channels that will result in the smallest average number of stuff bits per control bit per channel. (b) What is this smallest stuff ratio?

Chapter 4

Baseband Digital Systems

4.0 INTRODUCTION

In Chap. 3 most of the blocks involved in a typical communication system, as illustrated in Fig. 1.3-1, were discussed. In this chapter we continue the discussions by concentrating mainly on the operation of receiver demodulation.

The demodulator is a critical system element, for it is here that the principal performance capability of the system is established. It is the demodulator that examines the input stream of channel pulses, which are partially obscured by channel noise, and determines which code digits are being conveyed in the received waveform. Such determinations always involve occasional errors because of the noise, and we usually seek to find *optimum* demodulators that minimize the probability of occurrence of errors.

In this chapter several forms of baseband digital systems are discussed. Some of these forms do not involve channel encoding or decoding and are capable of directly converting an analog message to a digital waveform for transmission. Delta modulation (DM), delta-sigma (D-SM), and adaptive delta modulation (ADM) systems have this form. As noted in Chap. 3, these systems typically operate with sources having outputs that are correlated over time; they do, therefore, represent a class of source encoder not considered in Chap. 3.

Other forms of baseband systems to be discussed, such as pulse code modulation (PCM) and M-ary systems, utilize the quantizing and encoding functions of Chap. 3. In our discussions of these, the word *system* will refer primarily to the demodulation operation.

It is appropriate to begin by defining an *optimum* binary digital system, since we usually attempt to realize near optimum results in practice. This initial work is relatively mathematical but should still be rewarding to those readers interested in the details of how such systems are optimized. Readers more concerned with applications of results and having less interest in theoretical developments can omit the initial work and go directly to Sec. 4.4 but must be careful not to violate the assumptions and limitations that apply to the results.

4.1 REQUIREMENTS AND MODELS FOR SYSTEM OPTIMIZATION

The proper selection of an optimum system requires that at least three quantities be defined [1]:

1. System constraints,
2. Input specification,
3. Optimization criterion.

System constraints vary widely. The system might be constrained, for example, only to the class of linear physically realizable systems. In other cases realizability may be of no concern or a nonlinear system might be acceptable. In the present case we shall require only that the optimum system be linear. Even though a nonrealizable system can result, it can always be approximated in practice and it can be used as a basis of performance comparison for other designs.

Input specification requires that at least some knowledge be available about the receiver's input waveform, which is usually the sum of a desired signal and undesired noise. As a rough rule of thumb, the more that is known about the inputs to the receiver, the more optimum a system can be in some sense.

The optimization criterion is a measure of the goodness of a system. It is usually a quantity that can be either maximized or minimized; the optimum system then becomes that which produces the maximum or minimum measure. One type of optimum system is discussed in Appendix B and is called a *matched filter*. It is selected to maximize the ratio of output peak signal power, at some specified instant in time, to average noise power. The optimization criterion is therefore to maximize output peak signal-to-noise ratio.

For the present problem we require the optimum system to be linear. Input specification and optimization criterion are defined in detail in the following subsections.

Basic Binary System

Figure 4.1-1 is helpful in defining the overall system to be optimized. At the transmitter m represents a binary digital message sequence that can have either of two possible values in any one symbol interval. They are $m = m_1$ if a binary **0** occurs, or $m = m_2$ if a **1** is transmitted. The symbol duration is denoted by T_b, as usual, and is the same for all intervals of the sequence. In general, the corresponding transmitted waveform is represented by $s(t)$. However, in any one symbol interval $s(t)$ will always correspond to either the signal $s_1(t)$ when $m = m_1$ or $s_2(t)$ when $m = m_2$. Both $s_1(t)$ and $s_2(t)$ are assumed arbitrary, except that they are real and nonzero only in the symbol intervals in which they are transmitted; that is, their duration is T_b. The receiver (demodulator) is assumed to know the forms of $s_1(t)$ and $s_2(t)$, but it does not know *which* is transmitted in a given symbol interval because it has no knowledge of which message (m_1 or m_2) has been originated by the source.

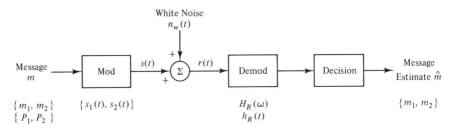

Figure 4.1-1. Basic binary digital system.

Noise Model

The input waveform to the receiver, denoted generally by $r(t)$, is assumed to be the sum of the transmitted signal and white, zero-mean, Gaussian noise, denoted by $n_w(t)$. The noise power density spectrum is denoted by $\mathcal{N}_0/2$ as usual, and applies to frequencies $-\infty < \omega < \infty$.

It is helpful to think of the white noise as resulting from the limit of ideally bandlimited white noise as its bandwidth becomes infinite. The bandlimited noise, denoted by $n_b(t)$, has a power spectrum

$$\mathcal{S}_{n_b}(\omega) = \left(\frac{\mathcal{N}_0}{2}\right) \text{rect} \left[\frac{\omega}{2W_N}\right] \tag{4.1-1}$$

where W_N is its bandwidth (radians per second). By inverse Fourier transformation of $\mathcal{S}_{n_b}(\omega)$ the corresponding autocorrelation function is

$$R_{n_b}(\tau) = \frac{\mathcal{N}_0 W_N}{2\pi} \text{Sa}(W_N \tau). \tag{4.1-2}$$

Figure 4.1-2 illustrates the behavior of $\mathcal{S}_{n_b}(\omega)$ and $R_{n_b}(\tau)$.

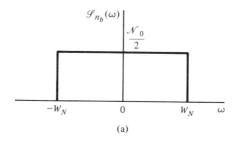

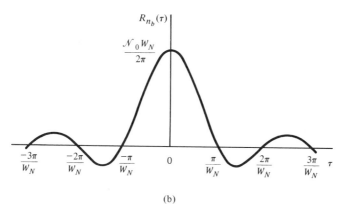

Figure 4.1-2. (a) Power spectrum and (b) autocorrelation function of band-limited white Gaussian noise.

Samples of $n_b(t)$ taken ΔT apart at times

$$t_k = k\Delta T, \qquad k = 1, 2, ..., K \tag{4.1-3}$$

are uncorrelated if ΔT is chosen as

$$\Delta T = \frac{\pi}{W_N}, \tag{4.1-4}$$

because the noise has zero mean and its autocorrelation is zero for sample time separations that are multiples of π/W_N. Furthermore, since the noise is assumed Gaussian, these samples are statistically independent. If the variance of noise samples is denoted by σ^2, then σ^2 is given by

$$\sigma^2 = E\,[n_b^2(t)] = R_{n_b}(0) = \frac{\mathcal{N}_0 W_N}{2\pi} = \frac{\mathcal{N}_0}{2\Delta T}. \tag{4.1-5}$$

Alternatively,

$$\sigma^2 \Delta T = \frac{\mathcal{N}_0}{2}. \tag{4.1-6}$$

Next, we observe that the joint probability density function of all K noise samples, denoted by $p_{n_b}(n_1, n_2, \ldots, n_K)$, is the product of the densities of the individual noise samples, denoted by $p_{n_k}(n_k)$, because samples are independent. Therefore, because samples are Gaussian, we have

$$p_{n_b}(n_1, n_2, \ldots, n_K) = \prod_{k=1}^{K} p_{n_k}(n_k) = (2\pi\sigma^2)^{-K/2} \exp\left[-\sum_{k=1}^{K} \frac{n_k^2}{(2\sigma^2)}\right]. \tag{4.1-7}$$

The above noise model constitutes our specification of the receiver's input noise. Choice of this noise model greatly facilitates the problem of finding optimum receivers. If the K noise samples occur during one symbol interval T_b, then

$$K = \frac{T_b}{\Delta T} = \frac{W_N T_b}{\pi} \tag{4.1-8}$$

independent noise samples are available. As $W_N \rightarrow \infty$ such that $n_b(t) \rightarrow n_w(t)$, white noise, the number of samples becomes infinite and $\Delta T \rightarrow 0$. Later we shall find that $K \rightarrow \infty$ allows some summations to be approximated as integrals that yield the optimum systems, which, of course, is why the noise model is used.

Signal Model

Let $r(t)$, $s_1(t)$, and $s_2(t)$, evaluated at sample times t_k, $k = 1, 2, \ldots$, K, be defined as follows

$$r_k \triangleq r(t_k) \tag{4.1-9}$$

$$s_{1k} \triangleq s_1(t_k) \tag{4.1-10}$$

$$s_{2k} \triangleq s_2(t_k). \tag{4.1-11}$$

Here $\{r_k\}$ is a set of random variables representing samples of the receiver's input waveform. They are random because of the channel noise:

$$r_k = \begin{cases} s_{1k} + n_k, & \text{if } m_1 \text{ sent} \\ s_{2k} + n_k, & \text{if } m_2 \text{ sent.} \end{cases} \tag{4.1-12}$$

Thus $n_k = r_k - s_{ik}$, $i = 1$ or 2. The joint probability density of the random variables $\{r_k\}$, conditional on m_1 or m_2 being sent, is

$$p_r(r_1, r_2, \ldots, r_K | m_i) = \prod_{k=1}^{K} p_{n_k}(r_k - s_{ik})$$

$$= (2\pi\sigma^2)^{-K/2} \exp\left\{-\sum_{k=1}^{K} \frac{(r_k - s_{ik})^2}{(2\sigma^2)}\right\}, \quad i = 1, 2. \tag{4.1-13}$$

This discussion constitutes the specification of the input signal. It remains to define the criterion of optimality.

Optimization Criterion

At appropriate times, once for each symbol interval, we shall require that the receiver demodulator decide which waveform, $s_1(t)$ or $s_2(t)$, was transmitted.† Of course, such decisions amount to deciding which of the two binary messages, m_1 or m_2, was transmitted. Thus the receiver is basically a decision device; if its decision is optimum in a typical interval, it will be optimum for other intervals as well. This fact allows us to examine the optimization problem for only one (typical) symbol interval. For convenience, we assume the interval exists from $t = 0$ to $t = T_b$.

The receiver can only work with its input waveform, $r(t)$. Its decisions must be made on the basis of observations of $r(t)$. We shall assume these observations are in the form of samples taken closely in time, ΔT apart. As the number of samples becomes very large ($\rightarrow \infty$) and ΔT becomes very small ($\rightarrow 0$), the samples are equivalent to observing the continuous waveform $r(t)$.

Let K samples be taken at times $t_k = k\Delta T$, $k = 1, 2, \ldots, K$, where $T_b = K\Delta T$. We choose ΔT according to (4.1-4). These choices allow the noise and signal models described above to apply. Now suppose the receiver observes a set of *specific values*, denoted by $\{\rho_k\}$, of the sample random variables $\{r_k\}$ during a given symbol interval. We seek to find how a receiver should process these values to optimally decide which message was transmitted. Next, suppose the receiver could somehow calculate two probabilities: first, the probability that message m_1 was sent given that the specific set of observations $\{\rho_k\}$ was obtained and second, the analogous probability that m_2 was sent.‡ Denote these probabilities by $P(m_1|\rho_1, \ldots, \rho_K)$ and $P(m_2|\rho_1, \ldots, \rho_K)$, respectively. They are called *a posteriori* probabilities. An obviously reasonable decision as to whether m_1 or m_2 was transmitted is: If

$$P(m_2|\rho_1, \ldots, \rho_K) > P(m_1|\rho_1, \ldots, \rho_K), \qquad \text{choose } m_2, \qquad (4.1\text{-}14)$$

and choose m_1 otherwise.

In fact, (4.1-14) becomes our optimization criterion. The system that satisfies (4.1-14) is the optimum system.

Our criterion can be put in a more convenient form by observing that [see (B.1-3), Bayes' rule]

$$P(m_i|\rho_1, \ldots, \rho_K)\, P(\rho_1, \ldots, \rho_K) = P(\rho_1, \ldots, \rho_K|m_i)\, P(m_i) \qquad (4.1\text{-}15)$$

for $i = 1$ and 2, where $P(m_i)$ is the probability that message m_i was transmitted.

---

† This is called *bit-by-bit signaling*.

‡ The receiver does not actually compute these probabilities; it is only helpful to imagine it does. The actual receiver is equivalent to having made such calculations.

Also, since

$$P(\rho_1, \ldots, \rho_K | m_i) = p_r(\rho_1, \ldots, \rho_K | m_i)\, d\rho_1\, d\rho_2 \cdots d\rho_K \qquad (4.1\text{-}16)$$

for $i = 1$ and 2, we can write the optimality criterion as

$$\frac{p_r(\rho_1, \ldots, \rho_K | m_2)}{p_r(\rho_1, \ldots, \rho_K | m_1)} > \frac{P(m_1)}{P(m_2)}, \qquad \text{choose } m_2,$$

$$(4.1\text{-}17)$$

and choose m_1 otherwise.

4.2 OPTIMUM BINARY SYSTEMS

The optimum binary system is one that satisfies the decision rule of (4.1-17). Define message probabilities (called *a priori* or *source* probabilities) by

$$P_1 \triangleq P(m_1) \qquad (4.2\text{-}1)$$

$$P_2 \triangleq P(m_2) \qquad (4.2\text{-}2)$$

and substitute (4.1-13) into (4.1-17). The inequality becomes

$$\exp\left\{\frac{1}{2\sigma^2} \sum_{k=1}^{K} [2\rho_k(s_{2k} - s_{1k}) - s_{2k}^2 + s_{1k}^2]\right\} > \frac{P_1}{P_2}. \qquad (4.2\text{-}3)$$

Because the exponential is monotonic in its exponent, we can take the natural logarithm in (4.2-3) to obtain an equivalent inequality:

$$\sum_{k=1}^{K} \rho(k\Delta T)\, [s_2(k\Delta T) - s_1(k\Delta T)]\Delta T - \sum_{k=1}^{K} [s_2^2(k\Delta T)$$

$$- s_1^2(k\Delta T)]\frac{\Delta T}{2} > \sigma^2 \Delta T \ln\left(\frac{P_1}{P_2}\right); \qquad (4.2\text{-}4)$$

where we substituted

$$\rho_k = \rho(t_k) = \rho(k\Delta T) \qquad (4.2\text{-}5)$$

$$s_{1k} = s_1(t_k) = s_1(k\Delta T) \qquad (4.2\text{-}6)$$

$$s_{2k} = s_2(t_k) = s_2(k\Delta T). \qquad (4.2\text{-}7)$$

In the limit as K bcomes large and ΔT becomes small,

$$\Delta T \rightarrow dt \qquad (4.2\text{-}8)$$

$$k\Delta T \rightarrow t, \qquad (4.2\text{-}9)$$

(4.2-4) can be written as

$$\int_0^{T_b} r(t)\, [s_2(t) - s_1(t)]\, dt > \frac{E_2 - E_1 + \mathcal{N}_0 \ln(P_1/P_2)}{2}. \qquad (4.2\text{-}10)$$

Here

$$E_i = \int_0^{T_b} s_i^2(t)\, dt, \qquad i = 1 \text{ and } 2, \qquad (4.2\text{-}11)$$

are the energies in signals $s_i(t)$. In writing (4.2-10) we have used (4.1-6) and replaced $\rho(t)$ by $r(t)$, since $\rho(t)$ is only a specific realization of $r(t)$. Equation (4.2-10) then represents the general operation on $r(t)$ that is performed by the optimum receiver.

By defining a *threshold*, V_T, as

$$V_T = \frac{E_2 - E_1}{2} + \frac{\mathcal{N}_0}{2} \ln\left(\frac{P_1}{P_2}\right), \qquad (4.2\text{-}12)$$

(4.2-10) allows the system's decision rule to be written as:

$$\text{If } \int_0^{T_b} r(t)[s_2(t) - s_1(t)]\, dt > V_T, \qquad \text{choose } m_2,$$

$$\qquad\qquad\qquad\qquad\qquad\qquad\qquad\qquad\qquad (4.2\text{-}13)$$

and choose m_1 otherwise.

Correlation Receiver Implementation

Since the receiver knows the forms of the waveforms $s_1(t)$ and $s_2(t)$ that are used by the transmitter, it can generate these waveforms and implement the test of (4.2-13), as shown in Fig. 4.2-1. The input waveform in a given symbol interval is multiplied by replicas of the two waveforms $s_1(t)$ and $s_2(t)$ and integrated over the symbol interval. The threshold V_T is subtracted from the difference of the integrals and the resulting signal is sampled at the end of the interval. If this sample, denoted by D, is greater than 0, the receiver decides m_2 was transmitted. If $D < 0$, it decides in favor of m_1, and if $D = 0$ a random choice between m_1 and m_2 is made. At the end of each symbol interval the integrator is discharged (reset to zero) in preparation for integration over a new interval of duration T_b. The process is repeated in each symbol interval.†

Matched Filter Implementation

Consider the responses of filters matched to $s_i(t)$, $i = 1$ and 2, when $r(t)$ is applied at their inputs. The matched filters' impulse responses are [see Sec. B.9]‡

$$h_i(t) = s_i(T_b - t), \qquad i = 1 \text{ and } 2, \qquad (4.2\text{-}14)$$

where $s_i(t)$ are nonzero only for $0 \leqslant t \leqslant T_b$. Filter responses, denoted by $r_{oi}(t)$, are

$$r_{oi}(t) = \int_{-\infty}^{\infty} r(\xi)\, h_i(t - \xi)\, d\xi$$

$$\qquad\qquad\qquad\qquad\qquad\qquad\qquad\qquad (4.2\text{-}15)$$

$$= \int_{t-T_b}^{t} r(\xi) s_i(T_b - t + \xi)\, d\xi, \qquad i = 1 \text{ and } 2.$$

---

† Note that symbol synchronization is necessary so that samples are properly timed in the optimum system.

‡ The constant t_0 in Sec. B.9 has been set equal to T_b to make $h_i(t)$ causal.

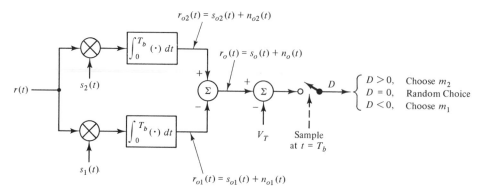

Figure 4.2-1. Optimum binary digital demodulator and decision in correlation form.

At the output's sample time $t = T_b$,

$$r_{oi}(T_b) = \int_0^{T_b} r(\xi) s_i(\xi)\, d\xi, \qquad i = 1 \text{ and } 2, \qquad (4.2\text{-}16)$$

which are the same as the integrator outputs of Fig. 4.2-1. Thus we conclude that the product-integrator cascades are equivalent to filters having the impulse responses of (4.2-14).† These matched filters are also defined by their transfer functions

$$H_i(\omega) = S_i^*(\omega) e^{-j\omega T_b}, \qquad i = 1 \text{ and } 2, \qquad (4.2\text{-}17)$$

where

$$s_i(t) \leftrightarrow S_i(\omega), \qquad i = 1 \text{ and } 2. \qquad (4.2\text{-}18)$$

Figure 4.2-2 illustrates the optimum receiver, using matched filters, that is equivalent to the correlation receiver of Fig. 4.2-1.

Optimum System Output Noise Power

Let $n_o(t)$ represent the noise of the optimum receiver's output, as shown in Fig. 4.2-1. If n_o is the random variable representing $n_o(t)$ at the sample (decision) time T_b, then

$$n_o = n_o(T_b) = \int_0^{T_b} n_w(t)\,[s_2(t) - s_1(t)]\, dt. \qquad (4.2\text{-}19)$$

---

† The reader should note that the matched filter and product-integrator (correlation) receivers are equivalent only at the sample time at the end of the bit interval.

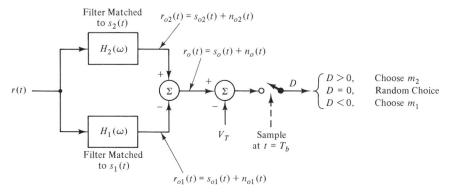

Figure 4.2-2.   Optimum binary digital demodulator and decision in matched filter form.

The output noise power (variance) on the decision sample, denoted by σ_o^2, is

$$\sigma_o^2 = E[n_o^2] = E\left\{\int_0^{T_b} n_w(t)[s_2(t) - s_1(t)]\,dt \int_0^{T_b} n_w(\alpha)\,[s_2(\alpha) - s_1(\alpha)]\,d\alpha\right\}$$

$$= \int_0^{T_b} \int_0^{T_b} \frac{\mathcal{N}_0}{2}\,\delta(\alpha - t)[s_2(t) - s_1(t)][s_2(\alpha) - s_1(\alpha)]\,d\alpha dt$$

$$= \frac{\mathcal{N}_0}{2}\int_0^{T_b} [s_2^2(t) + s_1^2(t) - 2s_1(t)\,s_2(t)]\,dt. \qquad (4.2\text{-}20)$$

The first two terms are the energies E_1 and E_2 in $s_1(t)$ and $s_2(t)$ given by

$$E_1 = \int_0^{T_b} s_1^2(t)\,dt \qquad (4.2\text{-}21)$$

$$E_2 = \int_0^{T_b} s_2^2(t)\,dt. \qquad (4.2\text{-}22)$$

If we define a quantity γ, which is a measure of the correlation between $s_1(t)$ and $s_2(t)$, by

$$\gamma \triangleq \frac{1}{\sqrt{E_1 E_2}}\int_0^{T_b} s_1(t)s_2(t)\,dt, \qquad (4.2\text{-}23)$$

then the output noise power of the optimum system becomes

$$\sigma_o^2 = \left(\frac{\mathcal{N}_0}{2}\right)[E_1 + E_2 - 2\gamma\sqrt{E_1 E_2}]. \qquad (4.2\text{-}24)$$

Optimum System Output Signal Levels

Let V_1 and V_2 represent the *signal* component, $s_o(t)$, in the output of the optimum system of Fig. 4.2-1 at the sample time $t = T_b$ when m_1 and m_2 are transmitted, respectively. These are

$$V_2 \triangleq s_o(T_b) = \int_0^{T_b} s_2(t)[s_2(t) - s_1(t)] \, dt$$

$$= E_2 - \gamma\sqrt{E_1 E_2}, \qquad m_2 \text{ sent}$$

(4.2-25)

$$V_1 \triangleq s_o(T_b) = \int_0^{T_b} s_1(t)[s_2(t) - s_1(t)] \, dt$$

$$= \gamma\sqrt{E_1 E_2} - E_1, \qquad m_1 \text{ sent},$$

(4.2-26)

where (4.2-21)–(4.2-23) have been used. These signal levels, taken at the sample time, are needed in computing error probabilities.

4.3 OPTIMUM BINARY SYSTEM ERROR PROBABILITIES

Because of noise, the output $r_o(t)$ of the optimum receiver is random. If message m_1 is transmitted, this output at the sample time $t = T_b$ has a Gaussian probability density with mean value V_1. Transmission of message m_2 leads to another Gaussian density, except this one is centered at a mean value V_2. These densities are illustrated in Fig. 4.3-1 where $p_{r_o}(r_o|m_i)$, $i = 1$ and 2, is the density of $r_o = r_o(T_b)$ given m_i is transmitted. It follows that

$$p_{r_o}(r_o|m_i) = p_{n_o}(r_o - V_i), \qquad i = 1 \text{ and } 2, \qquad (4.3\text{-}1)$$

where $p_{n_o}(\cdot)$ is the probability density function of the output (Gaussian, zero-mean) noise having variance σ_o^2 given by (4.2-24).

The receiver makes decisions by comparing the output r_o with V_T; if

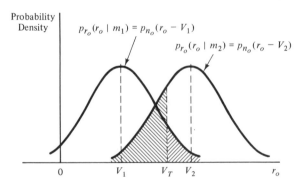

Figure 4.3-1. Probability density functions of optimum system output at the decision (sample) time for transmission of the two messages m_1 and m_2.

$r_o > V_T$, it decides m_2 was sent. However, if $r_o < V_T$ and m_2 is sent, an error is made; its probability is

$$P(e|m_2) = \int_{-\infty}^{V_T} p_{r_o}(r_o|m_2)\, dr_o$$

$$= \int_{-\infty}^{V_T} p_{n_o}(r_o - V_2)\, dr_o.$$

(4.3-2)

In a similar manner, if $r_o > V_T$ and m_1 is the correct message, an error is made having probability

$$P(e|m_1) = \int_{V_T}^{\infty} p_{r_o}(r_o|m_1)\, dr_o$$

$$= \int_{V_T}^{\infty} p_{n_o}(r_o - V_1)\, dr_o.$$

(4.3-3)

By averaging over the two possible errors, we obtain the average probability of error, denoted by P_e, for a given symbol interval (one bit in this case). After using the fact that output noise is Gaussian with variance σ_o^2, P_e can be written as

$$P_e = P(e|m_1)\, P_1 + P(e|m_2)\, P_2$$

$$= \frac{P_1}{2}\, \mathrm{erfc}\left(\frac{V_T - V_1}{\sqrt{2}\, \sigma_o}\right) + \frac{P_2}{2}\, \mathrm{erfc}\left(\frac{V_2 - V_T}{\sqrt{2}\, \sigma_o}\right),$$

(4.3-4)

where $\mathrm{erfc}(\cdot)$ is the complementary error function of (B.4-6). On use of (4.2-24)–(4.2-26) and (4.2-12) we obtain

$$P_e = \frac{P_1}{2}\, \mathrm{erfc}\left\{ \sqrt{\frac{E_1 + E_2 - 2\gamma\sqrt{E_1 E_2}}{4\mathcal{N}_0}} + \frac{\sqrt{\mathcal{N}_0}\, \ln(P_1/P_2)}{2\sqrt{E_1 + E_2 - 2\gamma\sqrt{E_1 E_2}}} \right\}$$

$$+ \frac{P_2}{2}\, \mathrm{erfc}\left\{ \sqrt{\frac{E_1 + E_2 - 2\gamma\sqrt{E_1 E_2}}{4\mathcal{N}_0}} - \frac{\sqrt{\mathcal{N}_0}\, \ln(P_1/P_2)}{2\sqrt{E_1 + E_2 - 2\gamma\sqrt{E_1 E_2}}} \right\}.$$

(4.3-5)

This expression for P_e is valid for arbitrary waveforms $s_1(t)$ and $s_2(t)$ and arbitrary choices of P_1 and P_2. Some simpler expressions follow some realistic assumptions, often true in practice. We take three cases.

Equal Probability Messages

If $P_1 = P_2 = \frac{1}{2}$, we have

$$P_e = \frac{1}{2}\, \mathrm{erfc}\left\{ \sqrt{\frac{E_1 + E_2 - 2\gamma\sqrt{E_1 E_2}}{4\mathcal{N}_0}} \right\}.$$

(4.3-6)

Because (4.3-6) still applies for arbitrary waveforms $s_1(t)$ and $s_2(t)$, it can be further minimized if we sacrifice this generality by selecting one, say $s_2(t)$, to be a suitable function of the other waveform $s_1(t)$. Since erfc(·) is a monotonic function of its argument, P_e is a minimum when

$$E_1 + E_2 - 2\gamma \sqrt{E_1 E_2} = E_1 + E_2 - 2 \int_0^{T_b} s_1(t)s_2(t)\, dt \qquad (4.3\text{-}7)$$

is maximum. Suppose we maximize (4.3-7) under a constraint of constant *total* energy

$$E_{\text{TOT}} = E_1 + E_2 = \text{constant}. \qquad (4.3\text{-}8)$$

The maximum will occur when the third (integral) term is maximum. From Schwarz's inequality

$$\left| -2 \int_0^{T_b} s_1(t)s_2(t)\, dt \right|^2 \leq 4 \int_0^{T_b} s_1^2(t)\, dt \int_0^{T_b} s_2^2(t)\, dt = 4E_1 E_2. \qquad (4.3\text{-}9)$$

The equality holds only when

$$s_2(t) = \Gamma s_1(t), \qquad (4.3\text{-}10)$$

with Γ a real constant. Squaring and integrating (4.3-10) gives

$$\Gamma = \pm \sqrt{\frac{E_2}{E_1}}. \qquad (4.3\text{-}11)$$

From use of (4.3-10) and (4.3-11), we write (4.3-7) as

$$E_{\text{TOT}} - 2 \int_0^{T_b} s_1(t)s_2(t)\, dt = E_{\text{TOT}} \mp 2 \sqrt{E_1 E_2}$$
$$= E_{\text{TOT}} \mp 2 \sqrt{E_1(E_{\text{TOT}} - E_1)}. \qquad (4.3\text{-}12)$$

The derivative of (4.3-12) with respect to E_1 is zero for $E_1 = E_{\text{TOT}}/2$. The second derivative of (4.3-12) with respect to E_1 is negative only for the lower signs in (4.3-12) when $E_1 = E_{\text{TOT}}/2$. Thus we conclude that (4.3-7) is maximum only when

$$E_1 = E_2 = \frac{E_{\text{TOT}}}{2} \qquad (4.3\text{-}13)$$

$$s_2(t) = \Gamma s_1(t) = -s_1(t). \qquad (4.3\text{-}14)$$

It is left for the reader as a simple exercise to show that the above developments are equivalent to the maximum of (4.3-7) occurring when $E_1 = E_2 = E_{\text{TOT}}/2$ and $\gamma = -1$. Our developments lead to two additional cases.

Equal Probability, Equal Energy Signals

Define E as the energy of each signal with

$$E \triangleq E_1 = E_2 = \frac{E_{\text{TOT}}}{2}. \qquad (4.3\text{-}15)$$

P_e of (4.3-6) reduces to

$$P_e = \frac{1}{2}\,\text{erfc}\left[\sqrt{\frac{E(1-\gamma)}{2\,\mathcal{N}_0}}\,\right].$$ (4.3-16)

Equal Probability, Antipodal Signals

When $s_2(t) = -s_1(t)$, the signals are called *antipodal*. This choice leads to the minimum value of (4.3-6) as discussed above:

$$P_e = \frac{1}{2}\,\text{erfc}\left[\sqrt{\frac{E}{\mathcal{N}_0}}\,\right].$$ (4.3-17)

Equation (4.3-17) is also obtainable from (4.3-16) by letting $\gamma = -1$.

4.4 BINARY PULSE CODE MODULATION

Binary *pulse code modulation* (PCM) is the name used to describe the process by which an analog message is converted to a binary digital waveform. The process consists of first sampling the analog signal according to the principles described in Chap. 2 so that the analog message can be recovered from its samples in the receiver. Each sample is then quantized to the nearest of a finite number of quantum levels and coded (often using the natural binary code). Finally, a waveform format is assigned to the code. The resulting PCM waveform is processed through the digital system. In this section we shall describe the basic functions performed by a PCM system and discuss its performance with several example waveform formats based on rectangular pulses.

Overall PCM System

Many ways exist for implementing PCM systems. Figure 4.4-1 illustrates one method that assumes N similar analog messages are time multiplexed. Each message is lowpass filtered to assure bandlimiting and is then sampled to generate N time-multiplexed sampled waveforms, each in its assigned time slot. The composite signal is compressed, if companding is used. Samples are next quantized, source and channel encoded, and finally converted into appropriate waveforms by the modulator. Although the transmitted waveform can have any of the formats discussed in Chap. 3, we shall limit our discussions to only unipolar, polar, Manchester, and differential formats, because these waveforms all have optimum receivers easily put in the form of those developed in Sec. 4.2.

Naturally, many variations are possible for the transmission system of Fig. 4.4-1. In some cases the channel encoder, the compressor, or both can be left out. If the composite message source is digital, such as one

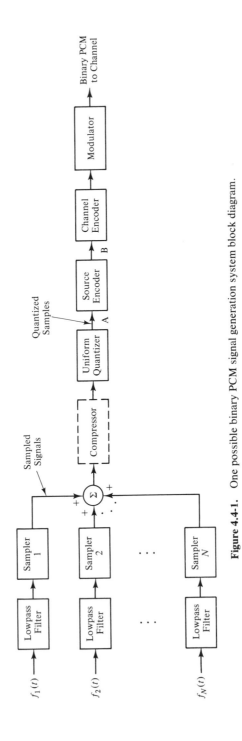

Figure 4.4-1. One possible binary PCM signal generation system block diagram.

with only discrete levels, for instance, it can be handled by accepting it as an input at point A. Digital sources already source encoded can be input directly to point B. By suitable multiplexing, combinations of these types of inputs can be handled.

The receiver necessary to recover messages from the PCM signal basically performs the inverse operations of the transmitter. Figure 4.4-2 illustrates the receiver based on the transmitter of Fig. 4.4-1. The receiver becomes optimum when the demodulator has the forms developed in Sec. 4.2.

Unipolar, Polar, and Manchester Formats

For example, a unipolar format corresponds to $s_1(t) = 0$, $s_2(t) = A$ for $0 \leq t \leq T_b$ and the optimum demodulator of Fig. 4.2-1 reduces to that shown in Fig. 4.4-3(a). From (4.2-12) the optimum threshold V_T is

$$V_T = \frac{A^2 T_b}{2} + \frac{\mathcal{N}_0}{2} \ln\left(\frac{P_1}{P_2}\right), \qquad \text{unipolar format,} \qquad (4.4\text{-}1)$$

since the energy, E_2, in $s_2(t)$ is

$$E_2 = A^2 T_b \qquad (4.4\text{-}2)$$

for the assumed rectangular pulses.

For a polar format where $s_1(t) = -A$ and $s_2(t) = A$, both for $0 \leq t \leq T_b$, the optimum demodulator has the form shown in Fig. 4.4-3(b), where

$$V_T = \left(\frac{\mathcal{N}_0}{2}\right) \ln\left(\frac{P_1}{P_2}\right), \qquad \text{polar format,} \qquad (4.4\text{-}3)$$

$$E_1 = E_2 = A^2 T_b. \qquad (4.4\text{-}4)$$

When the two binary symbols are equally probable, the threshold is zero. The polar waveform is a special case of antipodal signaling.

With a Manchester waveform the transmitted waveforms $s_1(t)$ and $s_2(t)$ are also antipodal. Here $s_2(t)$ is easily generated from a periodic square wave clock that generates a level A for $0 \leq t < T_b/2$ and $-A$ for $T_b/2 < t \leq T_b$. Its fundamental frequency is, therefore, $\omega_b = 2\pi/T_b$. By inversion, the clock generates $s_1(t) = -s_2(t)$ as well. The optimum demodulator becomes that shown in Fig. 4.4-3(c) where V_T is given by (4.4-3). In a practical realization a single integrator could be placed after the left summing junction for a hardware savings.

All the optimum demodulators of Fig. 4.4-3 have equivalent filter realizations that are easily derived from Fig. 4.2-2.

Error Probabilities

The error probabilities of Sec. 4.3 apply to the demodulators of Fig. 4.4-3 because they are optimum. If we define ε as the average energy per

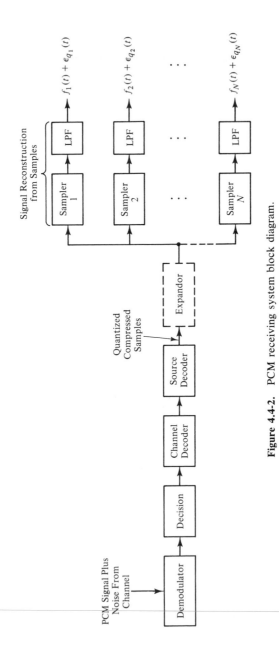

Figure 4.4-2. PCM receiving system block diagram.

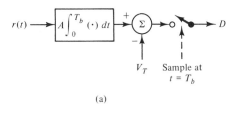

(a)

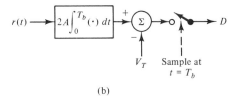

(b)

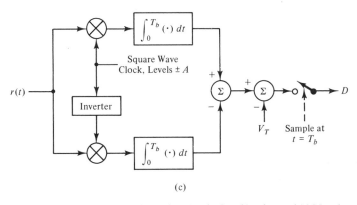

(c)

Figure 4.4-3. Optimum demodulators for (a) unipolar, (b) polar, and (c) Manchester waveforms.

symbol (bit) interval divided by twice the channel's white noise power density $[2(\mathcal{N}_0/2)]$, then

$$
\varepsilon = \begin{cases}
\dfrac{P_2 E}{\mathcal{N}_0} = \dfrac{P_2 A^2 T_b}{\mathcal{N}_0}, & \text{unipolar format} \\[2ex]
\dfrac{E}{\mathcal{N}_0} = \dfrac{A^2 T_b}{\mathcal{N}_0}, & \text{polar format} \\[2ex]
\dfrac{E}{\mathcal{N}_0} = \dfrac{A^2 T_b}{\mathcal{N}_0}, & \text{Manchester format}
\end{cases}
\tag{4.4-5}
$$

where E is the energy in a transmitted pulse of duration T_b and amplitude A. From (4.3-5), the average probability, P_e, of incorrectly decoding a bit (symbol) becomes

$$
P_e = \begin{cases}
\dfrac{P_1}{2}\operatorname{erfc}\left\{\sqrt{\dfrac{\varepsilon}{4P_2}}\left[1+\dfrac{\ln(P_1/P_2)}{\varepsilon/P_2}\right]\right\} \\
\qquad +\dfrac{P_2}{2}\operatorname{erfc}\left\{\sqrt{\dfrac{\varepsilon}{4P_2}}\left[1-\dfrac{\ln(P_1/P_2)}{\varepsilon/P_2}\right]\right\}, \quad \text{unipolar format} \\
\dfrac{P_1}{2}\operatorname{erfc}\left\{\sqrt{\varepsilon}\left[1+\dfrac{\ln(P_1/P_2)}{4\varepsilon}\right]\right\} \\
\qquad +\dfrac{P_2}{2}\operatorname{erfc}\left\{\sqrt{\varepsilon}\left[1-\dfrac{\ln(P_1/P_2)}{4\varepsilon}\right]\right\}, \quad \begin{array}{l}\text{polar or}\\\text{Manchester}\\\text{format.}\end{array}
\end{cases}
\tag{4.4-6}
$$

The quantity ε forms a fair basis on which to compare different systems. Figure 4.4-4 illustrates the behavior of (4.4-6) for various ratios P_1/P_2. For

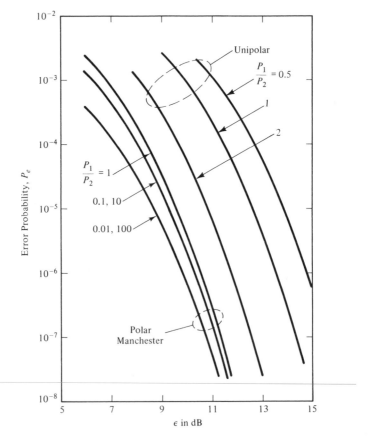

Figure 4.4-4. Error probability for unipolar, polar, and Manchester formats and unequal message probabilities.

the polar and Manchester cases P_e is not a sensitive function of P_1/P_2 for $0.1 \leq P_1/P_2 \leq 10$ and does not change greatly even over $0.01 \leq P_1/P_2 \leq 100$. For example, with $P_1/P_2 = 100$ and $P_e = 10^{-5}$, the increase in power required, relative to the $P_1/P_2 = 1$ case, is less than 1.0 dB. The difference becomes even smaller for smaller values of P_e. However, P_e for $P_1 \neq P_2$ is always smaller than when $P_1 = P_2 = \frac{1}{2}$. This behavior does not occur in the unipolar case where one waveform's energy is zero. For the often-satisfied case where $P_1 = P_2 = \frac{1}{2}$, we obtain

$$P_e = \begin{cases} \left(\dfrac{1}{2}\right) \mathrm{erfc}\left[\sqrt{\dfrac{\varepsilon}{2}}\right], & \text{unipolar format} \\[2em] \left(\dfrac{1}{2}\right) \mathrm{erfc}\,[\sqrt{\varepsilon}], & \text{polar or Manchester format.} \end{cases} \tag{4.4-7}$$

These two functions are plotted in Fig. 5.12-1 for comparison with the performance of various systems. For a given value of P_e the required value of ε is 3 dB larger for the unipolar format compared to the polar or Manchester format.

Effect of Differential Coding

If the binary source is differentially encoded, any of the binary waveform formats can be used, and the receiver is selected to match the format. For unipolar, polar, or Manchester formats, this means that either (4.4-6) or (4.4-7) still applies for the average probability of making an error in demodulating any given bit. However, because 2 bits (symbol intervals) are involved in a differential decoder (Fig. 3.6-3(c)), the decoding function degrades overall error probability.

Consider the decoder. First, suppose the currently arriving bit is correct; an error in output is made only if the adjacent earlier bit was in error. The probability of this joint event occurring is $P_e(1 - P_e)$. Next, suppose the current bit is in error; the output is in error only if the adjacent earlier bit is *correct*.† Again this joint event has probability $P_e(1 - P_e)$. The overall probability of a decoded bit being in error, denoted by $P_{e(\mathrm{diff})}$, is

$$P_{e(\mathrm{diff})} = 2\,P_e(1 - P_e). \tag{4.4-8}$$

If P_e is small,

$$P_{e(\mathrm{diff})} \approx 2\,P_e. \tag{4.4-9}$$

The approximate effect of differentially encoding data in a PCM system is, therefore, a doubling of the probability of bit errors. This increase in error probability is considered small in most applications and the advantages of differentially encoded data more than justify its use.

---

† Both bits in error will result in a correct output.

Finite Channel Bandwidth

The transmitted rectangular pulses arrive at the receiver undistorted only if the channel has infinite bandwidth. In a practical system a typical rectangular transmitted pulse of duration T_b will arrive at the receiver with nonzero rise and fall times. Because of the fall time, the pulse will spill over into the following symbol interval. The spillover, when integrated by the receiver (Fig. 4.4-3), will cause a change in the output level at the interval's sample time compared to its level without spillover. The rise time also affects the output level. Interference between symbol intervals in this fashion is termed *intersymbol interference,* a subject covered further in Sec. 4.6.

A rough idea of the effect of a real channel can be obtained by assuming it has a simple first-order lowpass transfer function

$$H_{ch}(\omega) = \left[1 + \left(\frac{j\omega}{W_{ch}}\right)\right]^{-1}, \qquad (4.4\text{-}10)$$

where W_{ch} is the channel's 3-dB bandwidth (radians per second). Now consider a typical level transition in a polar PCM transmitted waveform, for example, from $-A$ to A; the channel response is illustrated in Fig. 4.4-5.† A straightforward calculation of D in Fig. 4.4-3(b), when assuming $P_1 = P_2 = \frac{1}{2}$ and no noise, gives

$$D = 2A^2T_b\left[1 - \frac{2}{W_{ch}T_b}(1 - e^{-W_{ch}T_b})\right]$$

$$\approx 2A^2T_b\left[1 - \left(\frac{2}{W_{ch}T_b}\right)\right], \qquad W_{ch}T_b \gg 1. \qquad (4.4\text{-}11)$$

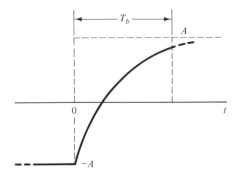

Figure 4.4-5. Transition between levels for polar PCM with restricted (lowpass) channel bandwith [2].

The ideal output is $2A^2T_b$. The bracketed factor represents degradation due to finite channel bandwidth. If the degradation factor is to be maintained

---

† For simplicity, level $-A$ is assumed to have occurred for several symbol intervals so the transition starts at $-A$.

at 0.9, or better, W_{ch} must not be less than $20/T_b = 10\omega_b/\pi \approx 3.2\ \omega_b$. From Fig. 3.7-2 this bandwidth corresponds to passing two full sidelobes in the power spectrum of the polar PCM signal. In Sec. 4.6 we show that transmitted pulses can purposely be shaped to reduce both intersymbol interference and the need for large channel bandwidth.

4.5 NOISE PERFORMANCE OF BINARY SYSTEMS

When a binary system operates with analog signals (PCM) where the receiver must reconstruct the original continuous message, the measure of noise performance is the ratio of system output signal power to noise power. When the system operates with digital messages, it is the average probability of making errors that is most important. We consider this second situation first.

Performance for Digital Messages with Coding

In many cases a group of k binary digits may represent a "word" of data, such as with natural binary source encoding when k equals N_b, the number of binary digits per word. Here no channel encoding is present, and we call this *uncoded* signaling. The probability of a word error, denoted by P_w, may be more important than P_e, the probability of a bit error. A word error occurs if any one or more of the k bits are in error; this probability is 1 minus the probability that all bits are demodulated correctly, as given by

$$P_w = 1 - (1 - P_e)^k \qquad (4.5\text{-}1)$$
$$\approx kP_e, \qquad P_e \ll 1.$$

In the case where the data word is coded using, for example, a t-error correcting (n, k) code, no error is made unless $t + 1$ or more of the n total bits in a word are in error. Probabilities involved here follow the binomial distribution. The probability of exactly i bit errors in n bits (any order) is $\binom{n}{i} P_e^i(1 - P_e)^{n-i}$, so the probability of $t + 1$ or more bit errors becomes

$$P_w = \sum_{i=t+1}^{n} \binom{n}{i} P_e^i (1 - P_e)^{n-i} \qquad (4.5\text{-}2)$$
$$\approx \binom{n}{t+1} P_e^{t+1} = \frac{n!\, P_e^{t+1}}{(t+1)!\,(n-t-1)!}, \qquad P_e \ll 1.$$

Of course, P_e here is the average probability of any one bit being in error.†

† Equation (4.5-2) approximates the sum by the first term. However, if n is large it may be necessary to include additional terms for the approximation to be accurate.

P_e in (4.5-2) does not necessarily have the same numerical value as P_e of (4.5-1).

To compare coded and uncoded systems, assume both use the same total time allocated to a codeword and the words have the same number (k) of data bits. These assumptions mean that both T_b and ε in the coded system are smaller than T_b and ε in the uncoded system, respectively, because the coded system must transmit its extra $n - k$ check digits. In the two systems P_w follows from (4.5-1) and (4.5-2) by using the appropriate ε (see (4.4-5)) with the appropriate expression for P_e. For a polar format with equal word probabilities, for example, we obtain

$$P_w \approx \left(\frac{k}{2}\right) \operatorname{erfc}(\sqrt{\varepsilon}), \qquad \text{no coding} \qquad (4.5\text{-}3)$$

$$P_w \approx \frac{n!}{(t+1)!(n-t-1)!}\left[\frac{1}{2}\operatorname{erfc}\left(\sqrt{\frac{k\varepsilon}{n}}\right)\right]^{t+1}, \qquad \text{with coding.} \qquad (4.5\text{-}4)$$

Here ε refers to the uncoded system and is assumed larger than about 4. System comparison usually entails plotting P_w versus ε for given n, k, and t. The difference in required values of ε for equal values of P_w is called *coding gain*. For large enough ε, coding gain is positive (ε required in coded system is less than ε required in uncoded system). For small enough ε, coding gain can be negative so that coding actually degrades performance. To illustrate these points we consider an example.

Example 4.5-1

A known perfect code is the (23, 12) *Golay code,* where $t = 3$ [3]. We use Appendix F to evaluate the complementary error functions of (4.5-3) and (4.5-4). Figure 4.5-1 plots the word probabilities. At $P_w = 10^{-4}$ the coding gain is approximately 2.6 dB.

Performance for Analog Messages Above Threshold (PCM)

Because of quantization, only the quantized version $f_q(t)$ of an analog message, $f(t)$, can be recovered by the receiver, even if there is no noise. As in Chap. 3 we represent $f_q(t)$ as $f(t)$ with quantization error $\varepsilon_q(t)$ as

$$f_q(t) = f(t) + \varepsilon_q(t). \qquad (4.5\text{-}5)$$

When noise is negligible, the system is said to be above threshold and performance is measured by the signal-to-quantization-noise power ratio

$$\left(\frac{S_o}{N_q}\right)_{\text{PCM}} = \frac{\overline{f^2(t)}}{\overline{\varepsilon_q^2(t)}}. \qquad (4.5\text{-}6)$$

We determined (S_o/N_q) in Chap. 3 for several forms of quantization. We shall not repeat those results here. However, as an example we summarize

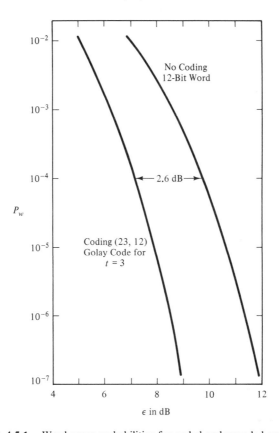

Figure 4.5-1. Word error probabilities for coded and uncoded systems.

one important case. For messages that have symmetrical fluctuations about zero and have an extreme value, denoted by $|f(t)|_{max}$, a uniform quantizer having a large number L of levels produces

$$\left(\frac{S_o}{N_q}\right)_{PCM} = \left(\frac{3L^2}{K_{cr}^2}\right)\frac{\overline{f^2(t)}}{\overline{f_r^2(t)}}. \tag{4.5-7}$$

Here $\overline{f_r^2(t)}$ is the message's power when at the *reference level* that establishes the quantizer's step size, δv, according to

$$\delta v = \frac{2|f_r(t)|_{max}}{L}. \tag{4.5-8}$$

The message's crest factor is K_{cr}, defined by

$$K_{cr}^2 = \frac{|f(t)|_{max}^2}{\overline{f^2(t)}}. \tag{4.5-9}$$

Equation (4.5-7) applies for $\overline{f^2(t)} \leqslant \overline{f_r^2(t)}$; when $\overline{f^2(t)} > \overline{f_r^2(t)}$, amplitude overload occurs (Chap. 3) and $(S_o/N_q)_{\text{PCM}}$ decreases.

Example 4.5-2

We calculate (4.5-7) for a sinusoidal message of peak amplitude 2.7 V when a 128-level uniform quantizer is designed for a maximum amplitude of 3.2 V. Here $\overline{f^2(t)} = (2.7)^2/2$ and $\overline{f_r^2(t)} = (3.2)^2/2$, whereas $K_{cr}^2 = 2$ for sinusoidal signals. Thus $(S_o/N_q)_{\text{PCM}} = [3(128)^2/2] (2.7/3.2)^2 = 17{,}496.0$ (or 42.4 dB).

Effect of Receiver Noise in PCM

When receiver noise is not negligible, we may write the recovered signal, denoted by $\hat{f}_q(t)$, as

$$\hat{f}_q(t) = f_q(t) + \varepsilon_n(t) = f(t) + \varepsilon_q(t) + \varepsilon_n(t). \qquad (4.5\text{-}10)$$

In other words, there is now an error $\varepsilon_n(t)$ in the reconstructed message $\hat{f}_q(t)$ due to noise. Because the quantization and receiver errors arise from different mechanisms, they may be taken as statistically independent and the total output waveform power is the sum of individual powers.

Define quantization and receiver error noise powers by

$$N_q = \overline{\varepsilon_q^2(t)} \qquad (4.5\text{-}11)$$

$$N_{rec} = \overline{\varepsilon_n^2(t)}. \qquad (4.5\text{-}12)$$

These allow output signal-to-noise power ratio to be defined by

$$\left(\frac{S_o}{N_o}\right)_{\text{PCM}} = \frac{\overline{f^2(t)}}{N_q + N_{rec}}. \qquad (4.5\text{-}13)$$

The presence of N_{rec} gives rise to a *threshold effect* in PCM.

Our basic problem is to find N_{rec}. To this end, we must relate input noise to how it causes errors in the output. The relationship is found from a close look at the behavior of the demodulation and decoding operations of Fig. 4.4-2. At the end of each word, the source decoder generates an exact quantum level corresponding to the recovered codeword (we consider only the case where no channel encoding is used). The recovered codeword, and therefore $\hat{f}_q(t)$, will contain error *only* if an error is made in individual pulse decisions. The effect of receiver noise is to cause occasional false decisions.

Thus $\hat{f}_q(t)$ will possess a noise error if a codeword is in error, and a codeword is in error if any one or more bit errors are made. Most PCM systems operate such that it is highly unlikely that more than one bit is in error in any one word. By assuming this to be the case, we examine the effect of bit errors on a bit-by-bit basis for an N_b-bit word.

If an error occurs in the least significant bit, an error in $\hat{f}_q(t)$ of δv

occurs. If such an error occurs in m words out of M possible words the average (mean-)squared error is $(\delta v)^2 m/M$. For a large number of words m/M is interpreted as the probability P_e of the least significant bit's being in error. Average-squared error in $\hat{f}_q(t)$ is then $(\delta v)^2 P_e$.

For the next least significant bit, a bit error causes an error in $\hat{f}_q(t)$ of $2(\delta v)$. The mean-squared error becomes $(2\delta v)^2 P_e$. Continuing the logic to the ith next least significant bit, the mean-squared error in $\hat{f}_q(t)$ is $(2^i \, \delta v)^2 P_e$. Now, recognizing that the total mean-squared error $\overline{\varepsilon_n^2(t)}$ in $\hat{f}_q(t)$ is the sum of the contributions from each bit in the code word, we have

$$N_{rec} = \overline{\varepsilon_n^2(t)} = (\delta v)^2 P_e \sum_{i=0}^{N_b-1} (2^2)^i = (\delta v)^2 P_e \frac{2^{2N_b}-1}{3}. \qquad (4.5\text{-}14)$$

Finally, since $2^{2N_b} \gg 1$, we substitute (4.5-8) and obtain approximately

$$N_{rec} = \frac{4|f_r(t)|_{\max}^2 \, 2^{2N_b} P_e}{3L^2}. \qquad (4.5\text{-}15)$$

To complete the calculation of N_{rec}, we may substitute the expression for P_e given by (4.4-7). For polar PCM with $P_1 = P_2 = \frac{1}{2}$ we get

$$N_{rec} = \frac{2|f_r(t)|_{\max}^2 \, 2^{2N_b}}{3L^2} \, \text{erfc} \left[\sqrt{\frac{A^2 T_b}{\mathcal{N}_0}} \right]. \qquad (4.5\text{-}16)$$

A similar result is achieved for unipolar PCM.

Performance Near Threshold in PCM

From (4.5-13), using (3.2-4) with $N_{q,ol} = 0$ since we assume $\overline{f^2(t)} \leqslant \overline{f_r^2(t)}$, (4.5-14), and the fact that $S_o = \overline{f^2(t)}$, we have

$$\left(\frac{S_o}{N_o} \right)_{\text{PCM}} = \frac{(S_o/N_q)_{\text{PCM}}}{1 + 2^{2N_b+1} \, \text{erfc} \left[\sqrt{A^2 \dfrac{T_b}{\mathcal{N}_0}} \right]}, \qquad \text{polar NRZ PCM.} \qquad (4.5\text{-}17)$$

This expression can be put in a more useful form by observing that the demodulator of Fig. 4.4-3(b) applies with $V_T = 0$ from (4.4-3). The peak power in the *signal* component of the waveform at the sampler, at the sample time, denoted by $\hat{S}_i$, is

$$\hat{S}_i \triangleq V_1^2 = V_2^2 = 4(A^2 T_b)^2 \qquad (4.5\text{-}18)$$

from (4.2-25) or (4.2-26) because $\gamma = -1$ and $E_1 = E_2 = A^2 T_b$. The average noise power at the input to the sampler, from (4.2-24), denoted by N_i, is

$$N_i = \sigma_o^2 = 2 \mathcal{N}_0 A^2 T_b. \qquad (4.5\text{-}19)$$

By defining sampler input *peak* signal-to-noise power ratio as

$$\left(\frac{\widehat{S}_i}{N_i}\right)_{\text{PCM}} = \frac{2\,A^2 T_b}{\mathcal{N}_0}, \qquad \text{polar NRZ PCM,} \qquad (4.5\text{-}20)$$

we may write (4.5-17) as

$$\left(\frac{S_o}{N_o}\right)_{\text{PCM}} = \frac{\left(\dfrac{S_o}{N_q}\right)_{\text{PCM}}}{1 + 2^{2N_b+1}\operatorname{erfc}\left[\sqrt{\dfrac{1}{2}\left(\dfrac{\widehat{S}_i}{N_i}\right)_{\text{PCM}}}\right]}, \qquad \text{polar NRZ PCM.}$$

$$(4.5\text{-}21)$$

Of course, $(S_o/N_q)_{\text{PCM}}$ is given by (4.5-7). A plot of (4.5-21) is shown in Fig. 4.5-2 for a message at its maximum no-overload level, $\overline{f^2(t)} = \overline{f_r^2(t)}$. The threshold effect is clearly evident and occurs for $(\widehat{S}_i/N_i)_{\text{PCM}}$ approximately from 10 to 15 dB for $N_b = 2$ to 10, respectively.

Another useful form of (4.5-17) can also be obtained. We write

$$\frac{A^2 T_b}{\mathcal{N}_0} = \frac{A^2 N_b T_b / T_s}{(\mathcal{N}_0 W_f / 2\pi)} \cdot \frac{W_f T_s}{2\pi N_b} = \left(\frac{2\pi S_i}{\mathcal{N}_0 W_f}\right)\frac{W_f T_s}{2\pi N_b}, \qquad (4.5\text{-}22)$$

where the input *average* signal power per message to the receiver (from the channel), denoted by S_i, is

$$S_i = \frac{A^2\, N_b T_b}{T_s}, \qquad (4.5\text{-}23)$$

and T_s is the time between message samples.† Finally, if we presume Nyquist sampling so $W_f T_s = \pi$, and use (1.4-5) so that (4.5-22) can be written in terms of a baseband system with the same average transmitted power, the same values of $\mathcal{N}_0$ and W_f, we have

$$\frac{A^2 T_b}{\mathcal{N}_0} = \frac{1}{2N_b}\left(\frac{S_o}{N_o}\right)_B. \qquad (4.5\text{-}24)$$

Hence (4.5-17) becomes

$$\left(\frac{S_o}{N_o}\right)_{\text{PCM}} = \frac{(S_o/N_q)_{\text{PCM}}}{1 + 2^{2N_b+1}\operatorname{erfc}\left[\sqrt{\dfrac{1}{2N_b}\left(\dfrac{S_o}{N_o}\right)_B}\right]}, \qquad \text{polar NRZ PCM.}$$

$$(4.5\text{-}25)$$

Figure 4.5-3 illustrates plots of (4.5-25) using (4.5-7) when $\overline{f^2(t)} = \overline{f_r^2(t)}$.

---

† In general, $T_s \neq N_b T_b$ because many PCM signals may be multiplexed in time. For similar messages using the same word lengths and number of bits, there are $N = T_s/N_b T_b$ time slots available in multiplexing.

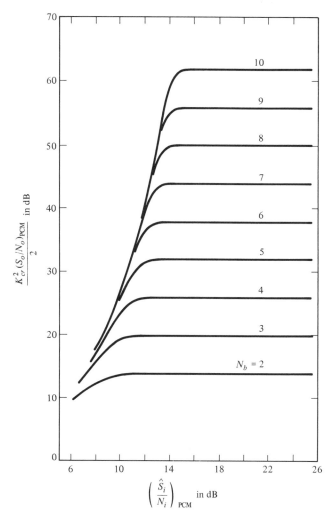

Figure 4.5-2.   Normalized performance of a binary polar NRZ PCM system when the analog message's samples are encoded with N_b-bit binary words. The abscissa is also equal to 2ε in decibels [2].

Threshold Calculation

The point of threshold will arbitrarily be defined as the point where $(S_o/N_o)_{PCM}$ drops 1 dB below the value of $(S_o/N_q)_{PCM}$. From (4.5-21) this occurs when the denominator becomes 1.26. Solving for the threshold

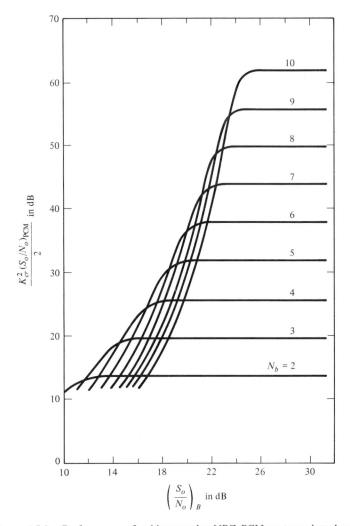

Figure 4.5-3. Performance of a binary polar NRZ PCM system plotted as a function of the performance of an equivalent baseband system. The abscissa is also equal to $2N_b\varepsilon$ in decibels [2].

signal-to-noise ratio $(\hat{S}_i/N_i)_{\text{PCM},th}$ gives

$$\left(\frac{\hat{S}_i}{N_i}\right)_{\text{PCM},th} = 2\left[\text{erfc}^{-1}\left(\frac{0.13}{2^{2N_b}}\right)\right]^2, \qquad \text{polar NRZ PCM.} \qquad (4.5\text{-}26)$$

Figure 4.5-4 illustrates the behavior of (4.5-26).

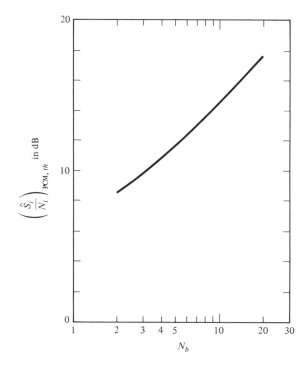

Figure 4.5-4.   Threshold peak signal-to-noise ratio of a binary polar NRZ PCM system [2].

4.6 INTERSYMBOL INTERFERENCE

In Sec. 4.4 we found that insufficient channel bandwidth caused pulses arriving at the receiver to spread into adjacent pulse intervals. This spillover caused errors in the signal component of the waveform sampled in the adjacent interval at the symbol decision time. This effect was termed *intersymbol interference*. In this section we shall discuss ways of reducing intersymbol interference.

Pulse Shaping to Reduce Intersymbol Interference

Consider a typical case where N equal-length time slots are available during a frame time, denoted by T_s. With N_b binary bits in a time slot the duration of a bit interval is $T_b = T_s/NN_b$. If, at the time of any one sample, the voltage is entirely due to the symbol corresponding to the sample, there is no intersymbol interference. This situation corresponds to waveforms from *all* other transmitted symbols passing through a null at the sample time. However, since symbols are transmitted every T_s/NN_b seconds, we must also require that the waveforms for all symbols pass through *periodic* nulls to prevent interference regardless of which sample is taken.

One example of a theoretically suitable waveshape, having the form $\sin(x)/x$, was encountered in sampling. It corresponded to the response of an ideal lowpass channel when excited by very narrow (impulsive) transmitted pulses. To employ this waveform in the more general problem, the *overall* product of the spectrum of the transmitted waveform, the channel transfer function, and the receiver matched filter would have to equal an ideal lowpass-type rectangular function of frequency. Such a result is not realizable and is difficult to approximate in practice. Even if it could be realized, there are still practical considerations in maintaining adequate synchronization of sample point timing that render the $\sin(x)/x$ waveform unsuitable [4].

An alternative approach was developed in 1928 when H. Nyquist (*Transactions of the AIEE*, Vol. 47, February, 1928, pp. 617–644) demonstrated how a pulsed signal having the desired distribution of nulls could be achieved, while at the same time its spectrum magnitude approximated a realizable filter transfer function. Nyquist's result has been called the *vestigial-symmetry theorem* [5] owing to the form of the spectrum. It may have any shape, as illustrated in Fig. 4.6-1(a), so long as it is real and odd symmetry exists about the points $\omega = \pm W$. Clearly, the form of the spectrum is one that may be approximated by a realistic system having a gradual roll-off characteristic. It remains to show that the location of nulls is as claimed.

Proofs of Nyquist's vestigial-symmetry theorem usually follow the asumption $W_1 \leq W$, the case of usual practical interest. On making this assumption, we may decompose the spectrum $S_o(\omega)$ into the sum of a rectangular component and components with odd symmetry about $\pm W$, as illustrated in Fig. 4.6-1(b) and (c). Inverse transforms of the components are easily shown to be

$$h(t) = \frac{W}{\pi} \frac{\sin(Wt)}{Wt} \tag{4.6-1}$$

$$h_1(t) = \frac{-2}{\pi} \sin(Wt) \int_0^{W_1} H_1(\omega + W) \sin(\omega t)\, d\omega. \tag{4.6-2}$$

The time function $s_o(t)$ corresponding to the spectrum $S_o(\omega)$ becomes

$$s_o(t) = \frac{W}{\pi} \frac{\sin(Wt)}{Wt} \left[1 - 2t \int_0^{W_1} H_1(\omega + W) \sin(\omega t)\, d\omega \right]. \tag{4.6-3}$$

Now regardless of the precise value of the term within the brackets, the factor $\sin(Wt)/Wt$ guarantees the existence of periodic nulls. Hence, for almost arbitrary choice of $H_1(\omega)$, intersymbol interference is zero if symbols in the PCM pulse train are conveyed in pulses having the shape of $s_o(t)$.

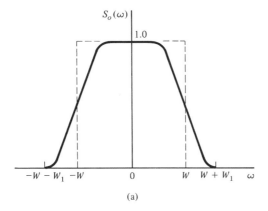

(a)

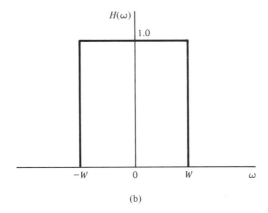

(b)

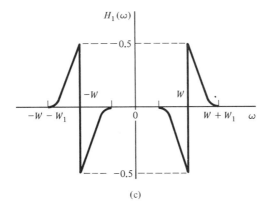

(c)

Figure 4.6-1.   Spectrums associated with Nyquist's vestigial-symmetry theorem. (a) A spectrum with the required symmetry, (b) its rectangular component, and (c) its components with odd symmetry about $\pm W$ [2].

As an example, consider the *raised-cosine* spectrum defined by

$$S_o(\omega) = \begin{cases} 1, & |\omega| < W - W_1 \\ \dfrac{1}{2} + \dfrac{1}{2}\cos\left[\dfrac{\pi}{2W_1}(|\omega| - W + W_1)\right], & W - W_1 \leqslant |\omega| \leqslant W + W_1 \\ 0, & W + W_1 < |\omega|. \end{cases}$$

$$(4.6\text{-}4)$$

Straightforward inverse Fourier transformation of (4.6-4) produces

$$s_o(t) = \frac{W}{\pi}\frac{\sin(Wt)}{Wt}\left[\frac{\cos(W_1 t)}{1 - (2W_1 t/\pi)^2}\right]. \qquad (4.6\text{-}5)$$

The above two functions are illustrated in Fig. 4.6-2 for $W_1/W = 0$, 0.5, and 1.0. Several important observations may be made about the pulses of (b). When $W_1/W = 1.0$, the sidelobes are very small (31.5 dB or more below the peak) indicating that intersymbol interference can be made small even in the presence of timing errors. Furthermore, additional nulls occur at $t = \pm(2n + 1)\pi/2W$, $n = 1, 2, \ldots$, which also tend to lower interference. Finally, the half-amplitude pulse duration for $W_1/W = 1.0$ is exactly π/W, the time duration of one bit. Some reflection on the reader's part will verify that a polar signal constructed from this basic pulse will have possible nulls precisely π/W seconds apart.† This is an advantage, since *synchronization* signals may be extracted from a received digital waveform having this property.

With pulse shaping we may still transmit code symbols at one symbol per π/W seconds. In other words, our *signaling rate* is W/π. However, observe that the *total* band required with pulse shaping becomes $W + W_1$ rather than W. Thus compared with an ideal signaling rate of $(W + W_1)/\pi$ symbols per second, we suffer a reduction in rate by the factor $W/(W + W_1)$.

Partial Response Signaling For Interference Control

Waveforms satisfying Nyquist's vestigial-symmetry theorem theoretically cause no intersymbol interference because of their periodic nulls in adjacent symbol intervals at those intervals' sample times. However, to achieve this condition bandwidth had to be increased from its minimum value $W = \pi/T_b = \omega_b/2$. If we relax our requirement of *no* intersymbol interference to a new condition of *controlled* interference, bandwidth increases are unnecessary.

For example, we could select a symbol waveform so that it interferes in a known manner with one adjacent interval following it but with no others, again because of periodic nulls. From knowledge of how the in-

---

† Actual null locations will correspond to the transition points in the binary sequence of ones and zeros.

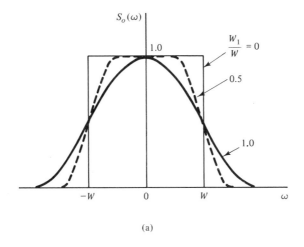

(a)

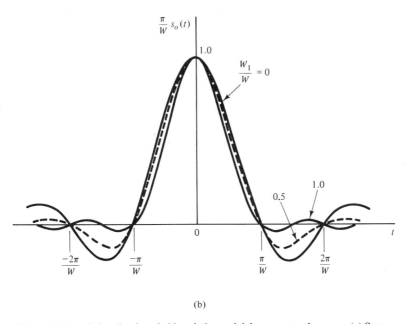

(b)

Figure 4.6-2. Pulse shaping via Nyquist's vestigial-symmetry theorem. (a) Spectrums and (b) shaped pulses [2].

terference occurs, we can conceivably correct (remove) it during demodulation. We construct a suitable waveform starting from the minimum-
bandwidth pulse described previously that satisfies Nyquist's vestigial-
symmetry theorem; it is defined here by†

$$p_1(t) = \text{Sa}\left[\frac{\pi(t - T_b)}{T_b}\right] \tag{4.6-6}$$

$$P_1(\omega) = T_b \, \text{rect}\left(\frac{\omega}{\omega_b}\right) \exp\left(\frac{-j2\pi\omega}{\omega_b}\right). \tag{4.6-7}$$

This waveform is illustrated in Fig. 4.6-3. If the waveform corresponds to,
for instance, the symbol interval from $t = 0$ to $t = T_b$, it clearly has nulls
at the sample times of all intervals, except its own at $t = T_b$. Thus it
causes no intersymbol interference, but it cannot be realized, as noted
previously.

Now we construct a new waveform, $p_2(t)$, by *adding* $p_1(t)$ with itself
delayed one symbol interval, as follows:

$$\begin{aligned}
p_2(t) &= p_1(t) + p_1(t - T_b) \\
&= \text{Sa}\left[\frac{\pi(t - T_b)}{T_b}\right] + \text{Sa}\left[\frac{\pi(t - 2T_b)}{T_b}\right] \\
&= \frac{\pi \cos\left[\left(\dfrac{\pi}{T_b}\right)\left(t - \dfrac{3T_b}{2}\right)\right]}{\left(\dfrac{\pi}{2}\right)^2 - \left[\left(\dfrac{\pi}{T_b}\right)\left(t - \dfrac{3T_b}{2}\right)\right]^2}
\end{aligned} \tag{4.6-8}$$

The spectrum $P_2(\omega)$, of $p_2(t)$, is

$$\begin{aligned}
P_2(\omega) &= P_1(\omega) + P_1(\omega)\exp(-j\omega T_b) \\
&= P_1(\omega) \, 2 \cos\left(\frac{\pi\omega}{\omega_b}\right)\exp\left(\frac{-j\pi\omega}{\omega_b}\right).
\end{aligned} \tag{4.6-9}$$

Both $p_2(t)$ and $|P_2(\omega)|$ are illustrated in Fig. 4.6-4. If $p_2(t)$ corresponds to
the bit interval from $t = 0$ to $t = T_b$, it contributes an amplitude 1.0 to
the interval's sample taken at $t = T_b$; it also contributes amplitude 1.0
to the *second* interval at its sample time $t = 2T_b$, which we call a *partial
response*. No other intervals have contributions from the first interval.
Thus the sample in any given interval contains a partial response from only
the one preceding interval and no other.

Consider interval 2, and assume a polar PCM case where digits **1** or

---

† All waveforms defined in the remainder of this section are based on $p_1(t)$, which is
defined so that it has unit maximum value. The unit-amplitude choice will allow waveforms
to be generated by a model like that of Fig. 3.6-2 of Chap. 3.

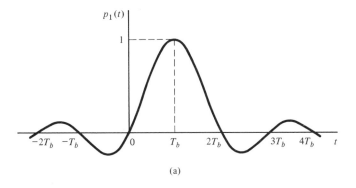

(a)

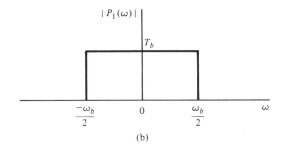

(b)

Figure 4.6-3. (a) A waveform that satisfies Nyquist's vestigial-symmetry theorem with minimum bandwidth. (b) The magnitude of the spectrum of the waveform of (a).

0 in any interval result in transmitting either $p_2(t)$ or $-p_2(t)$, respectively. Only four possibilities exist for intervals 1 and 2. They are sequences **00**, **01**, **10**, and **11**. Examination of the waveform sequences shows that the samples in interval 2 are either -2, 0, 0, or 2, respectively, for these four sequences. It follows that the decoding logic is: If the sample amplitude is 2, the digit is declared a **1** (the preceding digit is also a **1**); if the amplitude is -2 the digit is declared a **0** (preceding digit is also a **0**); if the sample is zero the digit is detected as a **1** if the preceding interval's decision was a **0**, and detected as **0** if the preceding decision was a **1**. The reader may recognize this decoding logic as being exactly the same as for the duobinary waveform of Sec. 3.6 that used rectangular pulses. Thus $p_2(t)$ of (4.6-8) used in partial response signaling is a duobinary waveform. As with the rectangular duobinary signal, precoding can also be used with the duobinary waveform using $p_2(t)$. Decoding logic is the same as described in Sec. 3.6.

The advantages of the above partial response waveform are (1) it transmits at data rate $f_b = 1/T_b$ bits/s while requiring only a bandwidth

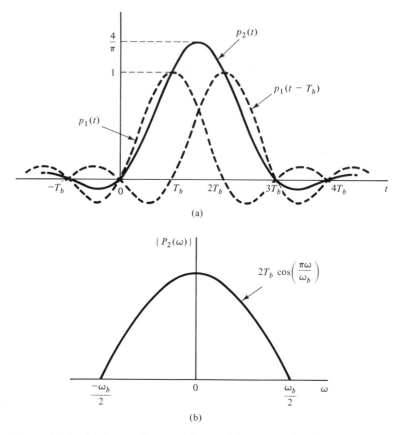

Figure 4.6-4. (a) A waveform used for partial response signaling and (b) the magnitude of the spectrum of $p_2(t)$.

$f_b/2$ Hz,† and (2) the spectrum of $p_2(t)$ is more easily approximated in practice than the ideal rectangular spectrum.

By a slight modification another useful partial response waveform can be designed that has only small spectral values at frequencies near dc. Construct $p_3(t)$ as follows:

$$p_3(t) = p_1(t) - p_1(t - 2T_b)$$

$$= \text{Sa}\left[\frac{\pi(t - T_b)}{T_b}\right] - \text{Sa}\left[\frac{\pi(t - 3T_b)}{T_b}\right] \qquad (4.6\text{-}10)$$

$$= \frac{2\pi \sin[(\pi/T_b)(t - 2T_b)]}{[(\pi/T_b)(t - 2T_b)]^2 - \pi^2}.$$

† *Bandwidth efficiency* is therefore 2 bits/s/Hz. The waveform of Fig. 4.6-3(a), corresponding to an ideal rectangular spectrum, also has this bandwidth efficiency but is more sensitive to pulse rate perturbations and its spectrum is less easily approximated.

The spectrum, $P_3(\omega)$, of $p_3(t)$ is

$$P_3(\omega) = P_1(\omega)\, 2j \sin\!\left(\frac{2\pi\omega}{\omega_b}\right)\exp\!\left(\frac{-j2\pi\omega}{\omega_b}\right). \qquad (4.6\text{-}11)$$

The waveform $p_3(t)$ and its spectrum magnitude are shown in Fig. 4.6-5. If a polar PCM waveform is constructed by generating $p_3(t)$ or $-p_3(t)$ in each digit interval having a **1** or **0**, respectively, the reader can readily verify that the waveform is modified duobinary. It can be decoded using the same logic described in Sec. 3.6 for either direct or precoded data sequences.

Generalized Partial Response Signaling

Duobinary and modified duobinary systems were introduced by Lender [6–8] and generalized by Kretzmer [9, 10]. The generalized systems are extensions of the models used in duobinary and modified duobinary systems, as illustrated in Fig. 4.6-6 [11]. As in Sec. 3.6, the sequence of (independent) data digits (**0s** and **1s**) determine the coefficients α_k. Coefficients w_0, w_1, $\ldots$, w_{N_p-1} have integer values and $N_p - 1$ denotes the number of adjacent intervals over which partial responses may be nonzero when $P_1(\omega)$ is defined by (4.6-7). For example,

$$w_0 = 1 \qquad (4.6\text{-}12)$$
$$w_1 = 1, \qquad (4.6\text{-}13)$$

and $N_p = 2$ in the duobinary case, whereas

$$w_0 = 1 \qquad (4.6\text{-}14)$$
$$w_1 = 0 \qquad (4.6\text{-}15)$$
$$w_2 = -1, \qquad (4.6\text{-}16)$$

and $N_p = 3$ in the modified duobinary case.†

From Fig. 4.6-6 the impulse response of the waveform generator, denoted as $p_{N_p}(t)$, is

$$p_{N_p}(t) = \sum_{i=0}^{N_p-1} w_i p_1(t - iT_b) \qquad (4.6\text{-}17)$$

which is the typical signal generated by a signaling interval. Its spectrum, $P_{N_p}(\omega)$, is

$$P_{N_p}(\omega) = P_1(\omega)\sum_{m=0}^{N_p-1} w_m e^{-jm\omega T_b} = P_1(\omega)\sum_{m=0}^{N_p-1} w_m D^m \qquad (4.6\text{-}18)$$

where

$$D \triangleq e^{-j\omega T_b}. \qquad (4.6\text{-}19)$$

† In duobinary and modified duobinary cases, $\alpha_k = A/2$ or $-A/2$ for binary **1** or **0**, respectively.

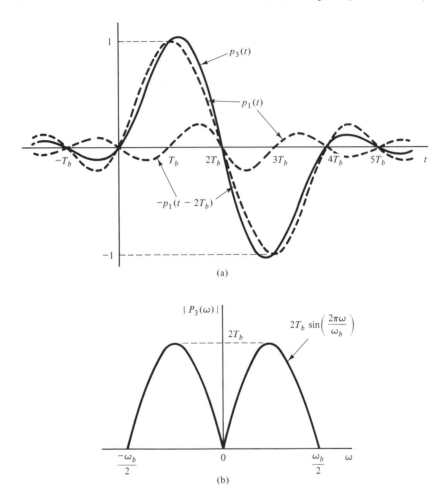

Figure 4.6-5. (a) A waveform used for a modified form of partial response signaling and (b) the magnitude of the spectrum of $p_3(t)$.

The sum multiplying $P_1(\omega)$ in (4.6-18) that involves D is called the partial response system's *polynomial;* because $P_1(\omega)$ has a value only for $-\omega_b/2 < \omega < \omega_b/2$, the system polynomial determines the *shape* of the spectrum of the generalized waveform but not the spectral extent. Many system polynomials have been studied and the most practical cases tabulated [11, 12].

Example 4.6-1

We determine $p_{N_p}(t)$ and $P_{N_p}(\omega)$ when $N_p = 3$, $w_0 = 1$, $w_1 = 2$, and $w_2 = 1$. From (4.6-17) using (4.6-6)

$$p_3(t) = \text{Sa}\left[\frac{\pi(t - T_b)}{T_b}\right] + 2\,\text{Sa}\left[\frac{\pi(t - 2T_b)}{T_b}\right] + \text{Sa}\left[\frac{\pi(t - 3T_b)}{T_b}\right],$$

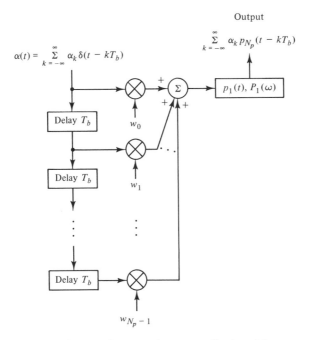

Figure 4.6-6.   Model system for generating a generalized partial response signal.

which reduces to

$$p_3(t) = \frac{2\pi^2 \operatorname{Sa}[\pi(t - 2T_b)/T_b]}{\pi^2 - [(\pi/T_b)(t - 2T_b)]^2}.$$

From (4.6-18) using (4.6-7)

$$P_3(\omega) = P_1(\omega) [1 + 2e^{-j\omega T_b} + e^{-j2\omega T_b}] = P_1(\omega)[1 + 2D + D^2]$$

$$= P_1(\omega)(1 + D)^2 = P_1(\omega) 4 \cos^2\!\left(\frac{\omega T_b}{2}\right) \exp(-j\omega T_b)$$

$$= P_1(\omega) 4 \cos^2\!\left(\frac{\pi\omega}{\omega_b}\right) \exp\!\left(\frac{-j2\pi\omega}{\omega_b}\right)$$

where the system's polynomial is $1 + 2D + D^2$. A polar PCM waveform with $\alpha_k = \pm 1$ using $p_3(t)$ will have five levels possible at sample times occurring at the ends of sampling intervals. These levels are 4, 2, 0, -2, and -4.

For other details on generalized partial response signaling the reader is referred to the literature. Pasupathy [13] gives a good summary of the concepts. Others have studied optimization of the receiver [14–16] and provide additional references.

Equalization Methods
to Control Intersymbol Interference

In two of the three preceding subsections we have explored ways of designing waveforms to have *controlled* intersymbol interference, which, by signal processing techniques, is removed in the receiver. The other subsection dealt with designing a waveform to have *no* intersymbol interference (ISI). In this subsection we introduce another approach to the control of ISI.

Suppose a pulse has been designed for transmission over a given channel so that the channel's response pulse, denoted by $p_{ch}(t)$, has no ISI. An example of $p_{ch}(t)$ would be a pulse satisfying Nyquist's vestigial-symmetry theorem. Thus samples of $p_{ch}(t)$, taken on either side of the maximum (at $t = 0$) at multiples of the bit interval T_b, are ideally zero. Next, suppose that the channel has a somewhat different transfer function than that expected in the design. The channel's actual response, denoted by $p_c(t)$, is different from $p_{ch}(t)$ and may contain ISI at sample times.† However, if the actual channel response is passed through a specially designed filter, called an *equalizing filter*, it is possible to restore the zero ISI condition at a finite number of sample times in the equalizing filter's output signal.

The structure of the equalizing filter, sometimes also called a *transversal equalizer*, is shown in Fig. 4.6-7. We assume a delay line with total delay $2NT_b$ that is tapped at intervals of T_b to produce $2N + 1$ taps. The signals at the taps are weighted by constants C_n, $n = 0, \pm 1, \pm 2, \ldots, \pm N$, prior to being summed to produce the final output, denoted by $p_{eq}(t)$. The response is

$$p_{eq}(t) = \sum_{n=-N}^{N} C_n p_c[t - (N + n)T_b]. \qquad (4.6\text{-}20)$$

The signal at the tap for $n = 0$ is delayed by NT_b, which we call the nominal delay of the filter. By choice of the constants C_n, it is possible to force $p_{eq}(t)$ to be zero at N sample intervals on either side of the desired peak output (which will occur at the nominal delay, as noted below).

To find the constants C_n that lead to the desired behavior, let samples of (4.6-20) be taken at times $t = (N + k)T_b$, $k = 0, \pm 1, \pm 2, \ldots, \pm N$. There will then be $2N + 1$ equations (one for each sample time) and $2N + 1$ unknown coefficients C_n. These equations can be placed in matrix

---

† Channel variations can occur, for example, in telephone systems where the actual signal path is determined by switchgear and destination. Clearly, a long distance call will involve a slightly different channel than a local call.

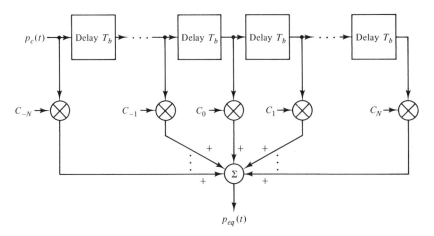

Figure 4.6-7.   Block diagram of an equalizing filter.

form as follows

$$
\begin{bmatrix} p_{eq}(0) \\ p_{eq}(T_b) \\ \vdots \\ p_{eq}(2NT_b) \end{bmatrix} =
$$

$$
\begin{bmatrix} p_c(0) & p_c(-T_b) & \cdots & p_c(-2NT_b) \\ p_c(T_b) & p_c(0) & \cdots & p_c[(-2N+1)T_b] \\ \vdots & \vdots & & \vdots \\ p_c(2NT_b) & p_c[(2N-1)T_b] & \cdots & p_c(0) \end{bmatrix} \begin{bmatrix} C_{-N} \\ C_{-N+1} \\ \vdots \\ C_N \end{bmatrix}.
$$

$$(4.6\text{-}21)$$

In general, we seek to choose the coefficients C_n such that the equalizer samples are the same as though the ideal channel response, $s_{ch}(t)$, was being sampled. Thus if we allow the ideal response to have the same nominal delay as the equalizer, we require $p_{eq}(t) = p_{ch}(t - NT_b)$ at all $2N + 1$ sample times. In other words,

$$
p_{eq}[(k+N)T_b] = p_{ch}(kT_b) = \begin{cases} p_{ch}(0), & k = 0 \\ 0, & k = \pm 1, \pm 2, \ldots, \pm N, \end{cases}
$$

$$(4.6\text{-}22)$$

and (4.6-21) becomes

$$
[p_{eq}] = [p_c][C],
$$

$$(4.6\text{-}23)$$

where we define matrices

$$
[p_{eq}] = \begin{bmatrix} p_{eq}(0) \\ p_{eq}(T_b) \\ \vdots \\ p_{eq}[(N-1)T_b] \\ p_{eq}(NT_b) \\ p_{eq}[(N+1)T_b] \\ \vdots \\ p_{eq}(2NT_b) \end{bmatrix} = \begin{bmatrix} 0 \\ 0 \\ \vdots \\ 0 \\ p_{ch}(0) \\ 0 \\ \vdots \\ 0 \end{bmatrix}, \qquad [C] = \begin{bmatrix} C_{-N} \\ C_{-N+1} \\ \vdots \\ C_{-1} \\ C_0 \\ C_1 \\ \vdots \\ C_N \end{bmatrix} \tag{4.6-24}
$$

and

$$
[p_c] = \begin{bmatrix} p_c(0) & p_c(-T_b) & \cdots & p_c(-2NT_b) \\ p_c(T_b) & p_c(0) & \cdots & p_c[(-2N+1)T_b] \\ \vdots & \vdots & & \vdots \\ p_c(2NT_b) & p_c[(2N-1)T_b] & \cdots & p_c(0) \end{bmatrix}. \tag{4.6-25}
$$

If the inverse of a square matrix is denoted by $[\cdot]^{-1}$, the solution of (4.6-23) is

$$
[C] = [p_c]^{-1}[p_{eq}]. \tag{4.6-26}
$$

From the form of $[p_{eq}]$ we easily find that $[C]$ is the middle column of $[p_c]^{-1}$ multiplied by $p_{ch}(0)$. The desired coefficients C_n are seen to be functions of samples of the unequalized channel response taken at times $t = 0, \pm T_b, \ldots, \pm 2NT_b$.

Example 4.6-2
Let the channel be ideally designed to produce a sampling pulse as shown in Fig. 4.6-8(a). Due to channel distortion the unequalized channel response of (b) is assumed to have ISI at only one sample time on each side of the maximum. Thus only $p_c(-T_b) = 0.1p_{ch}(0)$, $p_c(0) = p_{ch}(0)$, $p_c(T_b) = -0.2p_{ch}(0)$ are nonzero sample values. If we assume a three-tap equalizer $(N = 1)$ we have

$$
\begin{bmatrix} 0 \\ p_{ch}(0) \\ 0 \end{bmatrix} = \begin{bmatrix} 1.0 & 0.1 & 0 \\ -0.2 & 1.0 & 0.1 \\ 0 & -0.2 & 1.0 \end{bmatrix} \begin{bmatrix} C_{-1} \\ C_0 \\ C_1 \end{bmatrix} p_{ch}(0)
$$

from (4.6-23). Straightforward evaluation gives

$$
\begin{bmatrix} C_{-1} \\ C_0 \\ C_1 \end{bmatrix} = \begin{bmatrix} -0.1 \\ 1.0 \\ 0.2 \end{bmatrix} \frac{1}{1.04}.
$$

On using these coefficients in (4.6-20) the equalized channel responses at various

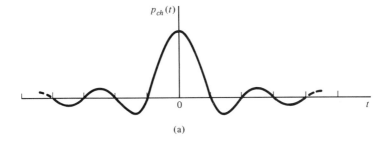

(a)

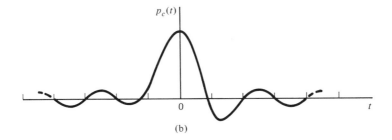

(b)

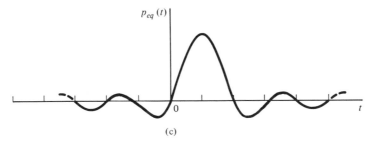

(c)

Figure 4.6-8.   Waveforms applicable to the equalized channel of Example 4.6-2.

sample times are

$$p_{eq}(-T_b) = \frac{-0.01}{1.04} p_{ch}(0)$$

$$p_{eq}(0) = 0$$

$$p_{eq}(T_b) = p_{ch}(0)$$

$$p_{eq}(2T_b) = 0$$

$$p_{eq}(3T_b) = \frac{-0.04}{1.04} p_{ch}(0).$$

Samples at other times are zero. The equalized channel response is illustrated in Fig. 4.6-8(c).

The transversal equalizer can force the equalized channel response to have nulls (zero ISI) at N sample intervals on either side of the desired maximum response. It does not guarantee that the equalized response will have nulls at sample times beyond $\pm NT_b$. In fact, if the unequalized channel has N_u side intervals with ISI and the equalizer uses $2N + 1$ taps with $N \leq N_u$, there will remain $2N_u$ sample times with ISI in the equalized response. N_u of these points fall below $t = 0$ and N_u fall above $2NT_b$. Only in special cases can the equalizer completely eliminate ISI. However, if N is large and ISI decreases as sample times become removed from the maximum, ISI can be made small.

Basically, two methods exist for setting equalizer tap coefficients. *Preset equalizers* use special pulse sequences to establish coefficients [17]. *Adaptive equalizers* use the data signal itself to continually adjust coefficients so that some error criterion, usually a mean-squared error, is minimized [18].

4.7 OPTIMUM DUOBINARY SYSTEM

To illustrate the application of partial response techniques we shall discuss a duobinary system and one approach to its optimization.

System

Proper optimization requires that we define the system constraints, input specifications, and optimization criteria involved. The system and its constraints are defined with the aid of Fig. 4.7-1(a). The models of Chap. 3 have been used to represent the transmitter. The transmitted waveform $s_T(t)$, having average power S_T, is the result of exciting a transmitter filter with transfer function $H_T(\omega)$ by a sequence of impulses with amplitudes α_k. For the duobinary system α_k has only values $A/2$ or $-A/2$ corresponding to binary data digits **1** or **0**, respectively. The binary digits are assumed to occur with equal probabilities, $P_1 = P_2 = 1/2$ and the data stream is assumed to have been precoded (Sec. 3.6). The input signal to the receiver, having average power S_i, is the output of the channel, assumed linear with transfer function $H_{ch}(\omega)$. The receiver is assumed linear with transfer function $H_R(\omega)$. The receiver's output signal $s_o(t)$ will be constrained to have the form discussed in Sec. 4.6; this form is defined by the pulse *shape* of (4.6-8). This choice is to eliminate intersymbol interference.

At the ends of symbol intervals the duobinary waveform has only three possible levels, which we define as A, 0, and $-A$. Thus the output signal $s_o(t)$ is generated according to Fig. 4.7-1(b), where $P_2(\omega)$ is given by (4.6-9). Due to our choice of output levels, the appropriate decision logic is to decide a symbol is a **1** if the receiver response $r_o(t)$ at the end of the

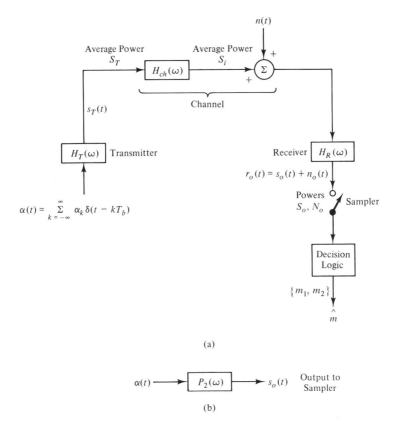

Figure 4.7-1. (a) A duobinary system and (b) the equivalent filter for transmitter, channel, and receiver operations applicable to the signal.

interval is near zero, and decide a 0 if it is near either $-A$ or A. We assume a sample is taken each symbol interval to enable these decisions.

For input specification we presume the channel noise $n(t)$ is stationary, zero mean and Gaussian with power spectrum $\mathscr{S}_n(\omega)$. We further assume the receiver is synchronized to the transmitter so that it knows the starting and ending times of symbol intervals.

The optimization criterion is basically to minimize average probability of making a symbol (bit) error, denoted by P_e. The minimization is done in two steps. First optimum thresholds are found that minimize P_e. Second, $H_T(\omega)$ and $H_R(\omega)$ are found that cause P_e to be smallest under the constraints of constant (fixed) transmitted power S_T and fixed output waveform shape [specified $P_2(\omega)$]. Thus $P_2(\omega)$, $H_{ch}(\omega)$, $\mathscr{S}_n(\omega)$ and S_T are specified in the minimization process.

Optimum Thresholds

Since the input noise has been assumed to be Gaussian, the output noise is Gaussian. Denote its probability density and variance (noise power) by $p_{n_o}(n_o)$ and σ_o^2, respectively. Then the probability densities applicable to the three possible output levels at the sample times are illustrated in Fig. 4.7-2. From Sec. 3.6, levels $-A$, 0, and A occur with probabilities $\frac{1}{4}$, $\frac{1}{2}$, and $\frac{1}{4}$, respectively. Since there are three amplitudes, the decision process generally involves two thresholds V_{T_1} and V_{T_2}.

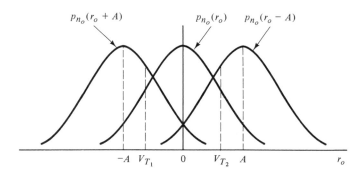

Figure 4.7-2. Probability density functions of duobinary receiver output r_o at sample times for three possible signal levels A, 0, and $-A$.

Now suppose level $-A$ occurs. A decision error occurs only if $V_{T_1} < r_o \leq V_{T_2}$.† An error on reception of level A also occurs when $V_{T_1} < r_o \leq V_{T_2}$, whereas a zero signal level results in an error only when $r_o \leq V_{T_1}$ or $V_{T_2} < r_o$. Average error probability then becomes

$$P_e = \frac{1}{4} \int_{V_{T_1}}^{V_{T_2}} p_{n_o}(r_o + A)\, dr_o$$

$$+ \frac{1}{4} \int_{V_{T_1}}^{V_{T_2}} p_{n_o}(r_o - A)\, dr_o + \frac{1}{2}\left[1 - \int_{V_{T_1}}^{V_{T_2}} p_{n_o}(r_o)\, dr_o \right]. \tag{4.7-1}$$

To find V_{T_1} and V_{T_2} that minimize P_e, we differentiate (4.7-1) with respect to these thresholds according to Leibniz's rule to get

$$\frac{\partial P_e}{\partial V_{T_1}} = -\frac{1}{4} p_{n_o}(V_{T_1} + A) - \frac{1}{4} p_{n_o}(V_{T_1} - A) + \frac{1}{2} p_{n_o}(V_{T_1}) = 0 \tag{4.7-2}$$

$$\frac{\partial P_e}{\partial V_{T_2}} = \frac{1}{4} p_{n_o}(V_{T_2} + A) + \frac{1}{4} p_{n_o}(V_{T_2} - A) - \frac{1}{2} p_{n_o}(V_{T_2}) = 0. \tag{4.7-3}$$

Typically $p_{n_o}(V_{T_1} - A)$ in (4.7-2) and $p_{n_o}(V_{T_2} + A)$ in (4.7-3) are small in

---

† If $r_o > V_{T_2}$, the decision would be that level A occurred. However a decision for level A corresponds to a data **0** as well as a decision for level $-A$, so no error is made.

relation to other terms so that we require

$$p_{n_o}(V_{T_1} + A) = 2 p_{n_o}(V_{T_1}) \qquad (4.7\text{-}4a)$$

$$p_{n_o}(V_{T_2} - A) = 2 p_{n_o}(V_{T_2}). \qquad (4.7\text{-}4b)$$

On substitution of the Gaussian function for $p_{n_o}(\cdot)$, we solve (4.7-4) to obtain the optimum thresholds

$$V_{T_2} = \frac{A}{2} + \frac{\sigma_o^2}{A} \ln(2) \triangleq V_T \qquad (4.7\text{-}5a)$$

$$V_{T_1} = -V_{T_2} = -V_T. \qquad (4.7\text{-}5b)$$

After substituting (4.7-5) into (4.7-1) and using the approximation erf $[(V_T + A)/\sqrt{2}\sigma_o] \approx 1$, we have

$$P_e = \frac{3}{4} - \frac{1}{4} \text{erf}\left[\frac{A^2 - \sigma_o^2 \ln(4)}{2\sqrt{2}\, A\, \sigma_o}\right] - \frac{1}{2} \text{erf}\left[\frac{A^2 + \sigma_o^2 \ln(4)}{2\sqrt{2}\, A\, \sigma_o}\right]. \qquad (4.7\text{-}6)$$

For the important case where $A^2 \gg \sigma_o^2$

$$P_e \approx \frac{3}{4} \text{erfc}\left(\frac{A}{2\sqrt{2}\,\sigma_o}\right). \qquad (4.7\text{-}7)$$

Close examination of (4.7-6) shows that P_e is further minimized by maximizing A^2/σ_o^2. Within our system constraints this maximization is achieved by proper choice of transmitting and receiving filters.

Optimum Transmit and Receive Filters

From Sec. 3.7 the power spectrum of $\alpha(t)$ in Fig. 4.7-1(a), denoted by $\mathcal{S}_\alpha(\omega)$, is

$$\mathcal{S}_\alpha(\omega) = \frac{A^2}{4T_b}. \qquad (4.7\text{-}8)$$

Average transmitted power becomes

$$S_T = \frac{1}{2\pi} \int_{-\infty}^{\infty} \mathcal{S}_\alpha(\omega)|H_T(\omega)|^2 \, d\omega = \frac{A^2}{8\pi T_b} \int_{-\infty}^{\infty} |H_T(\omega)|^2 \, d\omega. \qquad (4.7\text{-}9)$$

Our constraint that the output waveform $s_o(t)$ can have amplitudes A, 0, or $-A$ at sample times and have the pulse shape of (4.6-8) is equivalent to requiring

$$H_T(\omega) H_{ch}(\omega) H_R(\omega) = P_2(\omega). \qquad (4.7\text{-}10)$$

By solving (4.7-9) for A^2, substituting (4.7-10) to eliminate $H_T(\omega)$, and using the output noise power relationship

$$\sigma_o^2 = \frac{1}{2\pi} \int_{-\infty}^{\infty} \mathcal{S}_n(\omega)|H_R(\omega)|^2 \, d\omega, \qquad (4.7\text{-}11)$$

we have

$$\frac{A^2}{\sigma_o^2} = \frac{16\pi^2 \, S_T T_b}{\displaystyle\int_{-\infty}^{\infty} \frac{|P_2(\omega)|^2}{|H_{ch}(\omega)H_R(\omega)|^2} \, d\omega \int_{-\infty}^{\infty} \mathcal{S}_n(\omega)|H_R(\omega)|^2 \, d\omega}. \qquad (4.7\text{-}12)$$

Maximization of A^2/σ_o^2 is achieved by selecting $H_R(\omega)$ to minimize the denominator of (4.7-12). An easy application of Schwarz's inequality (see Sec. B.9 of Appendix B) with

$$A(\omega) = \sqrt{\mathcal{S}_n(\omega)} \, |H_R(\omega)| \qquad (4.7\text{-}13)$$

$$B(\omega) = \frac{|P_2(\omega)|}{|H_{ch}(\omega)H_R(\omega)|} \qquad (4.7\text{-}14)$$

gives the optimum receiver's power transfer function

$$|H_R(\omega)|^2 = \frac{|P_2(\omega)|}{C\sqrt{\mathcal{S}_n(\omega)} \, |H_{ch}(\omega)|}. \qquad (4.7\text{-}15)$$

Here C is an arbitrary real constant. By using (4.7-15) in (4.7-10) we have the optimum transmitting filter's power transfer function

$$|H_T(\omega)|^2 = \frac{C\sqrt{\mathcal{S}_n(\omega)} \, |P_2(\omega)|}{|H_{ch}(\omega)|}. \qquad (4.7\text{-}16)$$

The main results of (4.7-15) and (4.7-16) do not restrict the phase characteristics of $H_T(\omega)$, $H_{ch}(\omega)$, and $H_R(\omega)$, which can have any values so long as they combine in cascade according to (4.7-10).

For white noise where $\mathcal{S}_n(\omega) = \mathcal{N}_0/2$ the optimum filters become

$$|H_R(\omega)|^2 = \frac{\sqrt{2} \, |P_2(\omega)|}{C\sqrt{\mathcal{N}_0} \, |H_{ch}(\omega)|} \qquad (4.7\text{-}17)$$

and

$$|H_T(\omega)|^2 = \frac{C\sqrt{\mathcal{N}_0} \, |P_2(\omega)|}{\sqrt{2} \, |H_{ch}(\omega)|}. \qquad (4.7\text{-}18)$$

Except for the scale constant, the transmit and receive filters have the same frequency characteristics in this case.

In addition to white noise, if the channel is distortion-free as defined by

$$H_{ch}(\omega) = G_{ch} \, e^{-j\omega t_d}, \qquad (4.7\text{-}19)$$

where the voltage gain and delay are denoted as G_{ch} and t_d, respectively, then

$$|H_R(\omega)|^2 = \frac{|P_2(\omega)|}{G_{ch}^2} \qquad (4.7\text{-}20)$$

and

$$|H_T(\omega)|^2 = |P_2(\omega)|. \qquad (4.7\text{-}21)$$

In writing these two results the arbitrary constant C has been chosen as

$$C = \sqrt{2}\,\frac{G_{ch}}{\sqrt{\mathcal{N}_0}}. \tag{4.7-22}$$

From (4.7-20) we note that the receiver's gain is adjusted to compensate for any loss over the channel to keep the output levels fixed.

Example 4.7-1

A duobinary system uses a distortion-free channel with a power loss of 80 dB. If $A = 2$ V is required and the symbol duration is $T_b = 4$ ms, we find the average transmitted power and input average received power.

From (4.6-9) and (4.6-7)

$$|P_2(\omega)| = 2\,T_b \cos\!\left(\frac{\pi\omega}{\omega_b}\right) \mathrm{rect}\!\left(\frac{\omega}{\omega_b}\right).$$

By use of (4.7-21) in (4.7-9) we have

$$S_T = \frac{A^2}{4\pi} \int_{-\omega_b/2}^{\omega_b/2} \cos\!\left(\frac{\pi\omega}{\omega_b}\right) d\omega = \frac{A^2}{\pi T_b} = \frac{1000}{\pi} = 318.31 \text{ W}.$$

Because $G_{ch}^2 = 10^{-8}$ for a power loss of 80 dB, the input average power to the receiver is

$$S_i = 318.31\,(10^{-8}) = 3.18 \ \mu\text{W}.$$

Error Probability

After maximizing (4.7-12) using Schwarz's inequality, it is found that the maximum value is

$$\left(\frac{A^2}{8\sigma_o^2}\right)_{\max} = \frac{2\pi^2\,S_T T_b}{\left|\displaystyle\int_{-\infty}^{\infty} \frac{\sqrt{\mathcal{S}_n(\omega)}\,|P_2(\omega)|}{|H_{ch}(\omega)|}\,d\omega\right|^2}. \tag{4.7-23}$$

Error probability derives from using (4.7-23) in (4.7-6) or (4.7-7).

For a distortion-free channel with white noise, we readily find

$$\left(\frac{A}{2\sqrt{2}\,\sigma_o}\right)_{\max} = \frac{\pi}{4}\sqrt{\frac{G_{ch}^2\,S_T\,T_b}{\mathcal{N}_0}}, \tag{4.7-24}$$

so

$$P_e \approx \frac{3}{4}\,\mathrm{erfc}\!\left[\frac{\pi}{4}\sqrt{\frac{G_{ch}^2\,S_T\,T_b}{\mathcal{N}_0}}\right] \tag{4.7-25}$$

from (4.7-7). By recognizing that $S_T T_b$ is the average energy per bit, we can write (4.7-25) as

$$P_e \approx \frac{3}{4}\,\mathrm{erfc}\!\left[\frac{\pi}{4}\sqrt{G_{ch}^2\,\varepsilon}\right] \tag{4.7-26}$$

where

$$\varepsilon = \frac{S_T T_b}{\mathcal{N}_0}. \tag{4.7-27}$$

Figure 5.12-1 illustrates the behavior of (4.7-26) for $G_{ch} = 1$.

Example 4.7-2

Again assume the system of Example 4.7-1. For a white noise channel where $\mathcal{N}_0 = 10^{-8}/4\pi$ W/Hz, we find P_e. From the previous example, $S_T T_b = A^2/\pi = 4/\pi$ so $\varepsilon = 16(10^8)$ and $P_e \approx 0.75$ erfc $[(\pi/4)\sqrt{16}] = 6.97(10^{-6})$ after use of (F-5) in Appendix F.

4.8 OPTIMUM MODIFIED DUOBINARY SYSTEM

Suppose we are interested in constructing an optimum modified duobinary system using the waveform shape of (4.6-10) having the spectrum $P_3(\omega)$ of (4.6-11). If the receiver output waveform has possible values A, 0, and $-A$ at the sample times, taken at the ends of the symbol intervals, it is found that *all* the results given in Sec. 4.7 apply if $P_3(\omega)$ replaces $P_2(\omega)$. Thus Fig. 4.7-1(a) applies to the modified duobinary system with $\alpha_k = A/2$ or $-A/2$ when binary data digits are 1 or 0, respectively. It is necessary only to replace $P_2(\omega)$ with $P_3(\omega)$ in Fig. 4.7-1(b).

4.9 *M*-ARY PAM SYSTEM

When the transmitter can issue any one of several (M) pulse amplitudes during a symbol interval of duration T, we refer to the system as *M-ary pulse amplitude modulation*, (M-ary PAM). In this section we examine M-ary PAM under the assumption that each of the M possible waveform levels corresponds to one of M equally probable messages from a discrete source. The messages are also assumed statistically independent from one symbol interval to the next.† Thus the probability that a given message, and therefore a given pulse level, is transmitted is $1/M$.

Because levels are independent and equally probable, each represents $\log_2(M)$ bits of information occurring each T seconds. The rate of information transfer in M-ary PAM is then $(1/T)\log_2(M)$ bits/s. If M is a power of 2, such as $M = 2^{N_b}$, so $N_b = \log_2(M)$, a direct comparison to binary PCM can be made. If N_b is the number of binary bits per symbol, and if these N_b bits occur over T seconds, the same is in M-ary PAM, the two systems convey information at the same rate. However, in the process the binary PCM system must transmit N_b pulses while the M-ary system transmits

---

† More general systems can be developed, but the required mathematical detail makes the important results less obvious.

only one (during time T). The bandwidth required by the M-ary system is, therefore, smaller than that of the binary system by a factor $1/N_b$. Bandwidth saving is the main advantage of M-ary PAM; the saving is paid for by either decreased performance in noise, or a need for increased transmitter power.

Average Transmitted/Received Power

We shall assume that transmitted pulses are rectangular with amplitudes A_i, $i = 1, 2, \ldots, M$, that are uniformly separated by an amount A. Amplitudes A_i are assumed to be symmetrically displaced about zero. Hence,

$$A_i = \frac{A}{2}(2i - M - 1), \qquad i = 1, 2, \ldots, M. \tag{4.9-1}$$

If M is odd, levels are $0, \pm A, \pm 2A, \ldots, \pm A(M - 1)/2$. If M is even, levels are $\pm A/2, \pm 3A/2, \pm 5A/2, \ldots, \pm (M - 1)A/2$.

If no loss is assumed over the channel, the average power at the receiver, denoted by S_i, is the same as the transmitted power and is

$$S_i = \sum_{i=1}^{M} A_i^2 P(A_i) = \sum_{i=1}^{M} A_i^2 \left(\frac{1}{M}\right) \tag{4.9-2}$$

where $P(A_i) = 1/M$ is the probability that level A_i is transmitted. After substitution of (4.9-1) and use of known sums (Appendix D), (4.9-2) reduces to

$$S_i = \frac{A^2(M^2 - 1)}{12}. \tag{4.9-3}$$

Optimum Thresholds

At the ends of the symbol intervals, an optimum receiver will observe (sample) its output level and decide which of the M levels was most likely to have been transmitted. Now regardless of the exact form of the optimum receiver—as long as it is linear—the output levels, denoted by V_i, will be proportional to received pulse levels A_i and will be uniformly spaced by an amount denoted by V. Clearly, V is proportional to A, the received level separation.

Since channel noise is zero-mean, white, and Gaussian with power density $\mathcal{N}_0/2$, by assumption, the output noise at the sample time is Gaussian with mean V_i when the received pulse's amplitude is A_i. Thus the probability densities of Fig. 4.9-1 apply to the M possible output levels V_i. To establish which output level is actually received, we divide all possible receiver output levels, r_o, into regions separated by thresholds V_{T_i}, as shown. The first step in optimizing the system is finding optimum values of V_{T_i}.

Let $p_{n_o}(\cdot)$ denote the probability density of receiver output noise at

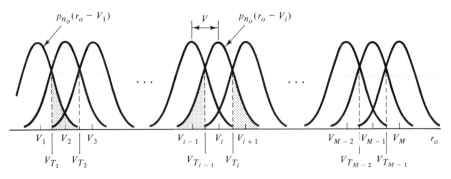

Figure 4.9-1. Probability density functions applicable to the output of an M-ary PAM system (at the sampler).

the sample time of an interval. If level V_1 is received the probability of error, given V_1, is

$$P(\varepsilon|V_1) = \int_{V_{T_1}}^{\infty} p_{n_o}(r_o - V_1)\, dr_o \qquad (4.9\text{-}4)$$

which is the leftmost shaded area in Fig. 4.9-1. Similarly, if V_i is received,

$$P(\varepsilon|V_i) = \int_{-\infty}^{V_{T_{i-1}}} p_{n_o}(r_o - V_i)\, dr_o + \int_{V_{T_i}}^{\infty} p_{n_o}(r_o - V_i)\, dr_o \qquad (4.9\text{-}5)$$

for $i = 2, 3, \ldots, M - 1$. This probability is the shaded area about V_i in Fig. 4.9-1. For level M,

$$P(\varepsilon|V_M) = \int_{-\infty}^{V_{T_{M-1}}} p_{n_o}(r_o - V_M)\, dr_o. \qquad (4.9\text{-}6)$$

The average probability of error, denoted by P_w, is then

$$P_w = \sum_{i=1}^{M} P(\varepsilon|V_i) P(V_i) = \frac{1}{M} \sum_{i=1}^{M} P(\varepsilon|V_i)$$

$$= \frac{1}{M} \sum_{i=1}^{M-1} \left[\int_{V_{T_i}}^{\infty} p_{n_o}(r_o - V_i)\, dr_o + \int_{-\infty}^{V_{T_i}} p_{n_o}(r_o - V_{i+1})\, dr_o \right]. \qquad (4.9\text{-}7)$$

By differentiation of (4.9-7) with respect to V_{T_i}, using Leibniz's rule, we establish a set of equations that, when set equal to zero, leads to the optimum thresholds V_{T_i}. It is found that

$$V_{T_i} = \frac{(V_{i+1} + V_i)}{2}, \qquad i = 1, 2, \ldots, M - 1, \qquad (4.9\text{-}8)$$

which fall midway between signal levels V_i. These values of V_{T_i} are illustrated in Fig. 4.9-1.

Probability of error given by (4.9-7) can now be found by substituting the optimum thresholds and the Gaussian form for $p_{n_o}(\cdot)$. If σ_o^2 is the input noise power at a sample time we readily find

$$P_w = \frac{M-1}{M} \, \text{erfc}\left[\frac{V}{2\sqrt{2}\,\sigma_o}\right]. \tag{4.9-9}$$

Optimum Receiver

Since the complementary error function, erfc(·), is a monotonically decreasing function of its argument, (4.9-9) is minimum when V^2/σ_o^2 is maximized. But V^2/σ_o^2 is maximum for a receiver that maximizes the output level peak power to average noise power ratio at the sample time, because V is the difference between adjacent signal levels ($V = V_{i+1} - V_i$). Thus the optimum receiver is a white-noise matched filter. For the rectangular pulse form assumed, its transfer function can be written as†

$$H_R(\omega) = e^{-j\omega T/2} \, \text{Sa}\left(\frac{\omega T}{2}\right). \tag{4.9-10}$$

During a symbol interval, the received signal pulse is

$$s_i(t) = A_i \, \text{rect}\left[\frac{t-(T/2)}{T}\right]. \tag{4.9-11}$$

By computing the response of the receiver at time T to this input, the reader can easily show (Prob. 4-48) that

$$V_i = A_i, \qquad i = 1, 2, ..., M. \tag{4.9-12}$$

so

$$V = A. \tag{4.9-13}$$

In a similar manner, when noise is considered we find

$$\sigma_o^2 = \frac{\mathcal{N}_0}{2T}. \tag{4.9-14}$$

Error Probability

By using (4.9-13) and (4.9-14) in (4.9-9) we have

$$P_w = \frac{M-1}{M} \, \text{erfc}\left[\sqrt{\frac{A^2 T}{4\,\mathcal{N}_0}}\right]. \tag{4.9-15}$$

Another useful form for word error probability follows the substitution of (4.9-3)

$$P_w = \frac{M-1}{M} \, \text{erfc}\left[\sqrt{\frac{3\,S_i T}{(M^2-1)\,\mathcal{N}_0}}\right]. \tag{4.9-16}$$

---

† This is an application of (B.9-7) with t_0 chosen to make the output level maximum occur at the sample time T.

A final form results from using the ratio of average energy per bit of information to twice the channel noise density, as given by

$$\varepsilon = \frac{S_i T}{\mathcal{N}_0 \log_2(M)}. \tag{4.9-17}$$

We then obtain

$$P_w = \frac{M-1}{M} \operatorname{erfc}\left[\sqrt{\frac{3 \log_2(M)}{M^2-1} \varepsilon}\right]. \tag{4.9-18}$$

Figure 4.9-2 illustrates the behavior of (4.9-18). Because the factor $(M-1)/M$ does not vary greatly ($\frac{1}{2}$ to 1), we find that average energy per information bit must increase by a factor $(M^2-1)/3 \log_2(M)$ as M increases above 2 to maintain P_w approximately constant.

Example 4.9-1

An eight-level PAM system transmits rectangular 50-μs pulses with amplitudes $A_i = \pm 1$ V, ± 3 V, ± 5 V and ± 7 V over a lossless channel for which $\mathcal{N}_0 = 5.2(10^{-6})$ W/Hz. We find average transmitted power and P_w. Here $A = 2$ V, so transmitted power, from (4.9-3), is

$$S_i = \frac{4(8^2-1)}{12} = \frac{63}{3} \text{ W}$$

(same as received power because of the lossless channel assumed). From (4.9-15)

$$P_w = \frac{8-1}{8} \operatorname{erfc}\left[\sqrt{\frac{4(50)10^{-6}}{4(5.2)10^{-6}}}\right] = \frac{7}{8} \operatorname{erfc}\left[\sqrt{\frac{50}{5.2}}\right] = 1.06(10^{-5}).$$

Performances of M-ary PAM and binary systems can be compared through their word error probabilities. In the latter case P_w derives from (4.5-3) using (4.4-5) for a polar format; it can be written as[†]

$$P_w = \frac{\log_2(M)}{2} \operatorname{erfc}\left[\sqrt{\frac{S_i T_b}{\mathcal{N}_0}}\right], \tag{4.9-19}$$

where S_i is the received average signal power. For values of $P_w < 10^{-4}$, the complementary error functions in (4.9-19) and (4.9-16) are so sensitive to their arguments that their coefficients have only small affect on the arguments that correspond to equal P_w. The two error probabilities are therefore approximately equal when

$$\frac{S_i T_b}{\mathcal{N}_0} \text{ (binary)} \approx \frac{3 S_i T}{(M^2-1) \mathcal{N}_0} \text{ (M-ary PAM)}. \tag{4.9-20}$$

Consider two cases.

First, assume the two systems have equal values of M, P_w, and $\mathcal{N}_0$,

---

[†] Since $k = N_b$, the number of binary bits, we have $k = \log_2(M)$ by assuming $M = 2^{N_b}$.

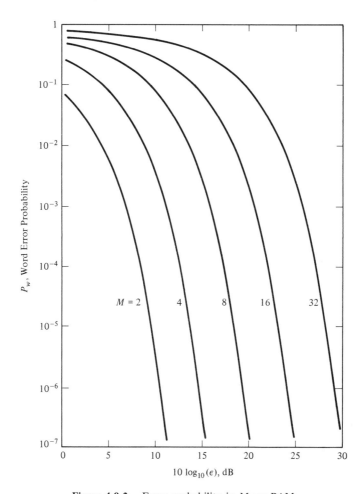

Figure 4.9-2. Error probability in *M*-ary PAM.

and transmit the same information rate (this requires $T = T_b \log_2(M)$). The *M*-ary system then requires less bandwidth (by a factor $1/\log_2(M)$) and more power (by a factor $(M^2 - 1)/3\log_2(M)$). In the second case the systems have equal values of M, P_w, $\mathcal{N}_0$, and bandwidth (this requires $T = T_b$). The *M*-ary system requires still more power (by a factor $(M^2 - 1)/3$) but transmits information at a higher rate (by a factor $\log_2(M)$) compared to the binary system. From these two cases we conclude, overall, that the main disadvantage to *M*-ary PAM is increased power required for a given performance. Its main advantage lies in its ability to use less bandwidth for a given information rate.

In Chaps. 1 and 3 we indicated that some systems are capable of direct digital encoding of the outputs of sources with memory—that is, sources with outputs that are correlated with time. In the next and the following three sections, several such systems are discussed in detail.

4.10 DELTA MODULATION

Delta modulation (DM), a European invention, was apparently first discussed in the English-language literature by de Jager [19]. It may be described as a type of PCM where only 1-bit encoding is used. One-bit coding is made possible by a feedback loop which is an integral part of the encoding process. In the ensuing development of DM principles, the simplest form of implementation is assumed. It has an ideal integrator in the loop feedback path and will be called a *basic* delta modulator. At the end of this section we consider some alternative configurations based on practical integrators.

System Block Diagram

Figure 4.10-1 illustrates the components of a DM system applicable to a single message. If time multiplexing of several messages is to be used, the separate pulse trains for the various messages may be added at point A. They are separated in the receiver by a suitable demultiplexer at point B. We shall assume in this section that N similar messages are multiplexed, each being bandlimited with maximum spectral extent W_f.

At the transmitter end of the channel, the pulse modulator accepts an input train of constant-amplitude clock pulses occurring at a rate $\omega_s/2\pi$ pulses per second. It converts these pulses to an output train of polar pulses with polarity determined by the "loop error" signal $-\varepsilon_q(t)$. If $-\varepsilon_q(t)$ is positive at the time of any given input pulse, the output pulse is positive. If $-\varepsilon_q(t)$ is negative, a negative output pulse is generated. The error $-\varepsilon_q(t)$

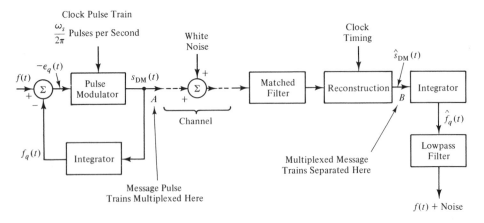

Figure 4.10-1. Block diagram of a delta modulation system [2].

is the difference $f(t) - f_q(t)$ between the input message $f(t)$ and $f_q(t)$, which is the integrated version of the generated pulse train $s_{DM}(t)$. We assume that these generated pulses have a duration τ which is short relative to the time $T_s = 2\pi/\omega_s$ between "sample" pulses. Thus, the (ideal) integrator output $f_q(t)$ will be a "stair-step" type of waveform with an up-going step for each positive pulse and a down-going step for a negative pulse. Each transmitted pulse is clearly a 1-bit encoded version of the signal $-\varepsilon_q(t)$. Now since the action of the loop will be to force $f_q(t)$ to be an approximation to $f(t)$, a fact made more obvious in the following paragraph, the overall action of the modulator is to transmit 1-bit information about the *changes* in $f(t)$ with time.

To better understand operation, assume $f(t)$ is applied, as shown in Fig. 4.10-2(a), at $t = 0$. Let the input clock pulse train timing be such that a pulse occurs at $t = 0$. At $t = 0$, and for a short time thereafter, $f(t) >$

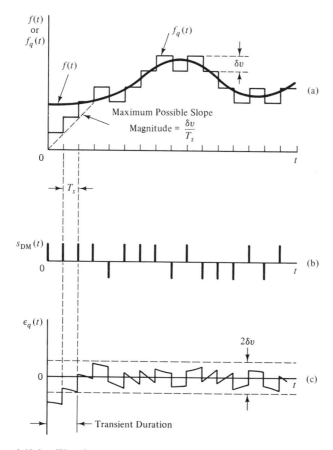

Figure 4.10-2. Waveforms applicable to delta modulation. (a) $f(t)$ and $f_q(t)$, (b) the DM signal, and (c) the error signal [2].

$f_q(t)$ and output pulses are all positive, as illustrated in (b). During this initial or *transient period* the signal $f_q(t)$ increases in steps of δv approaching the message until it finally exceeds it in voltage. After the transient period, $f_q(t)$ will become approximately equal to $f(t)$ and both positive and negative output pulses are generated with about equal regularity. The behavior of the error $\varepsilon_q(t)$ is illustrated in (c). Except during the transient period the error never exceeds $\pm \delta v$, where δv is called the *step size*.

At the receiver end of the channel a matched filter is shown in Fig. 4.10-1 to provide maximum noise immunity (see Appendix B). The next step involves a reconstruction of the transmitted pulse train to remove noise. The operation is similar to reconstruction in PCM where, at the times of the maximums of the pulses emerging from the matched filter, samples of the received signal are compared to a voltage threshold level (0 V here) to determine which polarity of pulse was received. Based on these pulse-to-pulse decisions, a version $\hat{s}_{DM}(t)$ of the original signal $s_{DM}(t)$ is reconstructed. In a practical receiver the reconstruction stage might be omitted (especially if only one message is involved). We include it here because it aids in both the understanding and the mathematical analysis of DM. If multiplexing is involved, the pulse trains are separated at point B, and each is applied to an integrator. For a typical message, the integrator output $\hat{f}_q(t)$ is approximately equal to the signal $f_q(t)$ in the transmitter (except for noise errors in reconstruction it would be exactly equal to $f_q(t)$ in principle). Thus $\hat{f}_q(t) \approx f_q(t)$. Finally, a lowpass filter, having a bandwidth W_f, is used to smooth out the sharp step-behavior of $\hat{f}_q(t)$. The final output becomes $f(t)$ plus errors which we may collectively consider to be noise.

Errors and Slope Overload

Even if receiver random noise is negligible, the reconstructed message $\hat{f}_q(t)$ in Fig. 4.10-1 can at best equal $f_q(t)$. It can never equal the original message $f(t)$ because of the 1-bit quantization involved. However, since

$$f_q(t) = f(t) + \varepsilon_q(t), \tag{4.10-1}$$

the receiver may be thought of as recovering $f(t)$ with an error $\varepsilon_q(t)$. This concept for handling errors was already introduced in the analysis of PCM, where $\varepsilon_q(t)$ was called the quantization error. It and amplitude overload errors were the only two nonnoise type errors in PCM. In DM there are also *two* types of nonnoise errors. One, called *granular error*, $\varepsilon_g(t)$, corresponds to the error due to quantization in the 1-bit coding process. The second is called *slope overload error, $\varepsilon_{so}(t)$*. The composite error due to both granular and slope overload errors is called *quantization error* in DM. Thus in terms of the component errors making up $\varepsilon_q(t)$ we may write (4.10-1) as

$$f_q(t) = f(t) + \varepsilon_g(t) + \varepsilon_{so}(t). \tag{4.10-2}$$

Granular error occurs when $f_q(t) \approx f(t) + \varepsilon_g(t)$—that is, during periods of time when there are no transient errors occurring. Slope overload errors occur when the input message at the delta modulator has a slope exceeding the slope capability of the feedback signal $f_q(t)$. From Fig. 4.10-2(a) the maximum generated slope capability is $\delta v / T_s$. Thus to prevent slope overload

$$\left| \frac{df(t)}{dt} \right|_{max} \leq \frac{\delta v}{T_s} \qquad (4.10\text{-}3)$$

is required. If, for short periods of time, (4.10-3) is not satisfied, slope overload errors occur. Slope overload may be corrected for a given message, by increasing the step size δv, or by decreasing T_s (sample at a higher rate). Both these solutions are later found to be undesirable, and slope overload is one of the fundamental limitations of DM.

To better appreciate the effects of slope overload consider the sinusoidal message

$$f(t) = A_f \cos(\omega_f t). \qquad (4.10\text{-}4)$$

From (4.10-3) the region of satisfactory performance corresponds to signal levels

$$\frac{A_f}{\delta v} \leq \frac{1}{2\pi} \left(\frac{\omega_s}{\omega_f} \right). \qquad (4.10\text{-}5)$$

The right side of this expression may be viewed as a kind of spectral characteristic for a basic delta modulator, since it is a function of frequency ω_f. Taken in this light, we expect that DM would be especially applicable to waveforms that have a power density spectrum that decreases as the reciprocal of the square of frequency. For such signals the various spectral components could be made to match, or operate in close proximity to the overload level, thereby optimizing performance. Indeed, this is the case, and basic DM has been found to perform well with speech or monochrome television video messages both of which possess average power spectra approximating the type described.

In order not to convey excessive granular noise $A_f / \delta v \gg 1$ is necessary in (4.10-5). This condition determines the sample rate as the following simple example points out.

Example 4.10-1

For an 800-Hz sinusoidal message we find the required sample rate to prevent overload when $A_f / \delta v = 20$ is to be realized. From (4.10-5)

$$f_s = \omega_s / 2\pi \geq 20 \omega_f = 40\pi(800) = 100.5 \text{ kHz}.$$

Note that this sample rate is 62.8 times as large as the Nyquist sample rate of 1.6 kHz.

Channel Bandwidth

Assuming N similar messages are multiplexed, the time per message sample is T_s/N. Within this time slot the sample pulse duration is $\tau \leqslant T_s/N$ seconds. Since the minimum channel bandwidth W_{ch} required to pass pulses of duration τ is π/τ, we have

$$W_{ch} \geqslant \frac{\pi}{\tau} \geqslant \frac{\pi N}{T_s} = \frac{N\omega_s}{2}. \tag{4.10-6}$$

In practice, one would probably design the pulse duration as large as possible (T_s/N) to convey maximum energy per sample, especially if receiver (thermal) noise was significant. Thus W_{ch} would not have to be radically larger than the minimum value $N\omega_s/2$.

Other Configurations

There are many variations of the basic delta modulator which uses an ideal integrator. In one case the integrator may be replaced by a time-varying network which adjusts step size with time to reduce slope overload. Such a modulator is called an *adaptive delta modulator*, and it is discussed in Sec. 4.13. In another variation, the message is integrated prior to modulation by the basic delta modulator. This variation, called *delta-sigma modulation*, is discussed in Sec. 4.12. It is a means of approximately matching the overload characteristics of DM to a message having a relatively flat power spectrum (owing to integration, the encoded message power spectrum decreases with frequency and therefore approximately matches the overload characteristic of the delta modulator).

Some variations of the basic delta modulator involve alteration of the feedback path to include more general linear networks that do not change with time. Clearly, from a practical standpoint, one cannot realize a perfect integrator. A simple lowpass one-section RC filter is often substituted instead. Johnson [20] has analyzed such a DM system and found that noise performance is unchanged over the basic modulator which has an ideal integrator. There is a change in the slope overload characteristic, however. If the filter time constant is $T_c = RC$ and its 3-dB frequency is $\omega_c = 1/T_c$, while channel pulses have amplitude A and duration T_s/N (maximum duration), then the slope overload characteristic for a sinusoidal message of peak amplitude A_f and frequency ω_f becomes

$$\frac{A_f}{A} = 1 \bigg/ \sqrt{1 + \left(\frac{\omega_f}{\omega_c}\right)^2}. \tag{4.10-7}$$

For frequencies above ω_c the behavior is similar to the basic delta modulator, since a $1/\omega_f$ characteristic is evident. The region $\omega_f > \omega_c$ again corresponds to slope overload. For frequencies $\omega_f < \omega_c$ the characteristic becomes flat, corresponding to *amplitude overload,* a condition not present in basic DM. For speech, this behavior is not serious, since the speech spectrum also tends to reach a maximum at lower frequencies.

A delta modulator may also contain two integrators in the feedback path. In this case message slope differences are encoded as opposed to amplitude differences. Instabilities may result, however, for two perfect integrators, and some practical modifications are required. The usual practical circuit is a two-section *RC* lowpass filter having an extra (stability-controlling) resistor placed in series with the output capacitor [19, 21]. Such modulators are called *delta modulators with double integration.* They may give improved noise performance compared to single-integration DM for speech. The article by Schindler [21] is a good summary of the various DM configurations. In a later paper [22] he includes a fairly extensive reference list which will serve those readers interested in additional depth on DM systems. Other references can be found in [23], and more detail on other feedback filters is available in [24, 25].

All the delta modulators discussed above use regularly spaced samples. It is also possible to transmit in the form of irregularly spaced samples. Such a system, called *asynchronous delta modulation,* is described by Hawkes and Simonpieri [26].

4.11 NOISE PERFORMANCE OF DELTA MODULATION

In this section we consider mainly the noise performance of a single (ideal) integrator form of DM system as depicted in Fig. 4.10-1.

Performance with Granular Noise Only

Let us first consider receiver input (channel) noise to be negligible. From Fig. 4.10-1 and (4.10-2) we have

$$\hat{f}_q(t) = f_q(t) = f(t) + \varepsilon_g(t) + \varepsilon_{so}(t). \qquad (4.11\text{-}1)$$

Here the recovered message $f(t)$ is corrupted by granular and slope overload errors $\varepsilon_g(t)$ and $\varepsilon_{so}(t)$. We shall treat the errors as noises and seek the amount of noise power due to each which reaches the final output through the lowpass filter. So far as signal power is concerned, the (ideal) filter bandwidth W_f is chosen just large enough to pass the signal undistorted. Output signal power is then

$$S_o = \overline{f^2(t)}. \qquad (4.11\text{-}2)$$

Granular noise is essentially uncorrelated with the signal [27] and is approximately uniformly distributed [27, 28], over the interval $2\delta v$. We may gain some intuitive agreement with this fact from the behavior of the error in Fig. 4.10-2(c) during the nontransient time interval. The granular noise power prior to lowpass filtering becomes

$$\overline{\varepsilon_g^2(t)} = \frac{(\delta v)^2}{3}. \tag{4.11-3}$$

For the filtered output granular noise power N_g, we adopt a heuristic argument given by Taub and Schilling [29]. By observing the form of $\varepsilon_g(t)$ from Fig. 4.10-2(c) we conclude that there is considerable high-frequency content in its spectrum. To make an estimate of its extent, imagine $\varepsilon_g(t)$ passed through a filter with a cutoff frequency ω_c chosen to pass $\varepsilon_g(t)$ with only reasonable distortion. The cutoff frequency is then a good estimate of the spectral extent of $\varepsilon_g(t)$. If we select the point of "reasonable" distortion as that where the filter rise time π/ω_c is equal to half the sample interval T_s, the spectrum of $\varepsilon_g(t)$ is found to extend to a cutoff frequency $\omega_c = \omega_s = 2\pi/T_s$. Now since the power spectrum of $\varepsilon_q(t)$ is more or less flat [27, p. 2120] at lower frequencies, we presume the total power to be uniformly distributed over $-\omega_c < \omega < \omega_c$. Filtered output power now easily calculates to

$$N_g = \frac{(\delta v)^2}{3}\left(\frac{W_f}{\omega_s}\right). \tag{4.11-4}$$

Thus in general, the signal-to-granular noise power ratio is

$$\left(\frac{S_o}{N_g}\right)_{DM} = \frac{3\overline{f^2(t)}}{(\delta v)^2}\left(\frac{\omega_s}{W_f}\right). \tag{4.11-5}$$

To prevent slope overload it is necessary that the maximum magnitude of the signal's slope, $|df(t)/dt|_{max}$, not exceed the slope capability of the system, which is $\delta v/T_s$. At some signal level $f_r(t)$, that we call the reference signal level, the two slopes become equal, and we choose step size according to

$$\delta v = \left|\frac{df_r(t)}{dt}\right|_{max} T_s. \tag{4.11-6}$$

With this choice slope overload will not occur for any message power $\overline{f^2(t)}$ $\leq \overline{f_r^2(t)}$. From (4.11-4) and (4.11-5) we have

$$N_g = \frac{4\pi^2}{3}\left|\frac{df_r(t)}{dt}\right|_{max}^2 \frac{W_f}{\omega_s^3} \tag{4.11-7}$$

$$\left(\frac{S_o}{N_g}\right)_{DM} = \frac{3\overline{f^2(t)}\,\omega_s^3}{4\pi^2\,|df_r(t)/dt|_{max}^2 W_f}. \tag{4.11-8}$$

This equation may be put in another useful form by defining a *slope crest factor* $\dot{K}_{cr}$, analogous to the amplitude crest factor of (4.5-9), by†

$$\dot{K}_{cr}^2 = \frac{|df(t)/dt|_{\max}^2}{\overline{[df(t)/dt]^2}} \qquad (4.11\text{-}9)$$

and observing that the rms bandwidth W_{rms} of the signal power density spectrum $\mathcal{S}_f(\omega)$ is

$$W_{\mathrm{rms}}^2 = \frac{\overline{[df(t)/dt]^2}}{\overline{f^2(t)}} = \frac{\displaystyle\int_{-\infty}^{\infty} \omega^2 \mathcal{S}_f(\omega)\, d\omega}{\displaystyle\int_{-\infty}^{\infty} \mathcal{S}_f(\omega)\, d\omega}. \qquad (4.11\text{-}10)$$

Since these two expressions must hold for any signal level, including the reference level, they are used in (4.11-8) to obtain

$$\left(\frac{S_o}{N_g}\right)_{\mathrm{DM}} = \frac{3}{4\pi^2\,\dot{K}_{cr}^2}\left(\frac{W_f}{W_{\mathrm{rms}}}\right)^2\left(\frac{\omega_s}{W_f}\right)^3\frac{\overline{f^2(t)}}{\overline{f_r^2(t)}}. \qquad (4.11\text{-}11)$$

Equation (4.11-11) gives the output signal-to-noise ratio for a single-integration DM system. The only system parameter appearing explicitly in the expression is the sample rate ω_s. By use of (4.10-6) we may put it in terms of channel bandwidth:

$$\left(\frac{S_o}{N_g}\right)_{\mathrm{DM}} \leq \frac{6}{\pi^2\dot{K}_{cr}^2}\left(\frac{W_f}{W_{\mathrm{rms}}}\right)^2\frac{1}{N^3}\left(\frac{W_{ch}}{W_f}\right)^3\frac{\overline{f^2(t)}}{\overline{f_r^2(t)}}. \qquad (4.11\text{-}12)$$

Thus performance in DM increases as the third power of channel bandwidth.

Response of a DM system to a sinusoidal message has special application to speech transmission. An example is next used to specialize (4.11-11) for a single message ($N = 1$).

Example 4.11-1

We obtain an expression for $(S_o/N_g)_{\mathrm{DM}}$ for a single sinusoidal message $f(t) = f_r(t) = A_f\cos(\omega_f t)$. By using the result of Prob. 4-53, we obtain $W_{\mathrm{rms}} = \omega_f$. For $\dot{K}_{cr}^2$ we have $|df(t)/dt|_{\max}^2 = (A_f\omega_f)^2$ and $\overline{[df(t)/dt]^2} = (A_f\omega_f)^2/2$, so $\dot{K}_{cr}^2 = 2$. Finally, (4.11-11) becomes

$$\left(\frac{S_o}{N_g}\right)_{\mathrm{DM}} = \frac{3}{8\pi^2}\left(\frac{W_f}{\omega_f}\right)^2\left(\frac{\omega_s}{W_f}\right)^3, \qquad \text{sinusoidal modulation.}$$

For speech signals it is experimentally found [19] that noticeable overload does not occur as long as signal amplitude does not exceed the amplitude

---

† Note that the dot does not imply the time derivative of K_{cr}. It is simply to denote a crest factor for a time-differentiated waveform.

of an 800-Hz sine wave set just at the slope overload value. If $W_f/2\pi = 3.5$ kHz is assumed and the typical spectral distribution of speech is considered [30], 800 Hz turns out to be $W_{rms}/2\pi$, the rms speech bandwidth. It is also known [19] that good speech transmission requires a sample rate of 40 kHz, while very good results are achieved for a 100-kHz sample rate. From Example 4.11-1 these values show that we must have $(S_o/N_g)_{DM} = 1086$ or 30.4 dB for good results or 42.3 dB for very good results.

For a DM system with double integration and operating with a sine wave at its reference level, $(S_o/N_g)_{DM}$ may be put in the form [19]

$$\left(\frac{S_o}{N_g}\right)_{DM} = \frac{0.0178(3)}{8\pi^2} \left(\frac{W_f}{\omega_f}\right)^2 \left(\frac{\omega_s}{W_f}\right)^5. \qquad (4.11\text{-}13)$$

The ratio of (4.11-13) to the expression in Example 4.11-1 shows that the double-integration system signal-to-granular noise power ratio is better by a factor of $0.0178(\omega_s/W_f)^2$. For $W_f/2\pi = 3.5$ kHz and $\omega_s/2\pi = 40$ kHz (good speech reproduction) or 100 kHz (very good speech) the factor evaluates to 3.7 dB and 11.6 dB, respectively. These improvements make the added complexity worth while for speech systems.

Performance with Slope Overload Noise Added

In general the slope overload error $\varepsilon_{so}(t)$ in (4.11-1) is correlated with the message. A number of analyses to find the slope overload noise power N_{so} at the receiver lowpass filter output have been conducted for a single-integrator DM system. These analyses [27, 31–33] are complicated and beyond the scope of this book. Hence we only state and use appropriate results. Greenstein [33] has given apparently the most general solution. It is difficult to apply, however, and in the important region of heavy overload Greenstein's work suggests (see his Fig. 7) that Abate's [32] empirical solution is reasonably accurate. Because Abate's results are easy to use, we present them here as taken from Greenstein's paper. In present notation, the slope overload noise power can be shown to be†

$$N_{so} = \left[\frac{(\delta v)^2}{3} \frac{W_f}{\omega_s}\right] \frac{2}{9} \left(\frac{\omega_s}{W_f}\right)^3 \left(\frac{1 + 3Z}{Z^2}\right) e^{-3Z}, \qquad (4.11\text{-}14)$$

where Z, as given by

$$Z^2 = \frac{(\delta v)^2}{T_s^2 \overline{[df(t)/dt]^2}}, \qquad (4.11\text{-}15)$$

is known as the *slope overload factor*. Equation (4.11-14) was derived for

---

† Steele [34] gives an approximation that is a simplification of the original result of Greenstein [33] but still not as simple as the result used here.

Gaussian random or noise-like messages. The bracketed factor is recognized as the granular noise power from (4.11-4).

The output signal-to-quantization noise power ratio is

$$\left(\frac{S_o}{N_q}\right)_{\text{DM}} = \frac{\overline{f^2(t)}}{N_g + N_{so}}.$$ (4.11-16)

To put this equation in a useful form we must observe that a Gaussianly distributed message has no well-defined maximum amplitude or maximum slope. This means that granular noise power cannot be expressed in the exact form of (4.11-7). However, for some reference power level $\overline{f_r^2(t)}$ of the message, let us set the step size according to

$$\delta v = \dot{K} T_s \sqrt{\overline{\dot{f}_r^2(t)}},$$ (4.11-17)

where we define

$$\dot{f}_r(t) = \frac{df_r(t)}{dt}.$$ (4.11-18)

Thus at the reference signal power level, $\delta v / T_s$ is proportional to the square root of the power in the derivative of the reference signal. The proportionality constant is $\dot{K}$. It is *analogous* to the slope crest factor $\dot{K}_{cr}$ for messages having a definite maximum slope but not equal to it.†

Now by substituting (4.11-17) into the expressions for N_g and N_{so}, as given by (4.11-4) and (4.11-14), we may find $(S_o/N_q)_{\text{DM}}$ for any signal power level different from the reference level:

$$\left(\frac{S_o}{N_q}\right)_{\text{DM}} = \frac{x^2 \dot{K}^2 (S_o/N_g)_{\text{DM},r}}{1 + \frac{2}{9}(\omega_s/W_f)^3 (3 + x)x \exp(-3/x)},$$ (4.11-19)

where

$$x = \frac{1}{\dot{K}} \sqrt{\frac{\overline{f^2(t)}}{\overline{f_r^2(t)}}},$$ (4.11-20)

$$\left(\frac{S_o}{N_g}\right)_{\text{DM},r} = \frac{3}{4\pi^2} \frac{1}{\dot{K}^2} \left(\frac{W_f}{W_{\text{rms}}}\right)^2 \left(\frac{\omega_s}{W_f}\right)^3,$$ (4.11-21)

and W_{rms} is the rms bandwidth of the power density spectrum of the signal $f(t)$. By substitution of minimum channel bandwidth we have

$$\left(\frac{S_o}{N_q}\right)_{\text{DM}} = \frac{x^2 \dot{K}^2 (S_o/N_g)_{\text{DM},r}}{1 + (16/9N^3)(W_{ch}/W_f)^3 (3 + x)x \exp(-3/x)}.$$ (4.11-22)

The quantity $(S_o/N_g)_{\text{DM},r}$ is the signal-to-granular noise ratio when the message is at its reference power level. The remaining factor represents the change

---

† A more exact interpretation follows a comparison of (4.11-17) with (4.11-15). $\dot{K}$ may be considered as a reference value of the slope overload factor.

in this ratio caused by the presence of slope overload noise and the fact that the message power may be different from the reference level.

Figure 4.11-1 provides several plots† of (4.11-22) for $N = 1$. To illustrate the practical application of the curves we take an example.

Example 4.11-2
In an assumed DM system $W_{ch}/W_f = 5$ and $\delta v/T_s$ is arbitrarily selected to be four times the rms signal derivative. Thus $\dot{K} = 4$ from (4.11-17). If the system transmits only one message, $\omega_s = 2W_{ch}$ and $\omega_s/W_f = 10$. Now suppose the message power density spectrum is uniform over $-W_f < \omega < W_f$. Then $W_{rms}^2 = W_f^2/3$. From (4.11-21)

$$\left(\frac{S_o}{N_g}\right)_{DM,r} = \frac{3(3)(10)^3}{4\pi^2(4)^2} = 14.25, \quad \text{or } 11.54 \text{ dB},$$

so $\dot{K}^2 (S_o/N_q)_{DM,r} = 228.0$, or 23.58 dB. Thus $(S_o/N_q)_{DM}$ will be less than 23.58 dB by the amount shown in Fig. 4.11-1 which depends on signal level. To see where the reference level falls on the $W_{ch}/W_f = 5$ curve, let $\overline{f^2(t)} = \overline{f_r^2(t)}$. Thus $x^2 = 1/\dot{K}^2 = 1/16$. The curve value is -12.05 dB at $x^2 = 1/16$, or -12.04 dB. Hence $(S_o/N_q)_{DM}$ becomes $23.58 - 12.05 = 11.53$ dB.

Note in example 4.11-2 that the operating point is not near the peak of the $W_{ch}/W_f = 5$ curve. The following example optimizes the system.

Example 4.11-3
Let us alter the choice of $\delta v/T_s$ in the last example to operate at the peak of the curve when $\overline{f^2(t)} = \overline{f_r^2(t)}$. We require x^2 to be -7.5 dB. Thus $x = 0.422$ or $\dot{K} = 2.37$. We now have

$$\left(\frac{S_o}{N_g}\right)_{DM,r} = 14.25 \left(\frac{4}{2.37}\right)^2 = 40.59, \quad \text{or } 16.08 \text{ dB},$$

and $\dot{K}^2(S_o/N_g)_{DM,r} = 228.0$, or 23.58 dB, the same as before. Now the loss, however, is only 8.51 dB. We obtain $(S_o/N_q)_{DM}$ equal to $23.58 - 8.51 = 15.07$ dB, an improvement of 3.54 dB over the previous example.

Effect of Receiver Noise on Performance

It only remains to determine the effect of receiver (channel) noise on the performance of a DM system with a single integrator. Presence of noise means that the reconstruction stage in Fig. 4.10-1 will occasionally make the wrong decision as to which polarity pulse is received. If a positive pulse is received and an error is made, the output generated is a negative pulse. We may consider the output as actually containing the true positive pulse plus a negative error pulse of *twice* the magnitude of the signal

---

† Problem 4-59 develops optimum values of x in terms of W_{ch}/W_f such that operation is at the curve peaks.

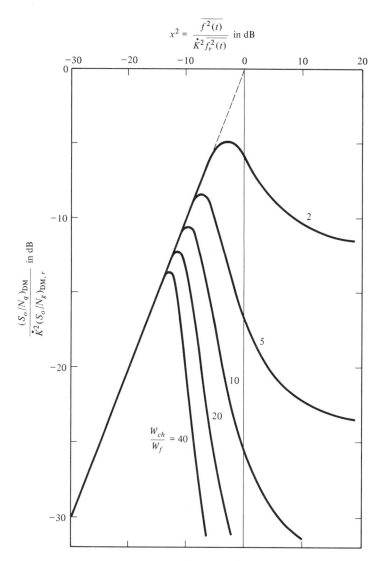

Figure 4.11-1. Loss in DM system signal-to-noise ratio due to message level variations and presence of slope overload noise. Parameter W_{ch}/W_f also equals $\omega_s/2W_f$[2].

pulse. In a similar manner, a negative received pulse erroneously reconstructed results in a negative signal pulse plus a positive double-amplitude error pulse. The overall effect at the input to the receiver integrator is an uncorrupted stream of signal pulses plus a stream of occasional double-amplitude noise pulses.

Since the pulses are short duration and the response of the integrators to any given noise pulse will be a step of amplitude $2\delta v$ or $-2\delta v$, we may consider the noise pulses as impulses with amplitudes $\pm 2\delta v$. The energy density spectrum for a single impulse is a constant level $4(\delta v)^2$. Now if the average time between noise pulses in any one message channel is $\overline{T}_{n1}$, then the power density spectrum of the noise pulse train may be taken approximately as $4(\delta v)^2/\overline{T}_{n1}$ at the integrator input. The effect of the integrator is to divide this power density by ω^2. Finally, we may integrate to obtain output power.

The integral depends on the lowpass filter characteristic. Because DM with ideal integrators does not transmit a dc signal well (the reader should verify this point for himself), there is no need for the filter to have very low-frequency response. If W_L is its low-frequency cutoff and $W_L \ll W_f$, the output noise power is

$$N_{rec} = \frac{4(\delta v)^2}{\pi \overline{T}_{n1} W_L}. \tag{4.11-23}$$

$\overline{T}_{n1}$ may be related to the probability P_e of *any* given received pulse's being erroneously reconstructed by approximately $P_e = N\tau/\overline{T}_{n1}$, where τ is the received pulse duration and N is the number of multiplexed messages. To determine P_e we may use the same methods as were applied to PCM in Sec. 4.4. For received pulses of amplitudes A or $-A$ and white input noise of spectral density $\mathcal{N}_0/2$ we obtain

$$P_e = \frac{1}{2}\operatorname{erfc}\left(\sqrt{\frac{A^2\tau}{\mathcal{N}_0}}\right), \tag{4.11-24}$$

where $\operatorname{erfc}(\cdot)$ is as defined in (B.4-6). Thus the receiver noise power becomes

$$N_{rec} = N_g \frac{6\omega_s}{\pi W_f W_L N\tau}\operatorname{erfc}\left(\sqrt{\frac{A^2\tau}{\mathcal{N}_0}}\right) \tag{4.11-25}$$

on use of (4.11-4).

A more convenient expression for N_{rec} results from examination of the input to the reconstruction stage of Fig. 4.10-1. For channel pulses of amplitude A and duration τ, the matched filter's transfer function is $\operatorname{Sa}(\omega\tau/2)$. At the time of a typical sample taken in the reconstruction stage, the filter's peak output signal amplitude is readily found to be A in response to the input pulse. Peak signal power at the sampler input, denoted by $\hat{S}_i$, is therefore

$$\hat{S}_i = A^2. \tag{4.11-26}$$

By integrating the noise power density spectrum at the sampler, which is given by $(\mathcal{N}_0/2)\operatorname{Sa}^2(\omega\tau/2)$, the input average noise power is found to be

$$N_i = \frac{\mathcal{N}_0}{2\tau}. \tag{4.11-27}$$

Thus peak signal-to-noise ratio at the sampler input is

$$\left(\frac{\hat{S}_i}{N_i}\right)_{\text{DM}} = \frac{2A^2\tau}{\mathcal{N}_0},$$ (4.11-28)

and (4.11-25) becomes

$$N_{rec} = N_g \frac{6\,\omega_s}{\pi\,W_f\,W_L\,N\tau}\,\text{erfc}\left[\sqrt{\frac{1}{2}\left(\frac{\hat{S}_i}{N_i}\right)_{\text{DM}}}\right].$$ (4.11-29)

Finally, we neglect slope overload noise for convenience and write

$$\left(\frac{S_o}{N_o}\right)_{\text{DM}} = \frac{\overline{f^2(t)}}{N_g + N_{rec}}.$$ (4.11-30)

On substitution of (4.11-29) we have

$$\left(\frac{S_o}{N_o}\right)_{\text{DM}} = \frac{(S_o/N_g)_{\text{DM}}}{1 + \dfrac{6\,\omega_s}{\pi W_f\,W_L\,N\,\tau}\,\text{erfc}\left[\sqrt{\dfrac{1}{2}\left(\dfrac{\hat{S}_i}{N_i}\right)_{\text{DM}}}\right]},$$ (4.11-31)

where $(S_o/N_g)_{\text{DM}}$ is given by either (4.11-11) or the numerator of (4.11-19), depending on how step size is defined. By substituting for ω_s and using the minimum channel bandwidth of (4.10-6), another useful form is

$$\left(\frac{S_o}{N_o}\right)_{\text{DM}} = \frac{(S_o/N_g)_{\text{DM}}}{1 + \dfrac{12}{\pi^2 N^2}\left(\dfrac{W_{ch}}{W_f}\right)^2\left(\dfrac{W_f}{W_L}\right)\text{erfc}\left[\sqrt{\dfrac{1}{2}\left(\dfrac{\hat{S}_i}{N_i}\right)_{\text{DM}}}\right]}$$ (4.11-32)

Typical evaluation of (4.11-32) ($N = 1$ and $W_f/W_L = 20$) shows that the 1-dB threshold signal-to-noise ratio $(\hat{S}_i/N_i)_{\text{DM},th}$ is not a sensitive function of W_{ch}/W_f. It varies from about 10 dB to about 14 dB for W_{ch}/W_f from 3 to 140, respectively.

Comparison of DM and PCM

When receiver noise is negligible and equal channel bandwidths are considered, PCM is always superior to DM for large values of W_{ch}/W_f. For small values of W_{ch}/W_f it is possible for DM to become superior to PCM. The point of changeover depends on signal conditions. The following example illustrates that a PCM system output signal-to-noise ratio is 3.3 dB larger than that of a DM system when $W_{ch}/W_f = 5$ and the message is sinusoidal. For $W_{ch}/W_f < 4$ it can be found that the DM system becomes slightly superior.

Example 4.11-4

We compare output signal-to-noise ratios of DM and PCM systems for a single sinusoidal message when $W_{ch}/W_f = 5$. For this message $K_{cr}^2 = 2$ and $\dot{K}_{cr}^2 = 2$.

Assume the message frequency is $\omega_f/2\pi = 800$ Hz and $W_f/2\pi = 3.5$ kHz, so that parameters approximate those of speech transmission in the DM system. Since $\omega_s/2W_f = W_{ch}/W_f$ for DM, the equation of example 4.11-1 gives

$$\left(\frac{S_o}{N_g}\right)_{DM} = \frac{3}{8\pi^2}\left(\frac{3.5}{0.8}\right)^2 (10)^3 = 727.3 \quad (\text{or } 28.6 \text{ dB}).$$

For PCM, assume $N_b = W_{ch}/W_f$. Equation (4.5-7) with $L = 2^{N_b}$ and $\overline{f^2(t)} = \overline{f_r^2(t)}$ gives

$$\left(\frac{S_o}{N_q}\right)_{PCM} = \frac{3(2^{10})}{2} = 1536 \quad (\text{or } 31.9 \text{ dB}).$$

Thus PCM is superior in performance to DM by 3.3 dB. Recalculation for $W_{ch}/W_f < 4$ shows DM slightly superior.

4.12 DELTA-SIGMA MODULATION

It was noted in Sec. 4.10 that a system using delta modulation performs most favorably with messages having a power density spectrum that decreases with increasing frequency (as $1/\omega^2$). It is not as well suited to messages having an approximately flat power spectrum. Furthermore, another difficulty arises if the message possesses a dc component, since it cannot easily be conveyed with DM having ideal integrators. Inose and Yasuda [35] have described a method whereby DM may be modified to perform well with messages having both a flat power spectrum and a dc component. The system is called *delta-sigma modulation* (D-SM).

A D-SM system is obtained, in concept, by adding one integrator and one differentiator to a basic DM system. With reference to Fig. 4.10-1, the integrator is added at the input so that an integrated version of the message $f(t)$ is applied to the summing point. The differentiator is provided in the receiver following the lowpass filter.†

Overload Characteristic

The D-SM system may be overloaded, just as the DM system, if the slope of the signal emerging from the input integrator is too large. This fact should be obvious, since from the summing point to the transmitter output we still have a basic DM system. Thus (4.10-3) applies. However, the maximum slope is now that of the *integral* of the message so that, in

---

† Our description will presume these two components are actually added, since this approach facilitates understanding of the operation of D-SM. In practice, no additional equipment is needed as compared to DM. On the contrary, D-SM can actually require one *fewer* component than DM. A practical implementation is discussed in reference [35]. A practical design methodology for D-SM is given in [36].

order to prevent overload, we require that

$$|f_r(t)|_{\max} \leqslant \delta v/T_s \tag{4.12-1}$$

be satisfied, where $f_r(t)$ is the reference signal level at which δv is chosen.

For a sinusoidal message, which may be used to infer the overload characteristic, (4.12-1) becomes

$$\frac{A_f}{\delta v} \leqslant \frac{\omega_s}{2\pi}. \tag{4.12-2}$$

Here A_f is the message peak value, δv is the step size, as before, and $T_s = 2\pi/\omega_s$ is the time between transmitted pulses. For transmitted pulses of short duration τ and amplitude A, step size is given by

$$\delta v = A\tau. \tag{4.12-3}$$

Equation (4.12-2) is independent of the frequency of the sinusoidal message. Thus although there is an upper bound to the amplitude that should be employed as a function of frequency, the bound is constant with frequency. We conclude, then, that D-SM will perform well with messages having a relatively flat spectrum.

Channel Bandwidth

The output pulse stream in D-SM has the same form as that of basic DM. Thus, minimum channel bandwidth is again given by (4.10-6).

Granular Noise Performance Limitation

As in DM, the final output of the D-SM receiver will be the recovered message $f(t)$ plus noise. The noise will again possess components of receiver (thermal) noise, granular noise, and overload noise. We treat only the first two and assume that overload noise is small. Even if receiver noise is small, granular noise is ever-present and ultimately limits the achievable output signal-to-noise ratio.

To determine the granular noise power reaching the output it is easiest to retrace some of the steps leading to (4.11-4) for a DM system. As noted earlier, the noise power spectrum at the lowpass filter input is approximately flat over $-W_f < \omega < W_f$, where W_f is the maximum frequency extent of the message. By working backward through the filter with the aid of (4.11-4), the power density is found to be $\pi(\delta v)^2/3\omega_s$. This power density also applies at the filter output, since it is presumed ideal. Adding a factor ω^2 to account for the effect of the differentiator the output granular noise power becomes

$$N_\rho = \frac{1}{2\pi} \int_{-W_f}^{W_f} \frac{\pi(\delta v)^2}{3\omega_s} \omega^2 \, d\omega \tag{4.12-4}$$

or

$$N_g = \frac{(\delta v)^2 W_f^3}{9\omega_s}. \tag{4.12-5}$$

Signal-to-noise ratio is obtained by dividing output signal power $\overline{f^2(t)}$ by (4.12-5). When we also substitute for the minimum value of δv, from (4.12-1), we have

$$\left(\frac{S_o}{N_g}\right)_{\text{D-SM}} = \frac{9}{4\pi^2} \frac{\overline{f_r^2(t)}}{|f_r(t)|_{\text{max}}^2} \left(\frac{\omega_s}{W_f}\right)^3 \frac{\overline{f^2(t)}}{\overline{f_r^2(t)}}. \tag{4.12-6}$$

In terms of the crest factor defined by (4.5-9), we have

$$\left(\frac{S_o}{N_g}\right)_{\text{D-SM}} = \frac{9}{4\pi^2} \frac{1}{K_{cr}^2} \left(\frac{\omega_s}{W_f}\right)^3 \frac{\overline{f^2(t)}}{\overline{f_r^2(t)}}. \tag{4.12-7}$$

For a sinusoidal message $K_{cr}^2 = 2$ and $\overline{f^2(t)} = \overline{f_r^2(t)}$, this expression reduces to

$$\left(\frac{S_o}{N_g}\right)_{\text{D-SM}} = \frac{9}{8\pi^2} \left(\frac{\omega_s}{W_f}\right)^3, \tag{4.12-8}$$

which is identical to that obtained by Johnson [20]. By comparing (4.12-7) with its delta modulation counterpart (4.11-11), we find that both depend on the cube of the sampling rate.

Johnson has also studied the case of practical integrators. That is, the integrators are simple, single-stage, RC lowpass filters having a 3-dB bandwidth ω_c. For a sinusoidal message, his result for $\overline{f^2(t)} = \overline{f_r^2(t)}$ is

$$\left(\frac{S_o}{N_g}\right)_{\text{D-SM}} = \frac{9}{8\pi^2} \left(\frac{\omega_s}{W_f}\right)^3 \frac{1}{[1 + 3(\omega_c/W_f)^2]}. \tag{4.12-9}$$

Finally, we note that Inose and Yasuda [35] have studied both single and double practical RC integrator forms of D-SM.

Example 4.12-1
DM and D-SM systems both operate with sinusoidal messages of frequency $\omega_f = W_f$. We compare their performances. From (4.12-8) and the result given in Example 4.11-1, we have (with $W_f = \omega_f$)

$$\frac{(S_o/N_o)_{\text{D-SM}}}{(S_o/N_o)_{\text{DM}}} = 3 \quad \text{(or 4.77 dB)}.$$

In this particular case D-SM performs better.

Performance with Receiver Noise Added

Receiver noise will occasionally cause the reconstruction stage in Fig. 4.10-1 to produce a pulse of incorrect polarity. The probability of such an incorrect decision is denoted as P_e. Just as in DM, the output of the reconstruction stage may be viewed as a stream of correctly recovered message pulses plus a stream of error or noise pulses having magnitude twice that of the message stream. By applying the previous discussion of the error pulse stream for DM, we may justify that the noise power density spectrum at the integrator input is $4(\delta v)^2 P_e/N\tau$. Here N is the number of multiplexed messages and τ is the duration of transmitted pulses.

Since the integrator cancels the effect of the differentiator (and both can actually be eliminated in practice), the output noise may easily be found by integration:

$$N_{rec} = \frac{1}{2\pi} \int_{-W_f}^{W_f} \frac{4(\delta v)^2 P_e}{N\tau} \, d\omega. \qquad (4.12\text{-}10)$$

Upon recognizing that P_e is again given by (4.11-24), N_{rec} evaluates to

$$N_{rec} = \frac{2(\delta v)^2 \, W_f}{\pi N \tau} \, \text{erfc}(\sqrt{A^2\tau/\mathcal{N}_0}). \qquad (4.12\text{-}11)$$

Because the D-SM receiver uses a matched filter as in DM, $A^2\tau/\mathcal{N}_0$ is given by (4.11-28):

$$\left(\frac{\hat{S}_i}{N_i}\right)_{\text{D-SM}} = \frac{2A^2\tau}{\mathcal{N}_0}. \qquad (4.12\text{-}12)$$

On using (4.12-5) in (4.12-11) output signal-to-noise ratio becomes

$$\left(\frac{S_o}{N_o}\right)_{\text{D-SM}} = \frac{(S_o/N_g)_{\text{D-SM}}}{1 + \dfrac{18\omega_s}{\pi \, N \, \tau \, W_f^2} \, \text{erfc}\left[\sqrt{\dfrac{1}{2}\left(\dfrac{\hat{S}_i}{N_i}\right)_{\text{D-SM}}}\right]}. \qquad (4.12\text{-}13)$$

Here $(S_o/N_g)_{\text{D-SM}}$ is given by (4.12-7).

4.13 ADAPTIVE DELTA MODULATION

In previous discussions we found that a delta modulation system became overloaded when the instantaneous (short-term) slope of the message was too large. Specifically, if the signal slope exceeded the ratio of step size δv to the time T_s between samples, overload occurred. For a given sample rate $\omega_s = 2\pi/T_s$, we could reduce slope overload by increasing step size or decreasing the message input power level (and thereby reducing signal slope). Both measures are undesirable, because they give decreased performance. Another disadvantage in DM is its sensitivity to signal level variations. Even if slope overload is absent, the signal-to-granular noise power ratio decreases when signal power level decreases [see (4.11.5)]. Thus in DM there exists only a finite range of signal-to-noise ratios, which exceeds whatever minimum one might establish as acceptable performance (for example, a good telephone link requires a signal-to-noise ratio of about 30 dB minimum). This finite interval is called the *dynamic range*.

Adaptive delta modulation (ADM) is a variation of DM which offers relief from the above disadvantages by adapting the step size to accommodate changing signal conditions. If the input signal's slope is large, step size is caused to increase, thereby reducing slope overload effects. On the other hand, step size is decreased if the message is changing slowly or decreases in power level, thereby reducing granular noise. This latter fact leads to one of the principal advantages of ADM, increased dynamic range. Clearly, from (4.11-5), if $(\delta v)^2$ decreases in proportion to decreases in message power level $\overline{f^2(t)}$, we may cause signal-to-noise ratio to become constant, or nearly so, over some range of signal power values. In practice, 40 dB to 50 dB of such *dynamic range extensions* may readily be achieved.

System Block Diagram

Step size in ADM can be changed by placing a variable-gain circuit in the feedback path of the basic DM generator (Fig. 4.10-1). The circuit may be made to give discrete [32, 37] or continuous [38–40] gain variations. Regardless of the type of gain variation, the main problem in ADM is the manner in which gain is controlled. At least two basic methods are possible. In the first, called *forward control*, a control voltage for step size adjustment is derived from the input message. In order for the receiver to have step size knowledge, this voltage must be transmitted over the channel separately from the ADM output pulse stream. A typical system [41] encodes the voltage in the form of a second DM pulse stream (at a low sample rate) and multiplexes the two streams in time. Disadvantages of the method are the need for 2-bit streams and the problems associated with separating the two streams at the receiver.

The second method of controlling gain, called *feedback control*, does not have the above disadvantages. The applicable block diagram is illustrated in Fig. 4.13-1(a). It is the same as in Fig. 4.10-1 for DM (up to point *A*) except for the addition of the variable-gain circuit and the step-size controller. The purpose of the controller is to sense the slope condition of the message conveyed in the transmitted pulse train $s_{ADM}(t)$. If the slope is large, the controller output causes the variable-gain circuit to have a large gain. Loop step size is then large. If the slope is small, the controller output causes a small gain and thereby a small step size. Because the loop will force $f_q(t)$ to approximately equal $f(t)$, all that is necessary in the receiver for message recovery is to reproduce the operations in the feedback path of the encoder (modulator). Thus, the receiver blocks shown in (b) are identical to their modulator counterparts. The receiver for an overall ADM system would be similar to that of Fig. 4.10-1 but with all the components to the right of point B replaced by the receiver of Fig. 4.13-1(b).

In the ADM system of the preceding paragraph, step size is varied to match signal conditions. Alternatively, step size can be kept constant by using a basic delta modulator and signal conditions at its input can be adjusted [42]. When signal level is large, a variable-gain circuit decreases the effective signal level at the delta modulator's input. Small signal levels result in large gains. The effect is a type of signal compression in the modulator. The inverse (expansion) operation is required in the receiver so the system uses a form of *companding*. The modulator's gain control signal is generated in a (second) feedback path around the modulator. Performance of this type of ADM is about equal to other ADM systems, but it has the disadvantage that the system requires a divider (for expansion), a device which is not always convenient in practice.

It is no doubt obvious that the central problem in ADM is the design

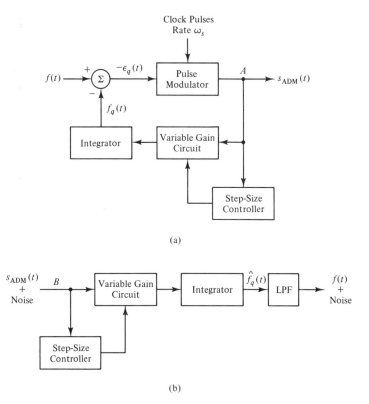

Figure 4.13-1.   (a) Adaptive delta modulator using feedback control and (b) the corresponding receiver decoder [2].

of the step-size controller. There are two important considerations. One is the functional law relating gain changes to signal slope changes. The other is the form of the implementation. In the following two subsections we discuss implementations. The natural course of discussion will involve the gain law as well. However, choice of a gain law is not inherent to any given implementation. The two can be separately chosen as far as principle goes. Some comments regarding an optimum gain law are included in the later consideration of noise performance of ADM.

Instantaneous Step-Size Control

When the variable-gain circuit is discrete, the usual method of implementing the step-size controller involves digital logic. Gain changes occur on a pulse-to-pulse basis and are said to be *instantaneous*. Several variations of design are possible [22, 32, 37, 43, 44]. In one approach [37] the controller has a 1-bit memory. By considering the fact that slope overload leads to a sequence of output pulses of identical polarity, it compares the polarity of each transmitted pulse with the immediately preceding one. If they have the same polarity the controller indexes the gain circuit to a larger gain value (typically by a factor P larger than the previous gain). If the polarities are not the same, a sign reversal has taken place and gain is indexed to a lower value (typically by a factor Q lower). This has sometimes been called *constant factor delta modulation* (CFDM). Jayant [37, 44] finds, from computer simulations with speech messages, that $PQ = 1$ and $P \approx 1.5$ represent optimum conditions to minimize quantization error power. Later simulations of Zetterberg and Uddenfeldt [43], however, indicate that $PQ = 1.05$ and $P = 1.1$ are more optimum, giving both minimum quantization error and maximum dynamic range. The difference in the two results apparently is due to the difference in the messages assumed in the simulations [43]. On the other hand, from reference [43], signal-to-noise ratio is not a sensitive function of P when optimum Q is used. It changes by about 3 dB as P changes† from 1.1 to 1.5. Optimum PQ remains in the vicinity of 1.1.

The above technique can be extended to controllers having more than a 1-bit memory [22, 32, 43]. Data of reference [43] indicate that, for up to 3 bits of memory, final receiver output signal-to-noise ratio does not change greatly (less than 1.5 dB for $1.1 < P < 1.5$ and the optimum Q in each case of P when ω_s/W_f is near 13.7). The decision logic chosen was to increase gain by the factor P if, for k bits of memory, all $k + 1$ pulses were of the same sign and to decrease gain by the factor Q otherwise.

Schindler [22] has shown that, even with no slope overload, there may be cases where two consecutive pulses transmitted may have the same

---

† These conclusions assume $7 \leqslant \omega_s/W_f \leqslant 13.7$.

polarity. On this basis he justifies that 2 (or more) bits of memory are better suited to control step size for nonstationary messages such as speech waveforms. A decision logic based on four total bits (3-bit memory) is given. It is most easily understood by assigning a 4-bit binary word to the four most recent pulse transmissions. If the most recent pulse corresponds to the least significant digit in the word, then gain is increased by 2 dB if the digital number is 0 or 15. Gain is increased by 1 dB if the number is 4, 7, 8, or 12. Gain is decreased by 0.125 dB for all other numbers. In a practical system, the decreases are accumulated until a total decrease of 1 dB is called for; the gain then is decreased by 1 dB.

Syllabic Step-Size Control

Many adaptive delta modulators do not change step size on a pulse-to-pulse basis. Rather, changes are made much more slowly. Since most research has been directed toward use of DM and ADM for speech encoding, such slow control is referred to as *syllabic*. The usual implementation involves a continuously-variable gain circuit controlled by an analog voltage from the step-size controller. This is often called *continuously variable slope delta modulation* (CVSD). The input to the controller is the ADM output pulse stream. This input may be processed either digitally [21, 39, 45, 46] or with analog circuitry [29, 38–41, 47] to obtain the analog control voltage.

The system of Greefkes and Riemens [46] is a good example of digital processing. Whenever the system output pulse stream contains four consecutive pulses of either state (on or off unipolar rather than polar pulses were used in reference [46]), a dc level is initiated. The level is maintained for a time equal to the total number of consecutive sampling intervals that the pulse state is preserved. The result is a sequence of "pulses" which are averaged by passing through a lowpass filter (bandwidth of $100/\pi$ Hz) to obtain the slowly-varying control voltage. Clearly, the control voltage is proportional to the average fraction of time that the system output pulse stream contains four or more consecutive pulses of the same state. However, we may show that the voltage is also proportional to the fraction of time that the slope (derivative) of the message exceeds half the average slope capability of the ADM system. The proof consists in recognizing that, for four pulses, half the average slope capability is exceeded when the *average* number of pulses of any one state exceeds two. Only the case where all four pulses have the same state satisfies this condition. Greefkes and Riemens define the ratio of the signal derivative to the maximum permissible value of this derivative as *modulation index*. Thus the analog voltage at the output of the step-size controller is proportional to the average fraction of time that the modulation index exceeds $\frac{1}{2}$.

A good example of an analog processor is the system of Tomozawa and Kaneko [38]. First, the system output pulse stream is passed through an integrator to obtain a signal representing the modulated signal component contained in the coded pulses. The integrator output is then envelope-detected. The detector feeds a lowpass filter which produces a slowly varying output voltage proportional to the modulating signal average level. The lowpass filter has a time constant of about 5 ms or a bandwidth (3 dB) of $100/\pi$ Hz. Thus the controller output is a voltage proportional to the average message voltage magnitude over a 5-ms time interval. The voltage, plus a small fixed bias, controls the gain of the variable gain circuit. The bias helps prevent instabilities [38] and prevents the variable gain from becoming zero when the detector output is zero.

It should be observed that the analog control voltages in the above two example systems are different. One is related to message *slope*, the other is related to message *magnitude*. In the following subsection, we discuss noise performance assuming syllabic step-size control, and show, at least for stationary random signals, that either approach leads to the same performance.

Quantization Noise Performance of ADM

The performance of a DM system may easily be extended to give the performance of ADM. We model the message as a zero-mean stationary Gaussian random signal in order to make use of previous work. The model will form a reasonable first approximation for other messages. The assumption of stationarity allows a fixed relationship to exist between signal power and the power in the signal derivative $\dot{f}(t) = df(t)/dt$. The relationship is

$$\overline{\dot{f}^2(t)} = W_{\text{rms}}^2 \overline{f^2(t)}, \tag{4.13-1}$$

where W_{rms} is the rms bandwidth of the power density spectrum of $f(t)$, as given by (4.11-10). Thus a certain average slope implies a certain signal power, and vice versa.

We treat ADM as DM with a variable step size δv that is a function of message slope, or the equivalent, message level. Receiver output signal-to-quantization noise power ratio must again be given by either (4.11-19) or (4.11-22) if suitable account is made for a variable step size. Other approaches, although less direct, are also possible [48]. Suppose we begin by placing an upper bound or "reference" value δv_r on δv by choosing the ratio of δv_r to sampling interval T_s to be proportional to the rms slope of the message when it is at some reference level $f_r(t)$. In other words,

$$\delta v_r = \dot{K} \sqrt{\overline{\dot{f}_r^2(t)}} \, T_s, \tag{4.13-2}$$

where $\dot{K}$ is a constant of proportionality. Equation (4.13-2) is just an extension of (4.11-17) for fixed δv in delta modulation.

Now the signal-to-quantization noise power ratio that we seek is given by

$$\left(\frac{S_o}{N_q}\right)_{\text{ADM}} = \frac{\overline{f^2(t)}}{N_g + N_{so}},$$

(4.13-3)

where the granular noise power N_g is given by (4.11-4) for any δv, and the slope overload noise power N_{so} is given by (4.11-14) for any δv and signal power. In order to find N_{so}, which is a function of Z as defined in (4.11-15), we first write Z^2 as

$$Z^2 = \frac{(\delta v)^2}{T_s^2 \overline{\dot{f}^2(t)}} = \left(\frac{\delta v}{\delta v_r}\right)^2 \frac{\dot{K}^2 \overline{\dot{f}_r^2(t)}}{\overline{\dot{f}^2(t)}} = \left(\frac{\delta v}{\delta v_r}\right)^2 \frac{\dot{K}^2 \overline{\dot{f}_r^2(t)}}{\overline{f^2(t)}}.$$

(4.13-4)

In obtaining this expression, T_s from (4.13-2) has been substituted and (4.13-1) was used. In terms of x, previously defined as

$$x^2 = \frac{\overline{f^2(t)}}{\dot{K}^2 \overline{f_r^2(t)}},$$

(4.13-5)

we have

$$Z^2 = \left(\frac{\delta v}{\delta v_r}\right)^2 \frac{1}{x^2}.$$

(4.13-6)

Substitution into (4.11-14) gives the desired expression for N_{so}. Performing somewhat similar operations on the ratio $\overline{f^2(t)}/N_g$ will allow us to write (4.13-3) as

$$\left(\frac{S_o}{N_q}\right)_{\text{ADM}} = \frac{\dot{K}^2 \left(\dfrac{S_o}{N_g}\right)_{\text{ADM},r} \left[\left(\dfrac{\delta v_r}{\delta v}\right) x\right]^2}{1 + \dfrac{2}{9}\left(\dfrac{\omega_s}{W_f}\right)^3 \left[3 + \left(\dfrac{\delta v_r}{\delta v}\right) x\right] \left(\dfrac{\delta v_r}{\delta v}\right) x \exp\left[-3\Big/\left(\dfrac{\delta v_r}{\delta v}\right) x\right]}.$$

(4.13-7)

Here

$$\left(\frac{S_o}{N_g}\right)_{\text{ADM},r} = \frac{3\overline{f_r^2(t)}}{(\delta v_r)^2} \frac{\omega_s}{W_f} = \frac{3}{4\pi^2 \dot{K}^2}\left(\frac{W_f}{W_{\text{rms}}}\right)^2 \left(\frac{\omega_s}{W_f}\right)^3$$

(4.13-8)

is the reference signal-to-granular noise power ratio. In terms of minimum channel bandwidth $N\omega_s/2$, which is the same as in DM, we write

$$\left(\frac{S_o}{N_q}\right)_{\text{ADM}} = \frac{\dot{K}^2 \left(\dfrac{S_o}{N_g}\right)_{\text{ADM},r} \left[\left(\dfrac{\delta v_r}{\delta v}\right) x\right]^2}{1 + \dfrac{16}{9N^3}\left(\dfrac{W_{ch}}{W_f}\right)^3 \left[3 + \left(\dfrac{\delta v_r}{\delta v}\right) x\right] \left(\dfrac{\delta v_r}{\delta v}\right) x \exp\left[-3\Big/\left(\dfrac{\delta v_r}{\delta v}\right) x\right]}.$$

(4.13-9)

We pause to examine our results. Observe that (4.13-7) is identical to (4.11-19) in form. In fact, x in the DM expression is replaced here by $x(\delta v_r/\delta v)$. Now since δv_r is the maximum value of δv, and it occurs whenever $\overline{f^2(t)} \geqslant \overline{f_r^2(t)}$, then, for this condition, $x^2(\delta v_r/\delta v)^2$ here is identical to x^2 for DM. Thus, whenever $x^2 \geqslant 1/\dot{K}^2$ we expect the normalized signal-to-noise ratio of ADM to be the same as in DM. When $x^2 < 1/\dot{K}^2$ the two are different. In DM, $(S_o/N_q)_{\text{DM}}$ decreases as x^2 decreases. In ADM, $(S_o/N_q)_{\text{ADM}}$ may be made nearly constant as x^2 decreases if $x^2(\delta v_r/\delta v)^2$ is nearly constant.

The condition required for $x^2(\delta v_r/\delta v)^2$ to be constant is readily found. When $\delta v = \delta v_r$, x^2 must equal $1/\dot{K}^2$. For $\delta v < \delta v_r$ we require

$$x^2 \left(\frac{\delta v_r}{\delta v} \right)^2 = \frac{1}{\dot{K}^2}, \qquad (4.13\text{-}10)$$

or, on substitution of (4.13-5),

$$\delta v = \delta v_r \sqrt{\frac{\overline{f^2(t)}}{\overline{f_r^2(t)}}}. \qquad (4.13\text{-}11)$$

In words, δv should be proportional to $[\overline{f^2(t)}]^{1/2}$, a condition which can be approximated in practice.

Unfortunately a normalized plot of (4.13-9) is not possible, as was done for its DM counterpart (4.11-22), because $\delta v_r/\delta v$ cannot readily be written in terms of x. We may, however, work toward developing some representative curves for a speech message to see the typical behavior of (4.13-9). First we assume the following approximation for (4.13-11):

$$\frac{\delta v}{\delta v_r} = \begin{cases} \dfrac{\delta v_0}{\delta v_r} + \left(1 - \dfrac{\delta v_o}{\delta v_r} \right) \sqrt{\dfrac{\overline{f^2(t)}}{\overline{f_r^2(t)}}}, & \overline{f^2(t)} < \overline{f_r^2(t)} \\ 1, & \overline{f^2(t)} \geqslant \overline{f_r^2(t)}. \end{cases} \qquad (4.13\text{-}12)$$

Here δv_0 is the smallest value that δv is allowed to have. The ratio $\delta v_r/\delta v_0$ when expressed in decibels may be called the *dynamic range factor* F_{dy}. That is,

$$F_{dy} = 20 \log_{10} \left(\frac{\delta v_r}{\delta v_0} \right). \qquad (4.13\text{-}13)$$

F_{dy} is the maximum amount that system dynamic range is increased over that of a basic DM system.

Next, we model the speech signal as a Gaussian random message having a simple lowpass power density spectrum. Below the 3-dB frequency $\omega = W_3$ it is constant. Above W_3 it decreases at -6 dB per octave frequency increase out to a point W_f where it abruptly becomes zero for $\omega > W_f$. For this model it is readily shown that

$$\left(\frac{W_{\text{rms}}}{W_f} \right)^2 = \left[\frac{(W_3/W_f)}{\tan^{-1}(W_f/W_3)} - \left(\frac{W_3}{W_f} \right)^2 \right]. \qquad (4.13\text{-}14)$$

For speech signals W_f/W_3 is approximately 4.35, corresponding to W_{rms} = $0.344W_f$ [30].

Figure 4.13-2 has been constructed from (4.13-9) for a single message using (4.13-12), W_{rms} = $0.344W_f$, W_{ch} = $10W_f$, and $\dot{K}$ = 2.985, which is the value that corresponds to the maximum point in Fig. 4.11-1. The effects of dynamic range factors of 10, 20, 30, and 40 dB are shown. For comparison, the response of a basic delta modulation system is shown dashed.

Example 4.13-1
For the system of Fig. 4.13-2 we calculate the increase in dynamic range of the ADM system relative to the DM system when the minimum required signal-to-noise ratio is 15 dB. F_{dy} = 40 dB is assumed.

In the slope overload region, (S_o/N_q) of 15 dB occurs for both systems at a relative signal strength of +6.8 dB. For smaller signal levels the DM and ADM systems reach 15 dB at relative signal strengths of −12.5 dB and −50.5 dB, respectively. Thus the dynamic range of DM is 6.8 + 12.5 = 19.3 dB. That of ADM is 6.8 + 50.5 = 57.3 dB. Dynamic range extension becomes 57.3 − 19.3 = 38.0 dB, or slightly below the maximum of 40 dB.

Hybrid Configurations

The principal configurations of ADM use either instantaneous or syllabic step-size control. It is possible, however, to combine both types of control to form what is called *hybrid companding delta modulation* (HCDM). The basic technique of Un and Magill [49] as modified by Un and Lee [50] has

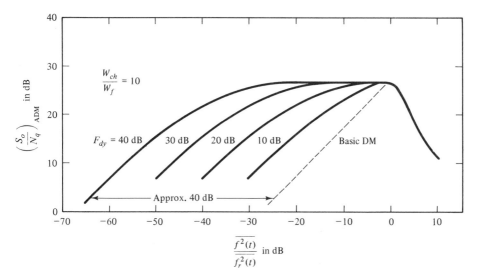

Figure 4.13-2. Receiver output signal-to-noise ratio in adaptive data modulation. A speech message is assumed for which W_{rms} = $0.344W_f$[2].

been compared to other ADM systems [51]; performance is claimed to be slightly superior (2 to 3 dB) to CVSD or CFDM systems.

The hybrid scheme can be modified to use more than one sample rate [52]. Buffers are used so that the channel pulse rate is maintained constant. Compared to fixed-rate HCDM, signal-to-quantization noise ratio can be improved by 3 dB to 4 dB [52].

4.14 DIFFERENTIAL PCM

From our earlier study of a delta modulator, we found that each transmitted pulse corresponded to a 1-bit encoding of the difference between the message $f(t)$ and an approximation $f_q(t)$ to it. The overall effect of the pulse modulator (Fig. 4.10-1) was to quantize the difference signal into two levels, sample the quantized difference, and generate a 1-bit coded pulse stream having one pulse per sample. If, instead, the difference signal is quantized into multiple levels, sampled, and each sample coded into an N_b-bit binary code, the result is called *differential pulse code modulation* (DPCM). The code is transmitted over the channel as binary pulses with N_b pulses per sample. DPCM is sometimes known as *delta PCM*.

DPCM System Block Diagram

DPCM systems may take on many forms. We shall discuss only the most elementary version and reserve comments on the variations for the end of this section. A simple DPCM modulator is illustrated in Fig. 4.14-1(a). The demodulator is shown in (b). When these components replace their counterparts in Fig. 4.10-1, a DPCM system is formed. Consider first the modulator. The error $-\varepsilon_q(t)$ is rounded off to the nearest of 2^{N_b} equally spaced levels where N_b is a positive integer. The quantized version of $-\varepsilon_q(t)$ acts as a modulating signal to the pulse amplitude modulator. By assuming the clock pulses have a short duration and constant amplitude, the modulator output is a train of amplitude-modulated pulses. Each pulse becomes a quantized sample of the error $-\varepsilon_q(t)$. Samples occur at the sample rate ω_s (every $T_s = 2\pi/\omega_s$ seconds). The PCM coder converts each sample pulse into an N_b-bit binary pulse sequence which is transmitted over the channel. The sample pulses are also applied to the integrator which closes the loop by producing $f_q(t)$ at its output.

Since the sample pulses are narrow, the integrator generates a staircase waveform as in a basic delta modulation system except there are now 2^{N_b} possible step sizes available (including sign). The steps are [53]

$$\pm \delta v_0, \pm 3\delta v_0, \ldots, \pm (2^{N_b} - 3)\delta v_0, \pm (2^{N_b} - 1)\delta v_0, \quad (4.14\text{-}1)$$

where δv_0 is the magnitude of the smallest step. Steps and pulse polarities are related through a simple binary code. Figure 4.14-2 sketches some typical waveforms of interest in generating a DPCM signal. A 2-bit code

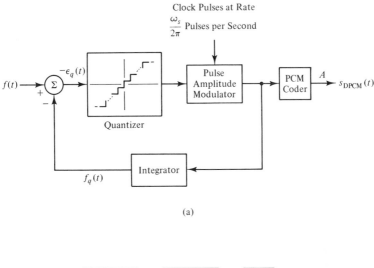

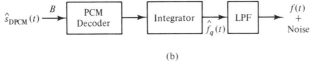

Figure 4.14-1.   Block diagrams applicable to a differential PCM system. (a) Modulator and (b) demodulator. The components fit into Fig. 4.10-1 to the left of point A and to the right of point B, respectively, to form a DPCM system [2].

is assumed; that is, $N_b = 2$. Quantization error magnitude $|\varepsilon_q(t)|$ may be as large as $3\delta v_0$, as shown in (c). The pulse stream of (b) assumes that steps of $3\delta v_0$, δv_0, $-\delta v_0$, and $-3\delta v_0$ are coded with pulses having polarities $++$, $+-$, $-+$, and $--$, respectively.

More generally, quantization error magnitude can be as large as $(2^{N_b} - 1)\delta v_0$, and $f_q(t)$ may still remain a good approximation to $f(t)$. Note, however, that periods of large error correspond to rapid changes occurring in the message. Thus if the signal's slope exceeds the maximum slope capability of the system, we may have *slope overload* just as in DM. To prevent slope overload

$$|\dot{f}(t)|_{\max} = |df(t)/dt|_{\max} \leqslant \frac{(2^{N_b} - 1)\delta v_0}{T_s} \qquad (4.14\text{-}2)$$

is required if the message has a maximum slope. If not, we may select a constant $\dot{K}$ such that

$$\dot{K}\sqrt{\overline{\dot{f}^2(t)}} \leqslant \frac{(2^{N_b} - 1)\delta v_0}{T_s}. \qquad (4.14\text{-}3)$$

These two conditions are analogous to (4.10-3) and (4.11-17) for DM.

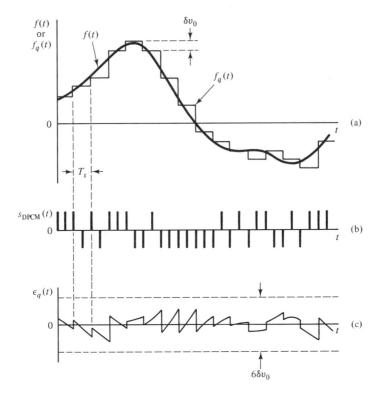

Figure 4.14-2. Waveforms applicable to DPCM. (a) $f(t)$ and $f_q(t)$, (b) DPCM pulse train, and (c) the error signal [2].

At the receiver end of the system, a PCM decoder recovers the amplitude-modulated sample pulses which, when integrated, become $\hat{f}_q(t)$, which is approximately equal to $f_q(t)$ back at the modulator. Except for receiver noise errors it would be the same. The lowpass filter smooths out the sharp steps in $\hat{f}_q(t)$, and the response is $f(t)$ pulse noise. As in DM the noise has granular, slope overload, and receiver noise components.

Channel Bandwidth

The channel bandwidth required in DPCM is that of an N_b-bit PCM system. Assuming N similar messages are time-multiplexed, we have

$$W_{ch} \geq \frac{\pi}{\tau} \geq \frac{\pi N N_b}{T_s} = \frac{N N_b \omega_s}{2}. \qquad (4.14\text{-}4)$$

Here the PCM pulses have duration τ.

Performance with Granular Noise Only

Several analyses may be cited which deal with the various noise components in the final receiver output signal [27, 28, 31, 53]. If only granular noise is present, which corresponds to times when the message is slowly varying such that the system easily follows the changes, van de Weg [53] has determined the output signal-to-granular noise power ratio. By using a random message model he obtains

$$\left(\frac{S_o}{N_g}\right)_{\text{DPCM}} = \frac{3}{2k}\left(\frac{\omega_s}{W_f}\right), \qquad N_b \geqslant 2, \qquad (4.14\text{-}5)$$

where W_f is the maximum spectral extent of the signal and

$$k = (\delta v_0)^2 / \overline{\dot{f}^2(t)}. \qquad (4.14\text{-}6)$$

Equation (4.14-5) is valid as a first approximation independent of the power spectrum of the message [53, p. 383], and van de Weg even justifies its application to a nonrandom sinusoidal signal [53, p. 384]. Thus we may apply (4.14-5) in a broad sense.

For signals having a definite maximum magnitude it is convenient to set the maximum step size $(2^{N_b} - 1)\delta v_0$ to prevent slope overload according to the maximum or reference slope value $\dot{f}_r(t)$ of $\dot{f}(t)$:

$$|\dot{f}_r(t)|^2_{\text{max}} = \frac{[(2^{N_b} - 1)\delta v_0]^2}{T_s^2}. \qquad (4.14\text{-}7)$$

For messages having no well-defined maximum magnitude, such as Gaussian random signals, we may choose the maximum step size such that system slope capability is larger than mean-squared signal slope by some factor $\dot{K}^2 > 1$:

$$\dot{K}^2 \overline{\dot{f}_r^2(t)} = \frac{[(2^{N_b} - 1)\delta v_0]^2}{T_s^2}. \qquad (4.14\text{-}8)$$

For this latter case k is easily evaluated. Substitution into (4.14-5) gives

$$\left(\frac{S_o}{N_g}\right)_{\text{DPCM}} = \frac{3(2^{N_b} - 1)^2}{8\pi^2 \dot{K}^2}\left(\frac{W_f}{W_{\text{rms}}}\right)^2\left(\frac{\omega_s}{W_f}\right)^3 \frac{\overline{f^2(t)}}{\overline{\dot{f}_r^2(t)}}, \qquad N_b \geqslant 2. \qquad (4.14\text{-}9)$$

For the former case, we only replace $\dot{K}^2$ by $\dot{K}^2_{cr}$, the slope crest factor defined in (4.11-9). W_{rms} is defined in (4.11-10).

We note that $(S_o/N_g)_{\text{DPCM}}$ increases as ω_s^3, which was also found for DM, D-SM, and ADM systems. If $N_b = 1$ we might expect that (4.14-9) would give results near those for DM. However, the value of (4.14-9) is actually 3 dB smaller than obtained in DM. This occurs because the derivation of (4.14-9) neglects correlation between successive steps when $N_b = 1$. For $N_b \geqslant 2$ the correlation becomes small [53]. Other important observations are that performance depends on the signal's power spectrum through the term $(W_f/W_{rms})^2$. The only system parameters† appearing explicitly in (4.14-9) are ω_s and N_b. Increasing either leads to better performance.

We now give three examples to illustrate use of (4.14-9) for three different information signals.

Example 4.14-1

Let us assume a random noiselike information signal having a uniform (constant) power density over $-W_f < \omega < W_f$ and a slope crest factor $\dot{K}_{cr} = \dot{K} = 4$. On using (4.11-10) we find $W_{rms}^2 = W_f^2/3$. From (4.14-9) with $\overline{f^2(t)} = \overline{f_r^2(t)}$

$$\left(\frac{S_o}{N_g}\right)_{\text{DPCM}} = \frac{9(2^{N_b} - 1)^2}{128\pi^2}\left(\frac{\omega_s}{W_f}\right)^3, \qquad N_b \geq 2.$$

If we now let $N_b = 2$ and $\omega_s/W_f = 10$, we have $(S_o/N_g)_{\text{DPCM}} = 64.1$, or 18.1 dB.

Example 4.14-2

Suppose $f(t)$ is a sinusoid of frequency $\omega_f \leq W_f$. From (4.11-9) we easily find that $\dot{K}_{cr}^2 = 2$. From Prob. 4-53 $W_{rms} = \omega_f$. From (4.14-9) with $\overline{f^2(t)} = \overline{f_r^2(t)}$

$$\left(\frac{S_o}{N_g}\right)_{\text{DPCM}} = \frac{3(2^{N_b} - 1)^2}{16\pi^2}\left(\frac{W_f}{\omega_f}\right)^2\left(\frac{\omega_s}{W_f}\right)^3, \qquad N_b \geq 2.$$

Example 4.14-3

As we have previously stated, results for delta modulation of speech signals may be found from the sinusoidal signal equation if $\omega_f/2\pi = 800$ Hz and $W_f/2\pi = 3.5$ kHz. By using these values in the equation of Example 4.14-2 we obtain

$$\left(\frac{S_o}{N_g}\right)_{\text{DPCM}} = \frac{3.59(2^{N_b} - 1)^2}{\pi^2}\left(\frac{\omega_s}{W_f}\right)^3, \qquad N_b \geq 2.$$

Again letting $N_b = 2$ and $\omega_s/W_f = 10$, we calculate $(S_o/N_g)_{\text{DPCM}} = 3274$, or 35.2 dB. By comparing with Example 4.14-1 we see that considerable difference in results can occur for different signal characteristics.

Performance Comparison with DM and PCM

DPCM is readily compared with DM. On the basis of equal channel bandwidth, the same values of $\dot{K}$, and equal message characteristics, we obtain

$$\left(\frac{S_o}{N_g}\right)_{\text{DPCM}} = \frac{(2^{N_b} - 1)^2}{2N_b^3}\left(\frac{S_o}{N_g}\right)_{\text{DM}}, \qquad (4.14\text{-}10)$$

which shows that DPCM is superior to DM if $N_b \geq 4$.

Van de Weg [53] has compared DPCM performance with that of PCM for a random message having a flat power spectrum and a slope crest factor of four. By assuming equal values of ω_s and the same number of bits per sample interval, it is found that DPCM performance is equal to or better

---

† $\dot{K}$ is a parameter related to both the system and the message.

than that of PCM when $\omega_s/W_f \geq 4$ and $N_b \geq 4$. On the other hand, for speech signals DPCM is superior to PCM for all values of ω_s/W_f and N_b when the comparison again assumes equal values of ω_s/W_f and N_b.

Performance with Slope Overload Noise Added

When receiver noise is negligible the total quantization noise power at the receiver output is the sum of granular and slope overload noise powers. O'Neal and Rice [31] have determined slope overload noise for a delta modulation system ($N_b = 1$). According to O'Neal [31], slope overload noise is determined from the delta modulation result by replacing the DM system step size by the maximum value $(2^{N_b} - 1)\delta v_0$ in the multilevel quantization (DPCM) case. Thus if we apply the same procedure to Abate's slope overload noise power which is given by (4.11-14), algebraic manipulation assuming random messages reveals

$$\left(\frac{S_o}{N_q}\right)_{\text{DPCM}} = \frac{x^2 \dot{K}^2 (S_o/N_g)_{\text{DPCM},.}}{1 + \frac{(2^{N_b} - 1)^2}{9}\left(\frac{\omega_s}{W_f}\right)^3 (3 + x)xe^{-3/x}}. \qquad (4.14\text{-}11)$$

Here $\dot{K}$ is determined by (4.14-8), x is given by

$$x^2 = \frac{\overline{f^2(t)}}{\dot{K}^2 \overline{f_r^2(t)}}, \qquad (4.14\text{-}12)$$

$\overline{f^2(t)}$ is the power in the message, which may be different from the reference level $\overline{f_r^2(t)}$, and $(S_o/N_g)_{\text{DPCM},r}$ is given by

$$\left(\frac{S_o}{N_g}\right)_{\text{DPCM},r} = \frac{3(2^{N_b} - 1)^2}{8\pi^2 \dot{K}^2}\left(\frac{W_f}{W_{\text{rms}}}\right)^2 \left(\frac{\omega_s}{W_f}\right)^3, \qquad N_b \geq 2. \qquad (4.14\text{-}13)$$

A plot of (4.14-11) for a fixed value of N_b shows a behavior versus x similar to that for DM given by (4.11-19).

Although we shall omit a discussion of receiver (channel) noise effects, it is known [54] that performance for a well-designed DPCM system is considerably better than that of a well-designed PCM system operating on the same digital channel, even if the channel is noisy.

Other System Configurations

As with most basic modulation techniques, additional complexity may lead to better performance. Such is the case with DPCM. Good discussions of the more advanced DPCM system configurations which result are given by Jayant [44] and Bayless *et al.* [55]. Our comments shall refer to data in reference [44]. For original sources and a good list of other references the reader is referred to Jayant.

We found that, by using a more complicated linear feedback network in a DM system, performance could be improved. The network could be either stationary (double versus single integration) or time-varying (adaptive DM). The analogous configurations also exist in DPCM. If the basic DPCM integrator is viewed as a prediction circuit, a more general predictor can be implemented that is either stationary or adaptive. Typical stationary or fixed predictors involve feedback networks which are specified by n constant parameters. The optimum values of these parameters are related to the correlation (and therefore spectral) properties of the message. The case $n = 1$ corresponds to the basic DPCM system. In practice $n \leq 3$ would be implemented because little performance improvement follows $n > 3$. For example, with $n = 3$ and speech messages, DPCM system performance (receiver output signal-to-noise ratio) is about 8.4 dB better than in a PCM system [44]. Only a 0.2-dB additional gain is obtained when n is increased to 5.

If the correlation properties of the message change with time, such as with speech waveforms, the optimum predictor must change with time. Thus the n parameters of the predictor may be made adaptive by performing periodic (typically on the order of every 4 ms for speech) analysis of short segments of the message to derive new optimum values. Such an *adaptive* DPCM system for $n = 3$ and speech signals is capable of about 1.6 dB improvement over a basic DPCM system with $n = 3$. For $n = 5$ and 10 the gains increase to 2.9 dB and 4.0 dB, respectively.

Finally, we note that configurations are possible which alter the quantizer characteristics. Because a DPCM system can slope overload similar to DM, the quantizer may be made adaptive such that a variable maximum step size (quantization level) occurs to accommodate signals which vary greatly in level (and therefore slope).

4.15 SUMMARY AND DISCUSSION

The principal theme of this chapter is the description of binary digicom systems that use baseband waveforms for channel transmission. Initially, the general optimum binary system is determined for a channel having white Gaussian noise and an infinite bandwidth. The correlation form of optimum receiver is shown in Fig. 4.2-1, where V_T is given by (4.2-12). An equivalent matched filter form of receiver is shown in Fig. 4.2-2. The general probability of bit error in these systems is given by (4.3-5).

Specializations are developed for the optimum binary system in Secs. 4.4 and 4.5 for specific unipolar, polar, and Manchester formats, as examples. The optimum demodulators (receivers) are shown in Fig. 4.4-3. The average probability of a bit error P_e is given by (4.4-6), which is plotted in Fig. 4.4-4. Differentially encoded data are also examined with P_e now given by (4.4-8). Effects of coding of digital data are discussed. In addition, the transmission of analog messages over the PCM digital system is developed in detail. Receiver output signal-to-noise ratio is given by (4.5-21) and plotted in Fig. 4.5-2.

The effects of a bandlimited channel are next discussed (intersymbol interference) and ways of either eliminating the effects (Nyquist's pulse shaping) or controlling it (partial response signaling and channel equalization) are developed. In the latter category, two optimum systems (duobinary and modified duobinary) are discussed in detail. The block diagram for both duobinary and modified duobinary systems is shown in Fig. 4.7-1(a) where the optimum duobinary filters are defined in (4.7-15) and (4.7-16). For the modified duobinary system it is necessary to replace $P_2(\omega)$ by $P_3(\omega)$ as defined in (4.6-11). Average probability of bit errors in these two systems is given by (4.7-7) using (4.7-23).

An example of a multi-level baseband system is included in Sec. 4.9, where M-ary PAM is discussed. Average probability of a word (symbol) error is found in (4.9-18) and plotted in Fig. 4.9-2. When compared with a binary system having the same word error probability, the M-ary system is found to require more average power; this fact is true when the systems have either the same information rate or have the same bandwidth. The main advantage of M-ary PAM is its smaller channel bandwidth required to transmit the same information rate (at same word error probability) as the binary system; its main disadvantage is the extra average power required to achieve the stated advantage.

In the remainder of this chapter, Secs. 4.10–4.14, several types of modulation systems are described that can convert an analog message directly to a digital waveform for channel transmission. These modulators combine in one step the functions of sampling, quantization, source encoding, and waveform formatting. They do not involve any channel encoding. The delta modulation (DM) system is best suited to messages with a power spectrum proportional to $1/\omega^2$, while the delta-sigma modulation (D-SM) system is best suited to signals with a constant power spectrum. The former spectrum form is a fair approximation for speech and some television signals. The adaptive delta modulation (ADM) system of Sec. 4.13 is a variation of DM that allows the receiver's output signal-to-noise ratio to be nearly constant over a range of input signal-to-noise ratios. It, therefore, improves the dynamic range achievable in the delta modulator. DM, D-SM, and ADM modulators all use a simple 1-bit encoding procedure (an error in a feedback loop is actually encoded). The differential pulse code modulation (DPCM) system of Sec. 4.14 is a variation of DM that uses a multi-bit encoding procedure.

Final receiver output noise powers in the DM, D-SM, ADM, and DPCM systems all contain component powers mainly of three types: (1) a granular noise power that is always present (it is due to the encoder's quantization noise), (2) another noise that is always present, called receiver noise (due to channel noise), and (3) a power component due to slope overload noise; this noise occurs only in time intervals where the transmitter's encoder is not capable of following rapid changes in the message's amplitude (above those for which the encoder was designed). Although basic procedures are given to compute all these noise components, only certain cases are given as examples. Output signal power-to-granular noise power ratio for DM, D-SM, and DPCM systems are given by (4.11-11), (4.12-7), and (4.14-9), respectively. Signal-to-noise power ratios based on output noise powers being the sums of granular and slope overload noise powers are given by (4.11-19), (4.13-7), and (4.14-11) for DM, ADM, and DPCM systems, respectively. Finally, the output ratio of signal powers to the sum of granular and receiver noise powers for DM and D-SM systems are, respectively, given by (4.11-31) and (4.12-13). Broadly speaking, the adaptive systems are preferred for their increased dynamic range while multi-bit encoding (DPCM) leads to improved performance. Thus recent emphasis in practical systems has been toward adaptive versions of DPCM.

Finally, it is noted that some recent books are recommended for additional reading on the topics of this chapter [56–58].

REFERENCES

[1] Thomas, J. B., *An Introduction to Statistical Communication Theory*, John Wiley & Sons, Inc., New York, 1969.

[2] Peebles, Jr., P. Z., *Communication System Principles,* Addison-Wesley Publishing Co., Inc., Reading, Massachusetts, 1976. (Figures 4.4-5, 4.5-2, 4.5-3, 4.5-4, 4.6-1, 4.6-2, 4.10-1, 4.10-2, 4.11-1, 4.13-1, 4.13-2, 4.14-1, and 4.14-2 have been adapted.)

[3] Golay, M. J. E., Notes on Digital Coding, *Proceedings IRE,* Vol. 37, No. 6, June 1949, p. 657.

[4] Schwartz, M., *Information Transmission, Modulation, and Noise,* 2nd ed., McGraw-Hill Book Co., New York, 1970.

[5] Carlson, A. B., *Communication Systems: An Introduction to Signals and Noise in Electrical Communication,* 2nd ed., McGraw-Hill Book Co., New York, 1975. (See also third edition, 1986.)

[6] Lender, A., The Duobinary Technique for High-Speed Data Transmission, *IEEE Transactions on Communication and Electronics,* Vol. 82, May 1963, pp. 214–218.

[7] Lender, A., Correlative Digital Communication Techniques, *IEEE Transactions on Communication Technology,* Vol. COM-12, December 1964, pp. 128–135.

[8] Lender, A., Correlative Level Coding for Binary-Data Transmission, *IEEE Spectrum,* Vol. 3, No. 2, February 1966, pp. 104–115.

[9] Kretzmer, E. R., An Efficient Binary Data Transmission System, *IEEE Transactions on Communications Systems,* Vol. CS-12, No. 2, June 1964, pp. 250–251.

[10] Kretzmer, E. R., Generalization of a Technique for Binary Data Communication, *IEEE Transactions on Communication Technology,* Vol. COM-14, No. 1, February 1966, pp. 67–68.

[11] Kabal, P., and Pasupathy, S., Partial Response Signaling, *IEEE Transactions on Communications,* Vol. COM-23, No. 9, September 1975, pp. 921–934.

[12] Huang, J., and Feher, K., On Partial Response Transmission Systems, *Conference Record 1977 International Conference on Communications,* Vol. 1, Chicago, Illinois, June 12–15, 1977, pp. 3.3-47 through 3.3-51.

[13] Pasupathy, S., Correlative Coding: A Bandwidth-Efficient Signaling Scheme, *IEEE Communications Society Magazine,* Vol. 15, No. 4, July 1977, pp. 4–11.

[14] Kobayashi, H., Correlative Level Coding and Maximum-Likelihood Decoding, *IEEE Transactions on Information Theory,* Vol. IT-17, No. 5, September 1971, pp. 586–594.

[15] Forney, Jr., G. D., Maximum-Likelihood Sequence Estimation of Digital Sequences in the Presence of Intersymbol Interference, *IEEE Transactions on Information Theory,* Vol. IT-18, No. 3, May 1972, pp. 363–378.

[16] Eggers, M. D., and Painter, J. H., Optimal Symbol-by-Symbol Detection for Duobinary Signaling, *IEEE Transactions on Communications,* Vol. COM-31, No. 9, September 1983, pp. 1077–1085.

[17] Lucky, R. W., Salz, J., and Weldon, Jr., E. J., *Principles of Data Communication,* McGraw-Hill Book Co., New York, 1968.

[18] Qureshi, S., Adaptive Equalization, *IEEE Communications Magazine,* Vol. 20, No. 2, March 1982, pp. 9–16.

[19] de Jager, F., Deltamodulation, A Method of P.C.M. Transmission Using the 1-Unit Code, *Phillips Research Reports,* Vol. 7, 1952, pp. 442–466.

[20] Johnson, F. B., Calculating Delta Modulator Performance, *IEEE Transactions on Audio and Electroacoustics,* Vol. AU-16, No. 1, March 1968, pp. 121–129.

[21] Schindler, H. R., Delta Modulation, *IEEE Spectrum,* Vol. 7, No. 10, October 1970, pp. 69–78.

[22] Schindler, H. R., Linear, Nonlinear, and Adaptive Delta Modulation, *IEEE Transactions on Communications,* Vol. COM-22, No. 11, November 1974, pp. 1807–1823.

[23] Steele, R., *Delta Modulation Systems,* John Wiley & Sons, New York, 1975.

[24] Uddenfeldt, J., and Zetterberg, L. H., Algorithms for Delayed Encoding in Delta Modulation with Speech-Like Signals, *IEEE Transactions on Communications,* Vol. COM-24, No. 6, June 1976, pp. 652–658.

[25] Uddenfeldt, J., Asymptotic Performance of Delayed Delta Coding with Single and Double Integration, *IEEE Transactions on Communications,* Vol. COM-26, No. 6, June 1978, pp. 907–913.

[26] Hawkes, T. A., and Simonpieri, P. A., Signal Coding Using Asynchronous Delta Modulation, *IEEE Transactions on Communications,* Vol. COM-22, No. 3, March 1974, pp. 346–348.

[27] Protonotarios, E. N., Slope Overload Noise in Differential Pulse Code Modulation

Systems, *Bell System Technical Journal,* Vol. 46, No. 9, November 1967, pp. 2119–2161.

[28] Goodman, D. J., Delta Modulation Granular Quantizing Noise, *Bell System Technical Journal,* Vol. 48, No. 5, May–June 1969, pp. 1197–1218.

[29] Taub, H., and Schilling, D. L., *Principles of Communiciation Systems,* McGraw-Hill Book Co., New York, 1971. (See also second edition, 1986.)

[30] French, N. R., and Steinberg, J. C., Factors Governing the Intelligibility of Speech Sounds, *Journal of the Acoustical Society of America,* Vol. 19, January 1947, pp. 90–119.

[31] O'Neal, Jr., J. B., Delta Modulation Quantizing Noise Analytical and Computer Simulation Results for Gaussian and Television Input Signals, *Bell System Technical Journal,* Vol. 45, No. 1, January 1966, pp. 117–141.

[32] Abate, J. E., Linear and Adaptive Delta Modulation, *Proceedings of the IEEE,* Vol. 55, No. 3, March 1967, pp. 298–308.

[33] Greenstein, L. J., Slope Overload Noise in Linear Delta Modulators with Gaussian Inputs, *Bell System Technical Journal,* Vol. 52, No. 3, March 1973, pp. 387–421.

[34] Steele, R., SNR Formula for Linear Delta Modulation with Band-Limited Flat and RC-shaped Gaussian Signals, *IEEE Transactions on Communications,* Vol. COM-28, No. 12, December 1980, pp. 1977–1984.

[35] Inose, H., and Yasuda, Y., A Unity Bit Encoding Method by Negative Feedback, *Proceedings of the IEEE,* Vol. 51, November 1963, pp. 1524–1535.

[36] Agrawal, B., and Shenoi, K., Design Methodology for $\Sigma\Delta M$, *IEEE Transactions on Communications,* Vol. COM-31, No. 3, March 1983, pp. 360–370.

[37] Jayant, N. S., Adaptive Delta Modulation with a One-Bit Memory, *Bell System Technical Journal,* Vol. 49, No. 3, March 1970, pp. 321–342.

[38] Tomozawa, A., and Kaneko, H., Companded Delta Modulation for Telephone Transmission, *IEEE Transactions on Communication Technology,* Vol. COM-16, No. 1, February 1968, pp. 149–157.

[39] Chakravarthy, C. V., and Faruqui, M. N., Two Loop Adaptive Delta Modulation Systems, *IEEE Transactions on Communications,* Vol. COM-22, No. 10, October 1974, pp. 1710–1713.

[40] Cartmale, A. A., and Steele, R., Calculating the Performance of Syllabically Companded Delta-Sigma Modulators, *Proceedings of the IEE* (London), Vol. 117, No. 10, October 1970, pp. 1915–1921.

[41] Brolin, S. J., and Brown, J. M., Companded Delta Modulation for Telephony, *IEEE Transactions on Communication Technology,* Vol. COM-16, No. 1, February 1968, pp. 157–162.

[42] Chakravarthy, C. V., A Class of Companded Unity Bit Coders, *IEEE Transactions on Communication Systems,* Vol. COM-30, No. 7, July 1982, pp. 1772–1775.

[43] Zetterberg, L. H., and Uddenfeldt, J., Adaptive Delta Modulation with Delayed Decision, *IEEE Transactions on Communications,* Vol. COM-22, No. 9, September 1974, pp. 1195–1198.

[44] Jayant, N. S., Digital Coding of Speech Waveforms: PCM, DPCM, and DM Quantizers, *Proceedings of the IEEE,* Vol. 62, No. 5, May 1974, pp. 611–632.

[45] Greefkes, J. A., A Digitally Controlled Delta Codec for Speech Transmission, *Conference Record, 1970 IEEE International Conference on Communications,* pp. 7-33 to 7-48.

[46] Greefkes, J. A., and Riemens, K., Code Modulation with Digitally Controlled Companding for Speech Transmission, *Philips Technical Review,* Vol. 31, No. 11/12, 1970, pp. 335–353.

[47] Greefkes, J. A., and de Jager, F., Continuous Delta Modulation, *Philips Research Reports,* R664, Vol. 23, 1968, pp. 233–246.

[48] Lee, H. S., and Un, C. K., Quantization Noise in Adaptive Delta Modulation Systems, *IEEE Transactions on Communications,* Vol. COM-28, No. 10, October 1980, pp. 1794–1802.

[49] Un, C. K., and Magill, D. T., The Residual-Excited Linear Prediction Vocoder with Transmission Rate Below 9.6 kbits/s, *IEEE Transactions on Communications,* Vol. COM-23, No. 12, December 1976, pp. 1466–1474.

[50] Un, C. K., Lee, H. S., and Song, J. S., Hybrid Companding Delta Modulation, *IEEE Transactions on Communications,* Vol. COM-29, No. 9, September 1981, pp. 1337–1344.

[51] Un, C. K., and Lee, H. S., A Study of the Comparative Performance of Adaptive Delta Modulation Systems, *IEEE Transactions on Communications,* Vol. COM-28, No. 1, January 1980, pp. 96–101.

[52] Un, C. K., and Cho, D. H., Hybrid Companding Delta Modulation with Variable-Rate Sampling, *IEEE Transactions on Communications,* Vol. COM-30, No. 4, April 1982, pp. 593–599.

[53] van de Weg, H., Quantizing Noise of a Single Integration Delta Modulation System with an N-Digit Code, *Phillips Research Reports,* Vol. 8, 1953, pp. 367–385.

[54] Chang, K-Y, and Donaldson, R. W., Analysis, Optimization, and Sensitivity Study of Differential PCM Systems Operating on Noisy Communication Channels, *IEEE Transactions on Communications,* Vol. COM-20, No. 3, June 1972, pp. 338–350.

[55] Bayless, J. W., Campanella, S. J., and Goldberg, A. J., Voice Signals: Bit-by-Bit, *IEEE Spectrum,* Vol. 10, No. 10, October 1973, pp. 28–34.

[56] Kanefsky, M., *Communication Techniques for Digital and Analog Signals,* Harper & Row, Publishers, New York, 1985.

[57] Ziemer, R. E., and Tranter, W. H., *Principles of Communications Systems, Modulation, and Noise,* 2nd ed., Houghton Mifflin Co., Boston, 1985.

[58] Ziemer, R. E., and Peterson, R. L., *Digital Communications and Spread Spectrum Systems,* Macmillan Publishing Co., New York, 1985.

PROBLEMS

4-1. From (4.1-1) bandlimited white noise becomes white noise as $W_N \to \infty$. Use (4.1-2) with $W_N \to \infty$ and justify that the same result is achieved. That is, show that

$$\lim_{W_N \to \infty} R_{n_b}(\tau) = \left(\frac{\mathcal{N}_0}{2}\right) \delta(\tau).$$

4-2. Explain in words why (4.1-16) is true.

4-3. An optimum binary system uses a unipolar-NRZ format with rectangular pulses. Pulse amplitude at the receiver is 1.5 V. Pulse duration is 2 μs, and white channel noise density is $\mathcal{N}_0/2 = 5(10^{-7})$ W/Hz. If optimum threshold V_T, as given by (4.2-12), is $3(10^{-6})$ J, find message probabilities P_1 and P_2.

4-4. Show that (4.2-17) is the Fourier transform of (4.2-14).

★4-5. A certain binary system seeks to generate $s_2(t)$ by switching a continuously running oscillator $3 \cos (\omega_0 t)$ to the channel for each binary **1**. For a binary **0**, $s_1(t) = 0$. If bit intervals of duration T_b are defined such that one starts at $t = 0$, determine how ω_0 must be related to T_b if the system is to be optimum.

4-6. A binary transmitter for equally probable messages uses the waveform

$$s_1(t) = \begin{cases} 5 \sin^2 \left(\dfrac{\omega_b t}{2} \right), & 0 \le t \le T_b \\ 0, & \text{elsewhere} \end{cases}$$

(a single pulse) when a binary **0** occurs, and

$$s_2(t) = \begin{cases} 5 \sin^2(\omega_b t), & 0 \le t \le T_b \\ 0, & \text{elsewhere} \end{cases}$$

for a binary **1**. (a) Find the energies E_1 and E_2 in these waveforms. (b) Find γ. (c) Find the optimum receiver output levels V_1 and V_2. (d) What value of $\mathcal{N}_0$ for the channel will give a minimum peak output signal power (square of signal level at sampler at sample time) to average noise power ratio of 1000 (or 30 dB)?

4-7. Use Schwarz's inequality to show that γ of (4.2-23) satisfies $-1 \le \gamma \le 1$.

4-8. Use (4.2-24) through (4.2-26) and show that the optimum binary system's threshold of (4.2-12) can be written as

$$V_T = \frac{V_2 + V_1}{2} + \frac{\sigma_o^2 \ln(P_1/P_2)}{V_2 - V_1}.$$

4-9. In an optimum binary system $V_1 = -1$ V, $V_2 = 5$ V, $V_T = 1.9$ V, and $\sigma_o = 0.5$ V. (a) Find message probabilities P_1 and P_2. (b) Find average error probability P_e. (*Hint:* Use results of Prob. 4-8.)

4-10. In an optimum binary digital system the two possible transmitted waveforms occur with equal probabilities and have equal energies. The ratio of average energy per bit to $\mathcal{N}_0$ is 10.4 and the system's average bit error probability is $1.18(10^{-4})$. Find γ for the two waveforms.

4-11. What would the average probability of a bit error become for the system of Prob. 4-10 if the transmitted waveforms were adjusted to be antipodal (energies are kept the same)?

4-12. An optimum unipolar-NRZ binary system transmits 12-V rectangular pulses of duration 1 ms over a channel for which $\mathcal{N}_0 = 4(10^{-3})$ W/Hz. Messages occur with probabilities $P_1 = P(m_1) = 0.9$ and $P_2 = P(m_2) = 0.1$. (a) Find optimum receiver threshold V_T. (b) Find the average transmitted pulse's energy. (c) Find ε. (d) Find average bit error probability P_e.

4-13. Work Prob. 4-12 if the system uses a polar-NRZ format.

4-14. A binary system has equally probable messages and uses a Manchester waveform format where $\varepsilon = 20$. The binary data are differentially encoded at the modulator prior to waveform formatting. (a) Find the receiver demodulator's average bit error probability. (b) What is the probability of a bit error after differential decoding? (c) What is the receiver's optimum threshold?

4-15. In a differential decoder for binary systems the decoded output is correct if the current and earlier adjacent bits are both in error. Explain why.

4-16. Use (4.4-11) to determine the smallest channel bandwidth that will guarantee that the decision level D is within 99% of the value it would have if channel bandwidth were infinite.

4-17. A binary PCM system uses a 4-bit natural binary source encoder and no channel encoding. It has an average bit error probability of 0.01. (a) What is the system's word error probability? (b) If enough additional transmitter power is installed to reduce bit error probability to 0.001 what is the word error probability?

4-18. A binary polar-NRZ PCM system can transmit either 4-bit uncoded words during a time interval T or it can transmit a double-error correcting (10, 4) code during the same time T. In each case the peak transmitted powers and channel noise densities are the same. (a) If $P_e = 2.339(10^{-3})$ in the uncoded case, find the word error probabilities for both coded and uncoded transmissions. (b) Which of the two systems would you use?

4-19. Work Prob. 4-18 if $P_e = 1.18(10^{-4})$ in the uncoded case.

4-20. In writing (4.5-2) series terms above the first were assumed negligible and $(1 - P_e)^{n-i-1} \approx 1$ in the first term. (a) Assume a (7, 4) single-error correcting code and find an expression in terms of P_e for the ratio of the second term (first neglected term) to the exact first term. (b) Evaluate your expression for $P_e = 10^{-1}, 10^{-2}$, and 10^{-3}. (c) What can you conclude about the approximate formula's accuracy from your results?

4-21. In a polar-NRZ binary PCM system that uses a uniform quantizer with an analog message for which $|f(t)|_{max} = 0.7$ V and $\overline{f^2(t)} = 0.01$ W, the output signal-to-quantization noise power ratio is $256.8(10^3)$ (or 54.1 dB) when the message power is 6 dB below its reference level. Channel noise is negligible. (a) What is the message's crest factor? (b) How many levels does the quantizer implement? (c) What is the quantizer's step size?

4-22. A message having symmetrical fluctuations, zero mean, and a crest factor of 3.4 is uniformly quantized in a PCM system to quantum levels ± 0.005 V, ± 0.015 V, ± 0.025 V, ..., ± 1.275 V. (a) What maximum signal-to-quantization noise power ratio is possible if no amplitude overload is to occur? (b) Repeat part (a) if the quantizer's input is a sinusoidal message instead. (c) For the sinusoidal message let the peak amplitude be 1.6 V. What fraction of time will amplitude overload occur?

4-23. Follow the procedures of the text to show that expressions analogous to (4.5-16), (4.5-17), and (4.5-21) for unipolar PCM with equally probable messages are

$$N_{rec} = \frac{2|f_r(t)|^2_{max} \, 2^{2N_b}}{3L^2} \, \text{erfc}\left[\sqrt{\frac{A^2 T_b}{4 \, \mathcal{N}_0}}\right], \qquad \text{unipolar-NRZ PCM}$$

$$\left(\frac{S_o}{N_o}\right)_{PCM} = \frac{(S_o/N_q)_{PCM}}{1 + 2^{2N_b+1} \, \text{erfc}[\sqrt{A^2 T_b/4 \, \mathcal{N}_0}]}, \qquad \text{unipolar-NRZ PCM}$$

$$\left(\frac{S_o}{N_o}\right)_{PCM} = \frac{(S_o/N_q)_{PCM}}{1 + 2^{2N_b+1} \, \text{erfc}\left[\sqrt{\frac{1}{8}\left(\frac{\hat{S}_i}{N_i}\right)_{PCM}}\right]}, \qquad \text{unipolar-NRZ PCM.}$$

4-24. If channel noise in the PCM system of Prob. 4-21 is *not* negligible and $(\hat{S}_i/N_i)_{PCM} = 18.0$, find $(S_o/N_o)_{PCM}$. Justify that your result agrees with that found from Fig. 4.5-2 by suitable scaling.

4-25. Explain why binary polar PCM uses less average power than unipolar PCM when the two have the same peak-to-peak amplitude separation between their pulse levels. Assume equally probable messages.

4-26. Two binary PCM systems, one unipolar and the other polar, are to produce the same average bit error probability when using the same channel and having the same bit durations. Find the necessary relationship between transmitted pulse amplitudes in the two systems. Assume equally probable messages.

4-27. Use results from Prob. 4-23 to find an expression for the 1-dB threshold signal-to-noise ratio in a unipolar PCM system with equally probable messages. Plot your results versus N_b for $2 \leq N_b \leq 20$. [*Hint:* Calculate N_b for various $(\hat{S}_i/N_i)_{PCM,th}$.]

4-28. A single analog message is transmitted by a binary polar-NRZ PCM system with a uniform quantizer. The message has a crest factor of $\sqrt{2}$ and is sampled every 50 μs, which is its Nyquist period. (a) What is the largest number of bits per word that can be used that will guarantee performance at or above threshold for an input peak signal-to-noise ratio of 12 dB? (b) What will be the maximum output signal-to-noise ratio? (c) What is T_b?

4-29. Inverse Fourier transform (4.6-4) to show that (4.6-5) is true.

4-30. (a) Unit-amplitude pulses of the form shown in Fig. 4.6-2 for $W_1/W = 0$ are used to transmit bits in a polar PCM system. Sketch the waveform for a bit sequence ...**11010011100**.... Use at least one sidelobe for each pulse in the sketch. Assume $T_b = \pi/W$. (b) Repeat part (a) except use the waveform for $W_1/W = 1$.

4-31. Show that a PCM waveform comprised of pulses of the form of Fig. 4.6-2 with $W_1/W = 0$ can be passed through a filter matched to the pulse shape (white receiver noise) without causing intersymbol interference in samples taken every bit interval.

★4-32. Let a PCM waveform comprised of pulses of the form of Fig. 4.6-2 with $W_1/W = 1.0$ be applied to a receiver filter matched to the shape of the pulses (white channel noise). Show that samples of the filter's output taken every bit interval *do* have intersymbol interference. However, show also that interference occurs only between adjacent bit intervals.

4-33. (a) Use pulses of the form of (4.6-8) and sketch the polar PCM waveform for the digit sequence ...**11010011100**.... (b) Verify that samples of the waveform taken at the end of each bit interval give the correct sequence when decoded according to the logic given in the text (no precoding).

4-34. Work Prob. 4-33 for the waveform of (4.6-10). The waveform is modified duobinary.

4-35. Show that $p_3(t)$, as given in Example 4.6-1, is correct; that is, supply the missing steps in the development.

4-36. The channel response in a binary system has a format made up from Nyquist (sampling) pulses defined by

$$s_o(t) = \frac{1}{T_b} \operatorname{Sa}\left(\frac{\pi t}{T_b}\right) \leftrightarrow \operatorname{rect}\left(\frac{\omega}{\omega_b}\right) = H_o(\omega),$$

where $\omega_b = 2\pi/T_b$ and T_b is the bit interval's duration. The formatted waveform is free of intersymbol interference. If the channel has distortion such that its response pulses are characterized by the spectrum

$$H_c(\omega) = H_o(\omega)\left\{1 + \sum_{n=1}^{N} [a_n\cos(n\omega T_b) + jb_n\sin(n\omega T_b)]\right\},$$

where a_n and b_n are real constants, find the response $s_c(t)$ in terms of $s_o(t)$. Show that $s_c(t)$ contains intersymbol interference and determine which sample times (at multiples of T_b) contain the interference. Discuss how the constants a_n and b_n affect the interference.

4-37. A typical pulse $p_c(t)$ emerging from the channel of a binary system has sample values $p_c(-2T_b) = -0.05p_{ch}(0)$, $p_c(-T_b) = 0.1p_{ch}(0)$, $p_c(0) = 0.9 p_{ch}(0)$, $p_c(T_b) = -0.15p_{ch}(0)$, and $p_c(2T_b) = 0.02p_{ch}(0)$. $p_c(t)$ is zero at all other possible sample times. Here $p_{ch}(0)$ is the maximum amplitude of the desired channel response. Find the coefficients required in a 3-tap channel equalizer. Find the values of the equalized channel response waveform at all possible sample times.

4-38. From (4.7-5a), threshold in an optimum duobinary system is a function of both sample time, peak signal power A^2, and noise power σ_o^2. How large must A^2/σ_o^2 be in order that the threshold be within 10% of its limiting value that is independent of noise?

4-39. Use (4.7-5) with (4.7-1) to show that (4.7-6) is true.

4-40. In an optimum duobinary system the channel has a transfer function that can be approximated by

$$H_{ch}(\omega) = \frac{1}{1 + (\omega/W_{ch})^2}.$$

Noise from the channel can be taken as white while the receiver's output pulse shape is to be determined by the channel and identical filters at transmitter and receiver. If the system bit rate is $\omega_b/2\pi$, sketch the required receiver transfer function's magnitude squared, $|H_R(\omega)|^2$, for $W_{ch}/\omega_b = 0.25, 0.5, 1.0$, and 2.0.

★4-41. An optimum duobinary system, originally designed for identical transmit and receive filters with a distortion-free white noise channel, gave a bit error probability of $2.234(10^{-3})$. It is modified to operate with the filters and channel defined in Prob. 4-40 with $W_{ch} = \omega_b$. If bit duration, average transmitted power, and channel noise level remain the same, what is the new bit error probability?

4-42. In an optimum duobinary system the transmitted average energy per bit divided by $\mathcal{N}_0$ (which is ε) is $1.459(10^7)$. Average bit error probability is $1.6568(10^{-5})$. The system operates on a lossy but distortion-free white noise channel. What is the gain of the channel?

4-43. Reduce (4.7-23) for an optimum modified duobinary system and show that (4.7-25) applies without change.

4-44. Work Prob. 4-42 for an optimum modified duobinary system.

4-45. Expand (4.9-2) and show that (4.9-3) is valid.

★4-46. When messages are not equally probable in an M-ary PAM system, derive expressions for the optimum thresholds that are now functions of noise power, message probabilities $P(V_i)$ as well as levels V_i.

4-47. An M-ary PAM system uses $M = 32$ levels uniformly and symmetrically displaced about zero. The largest possible receiver level at a sample time is 2.48 V while the average noise power level is $\sigma_o^2 = 4.3896(10^{-4})$ W. (a) What is the separation between the possible received signal levels? (b) What is the system's average probability of error P_w?

4-48. For an optimum M-ary PAM receiver that uses a white noise matched filter, as defined by (4.9-10), show that the filter's output levels V_i at the sample times are equal to the input pulse peak amplitudes A_i. (b) Show that the output noise power is given by (4.9-14).

4-49. If N_i is defined as the "input" average noise power to the sampler in an optimum M-ary PAM receiver, show that word error probability can be written as

$$P_w = \frac{M-1}{M} \text{erfc}\left[\sqrt{\frac{3}{2(M^2-1)}\left(\frac{S_i}{N_i}\right)_{M\text{-PAM}}}\right]$$

where we define

$$\left(\frac{S_i}{N_i}\right)_{M\text{-PAM}} \triangleq \frac{S_i}{N_i}.$$

4-50. A large power plant uses a single word-interleaved time-multiplexed line to send many control and monitoring signals to a main control console. The line has 500 equal-length time slots, each carrying 4-bit natural binary words that are Nyquist-rate samples of similar narrowband source messages of 10-Hz spectral extent. Word error probability is $1.4862(10^{-6})$. The multiplexer's transmitter is operating linearly with polar format with pulses at peak power capability. All available channel bandwidth is being used. The multiplexer is to be modified to handle twice as many similar messages, and the modulation

is changed to 16-level PAM (one level per 4-bit word) with its peak power
level raised to 20 times that of the original system. (a) What is the frame
duration of the 1000 multiplexed signals? (b) What is the duration of an
M-ary symbol? (c) What is the M-ary PAM system word error probability?
(d) Why is it necessary to convert to an M-ary PAM system?

4-51. In a basic DM system, $\delta v = 0.1$ V and $T_s = 20$ μs. (a) Based on examination
of the message

$$f(t) = A[1.2 \cos(400\pi t + \theta_1) + 1.3 \cos(600\pi t + \theta_2)$$
$$+ 0.5 \cos(1200\pi t + \theta_3)],$$

where θ_1, θ_2, and θ_3 are phase angles that can have any values, find the
largest value of A that will not cause slope overload. (b) Which frequency
is most important in overload control for this message?

4-52. Show that $\overline{[df(t)/dt]^2}/\overline{f^2(t)} = W_{rms}^2$ if W_{rms}^2 is given by the right side of (4.11-
10). (*Hint:* Consider the power into and out of a differentiation network.)

4-53. Using (4.11-10) show that the rms bandwidth for a sinusoidal signal of frequency
ω_f is ω_f.

4-54. A message $f(t)$ has a power spectrum

$$\mathcal{S}_f(\omega) = \begin{cases} \dfrac{B}{\omega^2} & W_L < |\omega| < W_f \\ 0, & \text{elsewhere,} \end{cases}$$

where B, W_L, and W_f are positive constants. (a) Find the power $\overline{f^2(t)}$ in the
message. (b) Find the message's rms bandwidth. (c) If the message, for
which slope crest factor is 5, $W_L/2\pi = 100$ Hz, and $W_f/2\pi = 3.5$ kHz, is
transmitted over a basic DM system with a relative pulse rate $\omega_s/W_f = 40$,
what maximum value of $(S_o/N_g)_{DM}$ is achievable without slope overload?

4-55. Work Prob. 4-54 for the power spectrum

$$\mathcal{S}_f(\omega) = B \operatorname{rect}\left(\frac{\omega}{2W_f}\right) - B \operatorname{rect}\left(\frac{\omega}{2W_L}\right).$$

4-56. A delta modulation system uses a nearly ideal single integrator that is replaced
by a double integrator. Determine the improvement in output signal-to-
granular noise power ratio that can be expected for sampling rates of $\omega_s = 5W_f$, $10W_f$, $20W_f$, and $40W_f$ when a sinusoidal message is transmitted.

4-57. By assuming operation above threshold in both PCM and DM systems, show
that

$$\frac{(S_o/N_q)_{PCM}}{(S_o/N_g)_{DM}} = \frac{\pi^2 2^{2N_b}}{2N_b^3}\left(\frac{\dot{K}_{cr}}{K_{cr}}\right)^2\left(\frac{W_{rms}}{W_f}\right)^2,$$

where N_b is the number of pulses in a PCM system sample. Assume Nyquist
sampling in the PCM case and equal channel bandwidths that are a common
factor $A_{ch} > 1$ larger than the minimums. That is, assume $W_{ch} = A_{ch}\pi/\tau$,
where τ is the largest allowable pulse duration in each system (no guard
times, NRZ PCM format).

4-58. Assume sinusoidal modulation and determine how much poorer a DM system
is than an 8-bit PCM system using the result of Prob. 4-57.

4-59. By use of (4.11-22) show that the (optimum) values of x which define the maximum points in Fig. 4.11-1 are given by

$$\frac{1}{x_{opt}} = \dot{K} = \ln\left[\frac{2}{N}\left(\frac{W_{ch}}{W_f}\right)\right]$$

for multiplexing of N similar signals using delta modulation.

4-60. A delta modulation system is designed to transmit three similar messages by time multiplexing. What is the optimum value of $\dot{K}$ to be used if $W_{ch}/W_f = 81.9$? For messages characterized by $W_f/W_{rms} = 2$, what optimum output signal-to-noise ratio can be achieved? Assume equalities in (4.10-6) apply.

4-61. Work Prob. 4-60 for $W_{ch}/W_f = 30.13$.

4-62. By using (4.11-32) show that the threshold value (1-dB definition) of input peak signal-to-noise ratio in DM is

$$\left(\frac{\hat{S_i}}{N_i}\right)_{DM,th} = 2\left\{erfc^{-1}\left[0.214N^2\left(\frac{W_f}{W_{ch}}\right)^2\left(\frac{W_L}{W_f}\right)\right]\right\}^2.$$

4-63. A single-integrator delta modulation system is to be optimally designed (Prob. 4-59) for transmitting a single message. If $W_{ch}/W_f = 27$ for the system, what must W_f/W_{rms} be for the message in order to achieve a 30-dB output signal-to-quantization noise power ratio? Assume the minimum channel bandwidth.

4-64. For the system of Prob. 4-63, at what input peak signal-to-noise power ratio does threshold occur if $W_L/W_f = 0.05$?

4-65. For a sinusoidal message of frequency ω_f show that, for equal sampling rates, a delta-sigma modulation system's performance can be no larger than three times that of a delta modulation system, assuming both use a single integrator. That is, show that

$$\left(\frac{S_o}{N_g}\right)_{D\text{-}SM} = 3\left(\frac{\omega_f}{W_f}\right)^2\left(\frac{S_o}{N_g}\right)_{DM} \leq 3\left(\frac{S_o}{N_g}\right)_{DM}.$$

4-66. Determine and plot (in decibels) how much poorer a practical D-SM system performs when it uses a simple lowpass filter of 3-dB bandwidth ω_c as compared to a system with an ideal integrator. Plot versus $\omega_c/W_f \leq 1$ when the message is a sinusoid of frequency $\omega_f = W_f$.

4-67. Show that the threshold input peak signal-to-noise ratio (1-dB definition) in a delta-sigma modulation system is given by

$$\left(\frac{\hat{S_i}}{N_i}\right)_{D\text{-}SM,th} = 2\left\{erfc^{-1}\left[\frac{0.0144\pi N\tau W_f^2}{\omega_s}\right]\right\}^2.$$

4-68. In a D-SM system a number of similar messages are time-multiplexed. Each message has a crest factor of 3, spectral extent of 3 kHz, and power level equal to its reference level. The system sample rate per message is 64 kHz and channel pulse duration is $\frac{1}{5}$ of a time slot duration. (a) Find $(S_o/N_g)_{D\text{-}SM}$. (b) Calculate $(S_o/N_o)_{D\text{-}SM}$ and determine if this system is operating above or below threshold when $(\hat{S_i}/N_i)_{D\text{-}SM} = 8$ at the receiver. Assume negligible overload noise.

4.69. What maximum dynamic range improvement can be achieved in an adaptive delta modulation system if the step size can be reduced from its maximum by a factor of 20? What is the improvement for factors of 40, 80, and 160?

4-70. Repeat the computations of the text that lead to Fig. 4.13-2 except assume step size is given by

$$
\frac{\delta v}{\delta v_r} =
\begin{cases}
\dfrac{\delta v_0}{\delta v_r} + \left(1 - \dfrac{\delta v_0}{\delta v_r}\right) \dfrac{\overline{f^2(t)}}{\overline{f_r^2(t)}}, & \overline{f^2(t)} \leqslant \overline{f_r^2(t)} \\[2ex]
1, & \overline{f^2(t)} > \overline{f_r^2(t)}.
\end{cases}
$$

Plot curves only for $\overline{f^2(t)} \leqslant \overline{f_r^2(t)}$ so that the slope overload term of (4.13-9) can be ignored.

4-71. Use (4.14-9) and develop a similar expression for a DM system to show that (4.14-10) is true if the two systems have equal channel bandwidths, values of $\dot{K}$, and use the same message. Assume both systems use the largest allowable pulse durations.

4-72. Performance of a DM system when a particular message is used is $(S_o/N_g)_{DM}$ = 6310.0 (or 38.0 dB). When the same message is sent over a DPCM system with the same channel bandwidth and value of $\dot{K}$ as the DM system, its performance is $(S_o/N_g)_{DPCM} = 400,691.2$. How many bits does the DPCM system use per sample?

4-73. By using (4.14-11) show that the value of x which optimizes $(S_o/N_q)_{DPCM}$ for a given ratio ω_s/W_f is given by

$$
\frac{1}{x_{opt}} = \ln\left[\left(\frac{2^{N_b} - 1}{\sqrt{2}}\right)^{2/3} \left(\frac{\omega_s}{W_f}\right)\right].
$$

Chapter 5

Bandpass Binary Digital Systems

5.0 INTRODUCTION

Communication systems often involve modulation of a carrier, which results, of course, in a bandpass waveform. Radio and television signals are good examples involving analog messages. A good example where the message is digital is the *modem*, a device used to connect a remote computer terminal to the main computer. This *modulation-dem*odulation apparatus modulates a carrier with the terminal's data stream for transmission to the computer (often over telephone lines) and recovers the data stream sent by the computer via a similar modulation.

In this chapter we shall examine several methods by which carriers are modulated by digital data. Attention is restricted only to binary systems. Various *M*-ary bandpass digital systems are discussed in Chap. 6. Initially we consider coherent systems. Some reflection on the reader's part will show that the theoretical developments of Secs. 4.1–4.3 apply also to coherent bandpass systems, and we shall make use of these facts through reference to the earlier work. Those readers who may have omitted the earlier theory (as indicated in Sec. 4.0) can still profit from the results to be given but may wish to review the theory if a more detailed understanding is needed of how the results are achieved.

5.1 OPTIMUM COHERENT BANDPASS SYSTEMS

As noted above, the developments of Secs. 4.1–4.3 for optimum binary systems also apply to bandpass systems, although they were applied only to baseband systems in Chap. 4. In this section we summarize the earlier work for convenient reference.

For two possible messages, labeled m_1 and m_2 that have probabilities $P_1 = P(m_1)$ and $P_2 = P(m_2)$, respectively, one of two possible waveforms, $s_1(t)$ or $s_2(t)$, is transmitted in a given symbol (bit) interval of duration T_b. Which is transmitted depends, respectively, on which message, m_1 or m_2, occurs in the symbol (bit) interval. The waveforms are arbitrary except they are assumed real and nonzero only in their symbol interval. They are transmitted over an infinite-bandwidth channel having white noise, $n_w(t)$, with two-sided power density $\mathcal{N}_0/2$ at the receiver. The composite waveform at the receiver becomes

$$r(t) = \begin{cases} s_1(t) + n_w(t), & m_1 \text{ sent} \\ s_2(t) + n_w(t), & m_2 \text{ sent.} \end{cases} \tag{5.1-1}$$

The optimum receiver is that which decides m_2 was transmitted if its probability of being sent [given the observed receiver input waveform $r(t)$ occurred] is larger than the probability that m_1 was sent (given the same observed waveform). The optimum receiver makes this decision during each and every symbol interval by performing the following test:

$$\text{If } \int_0^{T_b} r(t)[s_2(t) - s_1(t)]\, dt > V_T, \qquad \text{choose } m_2,$$
$$\text{otherwise choose } m_1. \tag{5.1-2}$$

Since the exact forms of $s_1(t)$ and $s_2(t)$ are known to the receiver, it is called *coherent*.† The receiver does not know *which* waveform is being received in any given symbol interval, however.

In (5.1-2) V_T is the optimum threshold that separates the regions of decision; it is given by

$$V_T = \frac{E_2 - E_1}{2} + \frac{\mathcal{N}_0}{2} \ln\left(\frac{P_1}{P_2}\right), \tag{5.1-3}$$

where the energies in the received waveforms are

$$E_i = \int_0^{T_b} s_i^2(t)\, dt, \qquad i = 1 \text{ and } 2. \tag{5.1-4}$$

The structure of the optimum coherent receiver follows the implementation of the left side of (5.1-2) as illustrated in Fig. 4.2-1 of Chap. 4. For convenience of reference the figure is reproduced here as Fig. 5.1-1(a). An equivalent matched filter form is shown in (b) as taken from Fig. 4.2-2.

---

† For simplicity of discussion we assume an idealized channel that does not act to cause the received waveforms to differ from those transmitted. Practical channels alter the transmitted signal's phase and time of arrival. A coherent system must also compensate for these effects through carrier and bit-timing synchronization. In this book we assume, for the most part, that synchronization has been accomplished, so our signal model is valid.

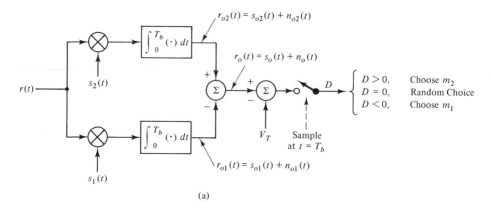

(a)

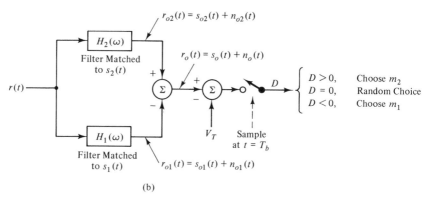

(b)

Figure 5.1-1.   Optimum coherent binary receivers. (a) Correlator form and (b) matched filter form.

Output Signal Levels and Noise Power

Let $r_o(T_b)$ denote the receiver's computation of the left side of (5.1-2) and let $s_o(T_b)$ and $n_o(T_b)$ represent, respectively, the output signal and noise at the symbol decision time (end of interval); then

$$r_o(T_b) = s_o(T_b) + n_o(T_b). \qquad (5.1-5)$$

In Chap. 4 we found that

$$s_o(T_b) = \begin{cases} V_1 = \gamma\sqrt{E_1 E_2} - E_1, & m_1 \text{ sent} \\ \\ V_2 = E_2 - \gamma\sqrt{E_1 E_2}, & m_2 \text{ sent} \end{cases} \qquad (5.1-6)$$

where

$$\gamma = \frac{1}{\sqrt{E_1 E_2}} \int_0^{T_b} s_1(t)s_2(t)\, dt \qquad (5.1-7)$$

represents the (normalized) correlation between the two waveforms $s_1(t)$ and $s_2(t)$. Output average noise power, denoted by σ_o^2, is

$$\sigma_o^2 = \overline{n_o^2(T_b)} = \left(\frac{N_0}{2}\right)[E_2 + E_1 - 2\gamma\sqrt{E_1 E_2}]. \tag{5.1-8}$$

Probability of Error

In general, (4.3-5) gives the average probability of bit errors, denoted by P_e. However, in many practical situations the simplifying assumption $P_1 = P_2 = \frac{1}{2}$ can be made, so that

$$P_e = \frac{1}{2}\,\text{erfc}\left[\sqrt{\frac{E_2 + E_1 - 2\gamma\sqrt{E_1 E_2}}{4N_0}}\right]. \tag{5.1-9}$$

If, in addition, energies are equal so $E = E_1 = E_2$, then

$$P_e = \frac{1}{2}\,\text{erfc}\left[\sqrt{\frac{E(1 - \gamma)}{2N_0}}\right]. \tag{5.1-10}$$

Finally, for antipodal signals where $s_1(t) = -s_2(t)$, we have $\gamma = -1$ and (5.1-10) reduces to

$$P_e = \left(\frac{1}{2}\right)\text{erfc}\left[\sqrt{\frac{E}{N_0}}\right]. \tag{5.1-11}$$

In the next few sections we consider several specific types of coherent bandpass systems for which the above optimum results apply.

5.2 COHERENT AMPLITUDE SHIFT KEYING

In binary *amplitude shift keying* (ASK), the amplitude of a carrier is switched between two levels, usually the extremes of full *on* and full *off*. As a result, ASK is sometimes referred to as *on-off keying* (OOK). The on condition might typically correspond to a code **1**, whereas off corresponds to a code **0**. In this section we examine the OOK version of ASK assuming carrier pulses of rectangular envelope with no spacing between pulses and equal-probability messages ($P_1 = P_2 = \frac{1}{2}$).

System Implementations

Figure 5.2-1(a) depicts an optimum binary ASK system (OOK variety) as derived from Fig. 5.1-1(a). The transmitter consists of a stream of data digits that modulates a carrier to generate the signals

$$s_{ASK}(t) = \begin{cases} s_1(t) = 0, & 0 \leq t \leq T_b, \text{ binary } 0 \\ s_2(t) = A\cos(\omega_0 t + \theta_0), & 0 \leq t \leq T_b, \text{ binary } 1. \end{cases} \tag{5.2-1}$$

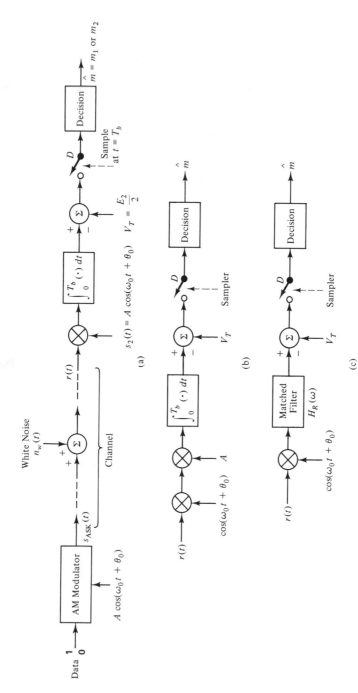

Figure 5.2-1. Block diagrams of (a) an optimum coherent ASK system, (b) an equivalent receiver, and (c) a matched filter receiver equivalent to that in (b).

The energy in $s_2(t)$ over one interval is

$$E_2 = \frac{A^2 T_b}{2} \tag{5.2-2}$$

whereas the average energy per bit divided by twice the channel noise density is

$$\varepsilon = \left(\frac{A^2 T_b}{2} \cdot \frac{1}{2} + 0 \cdot \frac{1}{2}\right)\frac{1}{\mathcal{N}_0} = \frac{A^2 T_b}{4\mathcal{N}_0}. \tag{5.2-3}$$

The receiver integrates the product $r(t)s_2(t)$ over each symbol interval. At the end of each interval, a sample is taken of the difference between the integrator's output and the threshold level V_T. If the difference D is positive, the decision is that m_2 was transmitted in the interval; if $D < 0$ the decision is in favor of m_1, but $D = 0$ leads to an arbitrary (random) choice of m_1 or m_2 (ideally, this event has zero probability of happening).

The receiver of Fig. 5.2-1(b) is clearly equivalent to that of (a). From earlier work, the product involving A followed by the integrator can be replaced by a filter matched to the *envelope* of $s_2(t)$, as shown in (c); this form of receiver can be viewed as a matched form of coherent detector. The required filter transfer function from (4.2-17) is (also see Prob. 5-7).

$$H_R(\omega) = AT_b\, e^{-j\omega T_b/2}\mathrm{Sa}\left(\frac{\omega T_b}{2}\right). \tag{5.2-4}$$

In the general theory leading to the system of Fig. 5.2-1(a), it was tacitly assumed that $s_2(t)$ was the same function regardless of what interval m_2 occurred. This condition is guaranteed if the coherent carrier's frequency $\omega_0/2\pi$ is a multiple of the bit rate $1/T_b$, regardless of the phase angle θ_0. However, the theory really requires only that the receiver know $s_2(t)$ in the interval it is used [$s_2(t)$ can vary from interval to interval as long as the receiver variations follow]. These facts mean that, if the local carrier is coherent, we only require $\omega_0/2\pi$ to be a multiple of *half* the bit rate. Even this condition can be ignored if $\omega_0/2\pi \gg 1/T_b$ and the local carrier is coherent. In the last case the receiver must still generate a locally coherent carrier, but the transmitter frequency can be arbitrary so long as it is much larger than the bit rate.

ASK Noise Performance

Since $s_1(t) = 0$, $E_1 = 0$ and $\gamma = 0$. From (5.1-9) error probability is

$$P_e = \left(\frac{1}{2}\right)\mathrm{erfc}\left[\sqrt{\frac{E_2}{4\mathcal{N}_0}}\right] = \left(\frac{1}{2}\right)\mathrm{erfc}\left[\sqrt{\frac{A^2 T_b}{8\mathcal{N}_0}}\right]. \tag{5.2-5}$$

In terms of ε, an alternative form is

$$P_e = \left(\frac{1}{2}\right)\mathrm{erfc}\left[\sqrt{\frac{\varepsilon}{2}}\right]. \tag{5.2-6}$$

This function is plotted in Fig. 5.12-1 for comparison with other systems. Another useful form for P_e results from defining peak signal power and average noise power at the sampler's input at the sample time by

$$\hat{S}_i \triangleq (V_2 - V_T)^2 = \frac{E_2^2}{4} \qquad (5.2\text{-}7)$$

$$N_i \triangleq \sigma_o^2 = \frac{\mathcal{N}_0 E_2}{2}, \qquad (5.2\text{-}8)$$

respectively. We have

$$P_e = \frac{1}{2}\,\text{erfc}\left[\sqrt{\frac{1}{2}\left(\frac{\hat{S}_i}{N_i}\right)_{\text{ASK}}}\right] \qquad (5.2\text{-}9)$$

where

$$\left(\frac{\hat{S}_i}{N_i}\right)_{\text{ASK}} = \frac{(V_2 - V_T)^2}{\sigma_o^2} = \frac{A^2 T_b}{4\mathcal{N}_0} = \varepsilon. \qquad (5.2\text{-}10)$$

Example 5.2-1

A binary ASK system for equally probable messages uses 100-μs bits and a channel for which $\mathcal{N}_0 = 1.338(10^{-5})$ W/Hz. We find the peak transmitter pulse amplitude that will produce a bit error probability $P_e = 2.055(10^{-5})$. From (5.2-6) erfc $[\sqrt{\varepsilon/2}] = 4.11(10^{-5})$, which occurs with $\sqrt{\varepsilon/2} = 2.9$ (from Appendix F). Thus $\varepsilon = 2(2.9)^2 = 16.82$. From (5.2-3)

$$A = 2\sqrt{\frac{\mathcal{N}_0\varepsilon}{T_b}} = 2\sqrt{1.338(10^{-5})\frac{16.82}{10^{-4}}} = 3.0 \text{ V}.$$

Note that this system has an input peak signal-to-noise ratio of

$$\left(\frac{\hat{S}_i}{N_i}\right)_{\text{ASK}} = \varepsilon = 16.82 \quad \text{(or 12.26 dB)}.$$

Signal Power Spectrum and Bandwidth

The transmitted ASK signal is equivalent to a carrier $\cos(\omega_0 t + \theta_0)$ multiplied by a unipolar pulse format. Since the power spectrum of the unipolar pulse stream was found in (3.7-10), we easily use the result of Prob. B-54 to obtain the power spectrum of the product:

$$\mathscr{S}_{\text{ASK}}(\omega) = \left(\frac{\pi A^2}{8}\right)[\delta(\omega - \omega_0) + \delta(\omega + \omega_0)]$$
$$+ \left(\frac{A^2 T_b}{16}\right)\left\{\text{Sa}^2\left[\frac{(\omega - \omega_0)T_b}{2}\right] + \text{Sa}^2\left[\frac{(\omega + \omega_0)T_b}{2}\right]\right\}. \qquad (5.2\text{-}11)$$

The bandwidth of this spectrum between first nulls on each side of the carrier is $4\pi/T_b = 2\omega_b$, twice the bit rate.

Local Carrier Generation for Coherent ASK

The coherent system requires a local carrier having the same phase as the incoming carrier wave. The need for generating this local signal is a main disadvantage of the coherent system. Conceptually, all we require to generate the local signal is a very narrowband bandpass filter centered at frequency ω_0 followed by an amplifier-limiter, as shown in Fig. 5.2-2(a). Because the ASK waveform will contain a large amount of carrier, a filter can select out the carrier and reject information sidebands if its bandwidth is small enough. An amplifier-limiter is then used only to provide a fixed-level output. Implementation of the method may be difficult in practice owing to the need for the very narrow bandwidth filter at large center frequencies. A more practical method would be to first mix down to a lower frequency, use a filter-amplifier-limiter, and then mix back up to the carrier frequency, as illustrated in (b). Still another scheme uses the phase-locked loop of (c); it behaves as a narrow filter at center frequency ω_0.

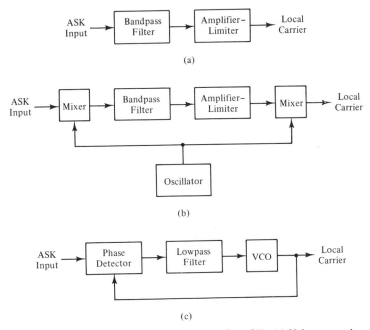

(a)

(b)

(c)

Figure 5.2-2.   Local carrier generation methods for ASK. (a) Using narrowband filter, (b) narrowband filter with mixers for center frequency shifting, and (c) the phase-locked loop [1].

5.3 PHASE SHIFT KEYING

In binary *phase shift keying* (PSK) the phase of a carrier is switched between two values according to the two possible messages m_1 and m_2.

The two phases are usually separated by π radians and this is the only case we consider; it is sometimes called *phase reversal keying* (PRK). Phase reversal keying corresponds to the two possible transmitter waveforms

$$s_{\text{PSK}}(t) = \begin{cases} s_1(t) = -A\cos(\omega_0 t + \theta_0), & 0 \leqslant t \leqslant T_b, & m_1 \text{ sent,} \\ s_2(t) = +A\cos(\omega_0 t + \theta_0), & 0 \leqslant t \leqslant T_b, & m_2 \text{ sent.} \end{cases} \quad (5.3\text{-}1)$$

The transmitted signal is equivalent to a double sideband suppressed-carrier amplitude-modulated waveform where the information signal is a digital waveform with polar format.

Optimum System

Figure 5.3-1 illustrates the block diagram of the optimum PSK system for $P_1 = P_2 = \frac{1}{2}$, as derived from Fig. 5.1-1(a). Because of the equal-energy condition $(E_1 = E_2)$ and the equal-message-probability assumption, threshold is zero from (5.1-3). From (5.3-1) we see that $s_1(t)$ and $s_2(t)$ are antipodal, so that $\gamma = -1$. The output signal levels at the sample time derive from (5.1-6):

$$s_o(T_b) = \begin{cases} V_1 = -2E_1 = -A^2 T_b, & m_1 \text{ sent} \\ V_2 = 2E_2 = A^2 T_b, & m_2 \text{ sent.} \end{cases} \quad (5.3\text{-}2)$$

From (5.1-8) the output noise power is

$$\sigma_o^2 = \mathcal{N}_0 A^2 T_b. \quad (5.3\text{-}3)$$

It is often assumed in the optimum PSK system that the transmitted carrier frequency is a multiple of the bit rate; that is, $\omega_0 = m\omega_b$, where $m = 1, 2, \ldots$. This assumption, as in ASK, makes the waveforms $s_1(t)$ and $s_2(t)$ identical in any intervals in which the respective messages m_1 and m_2 occur. However, it can be shown (Prob. 5-16) that the system of Fig. 5.3-1 remains optimum if ω_0 is a multiple of half the bit rate ($\omega_0 = m\omega_b/2$ for $m = 1, 2, \ldots$) provided the receiver's local carrier remains phase coherent with the transmitted carrier.

Spectral Properties of PSK

The binary PSK waveform of (5.3-1) can be considered to be a double sideband suppressed-carrier amplitude modulated signal (DSB) where the message is digital. The power spectrum of such a signal is comprised of scaled translations of the message's power spectrum, given by (3.7-11), to frequencies $\pm\omega_0$ according to Prob. B-54; we have

$$\mathcal{S}_{\text{PSK}}(\omega) = \frac{A^2 T_b}{4} \left\{ \text{Sa}^2 \left[(\omega - \omega_0)\frac{T_b}{2} \right] + \text{Sa}^2 \left[(\omega + \omega_0)\frac{T_b}{2} \right] \right\}. \quad (5.3\text{-}4)$$

Bandwidth between first nulls about the carrier is $2\omega_b$.

The spectrum of (5.3-4) contains no discrete components (impulses) at $\pm\omega_0$, which makes it difficult for the receiver to generate a coherent

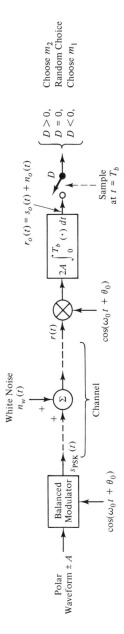

Figure 5.3-1. Block diagram of an optimum coherent binary PSK system for equal-probability messages.

253

local carrier. This problem is identical to that encountered in DSB systems, where Costas loops and squaring circuits are used to generate the local carrier. These techniques produce phase ambiguities of π radians, however, which must be resolved in the PSK application. The resolution can be accomplished by periodically transmitting a known signal for sign synchronization. Another technique differentially encodes the binary data prior to transmission; on reception the demodulated bit stream is then decoded [decoder shown in Fig. 3.6-3(c)].

Still another approach to local carrier generation is to carrier-modulate with phases not quite 180° apart. The transmitted spectrum would then contain a small discrete component that could be recovered by the methods used in ASK (Fig. 5.2-2). The disadvantage of this technique is an attendant increase in bit error probability.

Error Probability of PSK

Average probability of making a bit (symbol) error in the optimum PSK system is given by (5.1-11) with $E = A^2 T_b/2$:

$$P_e = \frac{1}{2}\text{erfc}\left(\sqrt{\frac{A^2 T_b}{2 \mathcal{N}_0}}\right). \tag{5.3-5}$$

Other forms result from substitution of

$$\varepsilon = \frac{A^2 T_b}{2 \mathcal{N}_0} \tag{5.3-6}$$

or

$$\left(\frac{\hat{S}_i}{N_i}\right)_{\text{PSK}} = \frac{V_1^2}{\sigma_o^2} = \frac{V_2^2}{\sigma_o^2} = \frac{A^2 T_b}{\mathcal{N}_0} = 2\varepsilon. \tag{5.3-7}$$

Thus,

$$P_e = \frac{1}{2}\text{erfc}\left(\sqrt{\varepsilon}\right) = \frac{1}{2}\text{erfc}\left[\sqrt{\frac{(\hat{S}_i/N_i)_{\text{PSK}}}{2}}\right]. \tag{5.3-8}$$

The behavior of (5.3-8) is illustrated in Fig. 5.12-1.

Example 5.3-1
Suppose we examine a PSK system that uses the same bit duration (100 μs), the same channel noise level [$\mathcal{N}_0 = 1.338(10^{-5})$ W/Hz], and the same peak amplitude of transmitter pulses ($A = 3.0$ V) as the ASK system of Example 5.2-1. Here

$$\varepsilon = \frac{(3)^2\, 10^{-4}}{2(1.338)10^{-5}} \approx 33.6323.$$

Since ε is large, we may use the approximation (F-5) for the complementary error function to get

$$P_e = \frac{e^{-\varepsilon}}{2\sqrt{\pi\varepsilon}} = 1.204(10^{-16}).$$

When this value of P_e is compared to $P_e = 2.055(10^{-5})$ for the ASK system, we find a vast improvement. However, if the PSK system had used the same *peak-to-peak* separation in the transmitted signal as the ASK system, we would have obtained *equal* values of P_e. The reader is urged to prove this fact as an exercise.

5.4 QUADRATURE PSK

System for Two Message Sources

Consider two PSK systems, one exactly as shown in Fig. 5.3-1 and a second that is identical to the first except that it uses a carrier $\sin(\omega_0 t + \theta_0)$ instead of $\cos(\omega_0 t + \theta_0)$. It can be shown (Prob. 5-22) that the two systems operate independently as long as ω_0 is an integral multiple of half the bit rate ($\omega_0 = m\omega_b/2$, $m = 1, 2, \ldots$). By combining the two into a single equivalent system, we form a means of doubling the rate at which bits are transmitted over the channel (on the same carrier). The resulting system is called *quadrature* PSK (QPSK).

The equivalent transmitted signal is the sum of the two separate waveforms†

$$s_{\text{QPSK}}(t) = \mp A \cos(\omega_0 t + \theta_0) - (\mp A)\sin(\omega_0 t + \theta_0) \qquad (5.4\text{-}1)$$

for $0 \le t \le T_b$. The factor $\mp A$ of the $\cos(\omega_0 t + \theta_0)$ term represents the data from one source, whereas the $\mp A$ on $\sin(\omega_0 t + \theta_0)$ represents data from the second source. In each case the upper sign goes with a binary **0**, and the lower sign corresponds to a **1**. It is readily shown that (5.4-1) can be written as

$$s_{\text{QPSK}}(t) = \sqrt{2}\, A \cos(\omega_0 t + \theta_0 + \theta_k), \qquad (5.4\text{-}2)$$

where θ_k can have values $\pm \pi/4$, and $\pm 3\pi/4$ according to the messages. The possible amplitudes and phases (θ_k) of $s_{\text{QPSK}}(t)$ are shown as points in Fig. 5.4-1. For example, if the message source modulating the cosine carrier is a binary **0** and that of the second source modulating the sine carrier is a **1**, then the corresponding polar waveform levels are $-A$ and A, respectively; the transmitted composite waveform peak amplitude is $\sqrt{2}\, A$, and the value of θ_k is $3\pi/4$. Diagrams like Fig. 5.4-1 are called *signal constellations*; more will be said about these in Chap. 6.

Figure 5.4-2 illustrates the QPSK system just described. Because the channel is actually supporting two PSK signals at the same carrier frequency and bandwidth, the spectral bandwidth of QPSK is the same as a single PSK system. Average probability of a bit's being in error in either of the two receiver channels is also the same as in PSK. From (5.3-5) we have

---

† For convenience we add the negative of the system using the sine carrier. The sign can be accounted for in the receiver.

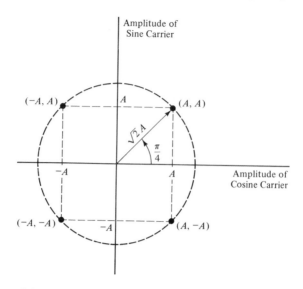

Figure 5.4-1. Possible amplitudes and phases of a QPSK waveform.

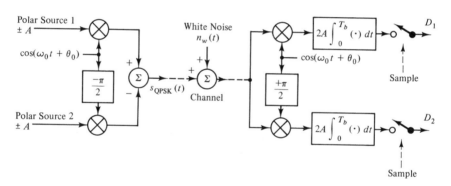

Figure 5.4-2. A QPSK system having two polar formatted message sources sharing the same channel and carrier frequency.

$$P_e = \frac{1}{2}\,\mathrm{erfc}\!\left[\sqrt{\frac{A^2 T_b}{2\,\mathcal{N}_0}}\right], \qquad \text{either message channel,} \qquad (5.4\text{-}3)$$

where T_b is, of course, the bit duration of the message polar format.

System for Single Message Source

Although the QPSK system just described assumed separate message sources at the same symbol (bit) rate, it is possible to have the system work with a single source at *twice* the bit rate† by using a *serial-to-parallel*

---

† T_b now is half T_b used before for each of the two sources.

converter. The functions performed by the transmitter modulator are shown in Fig. 5.4-3(a). The converter accepts the binary polar-NRZ formatted waveform with bit rate $f_b = 1/T_b$ and converts two bits at a time† into two new polar waveforms with "symbol" duration T, where

$$T = 2T_b. \quad\quad\quad (5.4\text{-}4)$$

Example waveforms are shown in Fig. 5.4-3(b). It is helpful to associate polar waveform 1 with the first bit of the 2-bit "words" and polar waveform 2 with the second bit; this convention was used in the example waveforms.

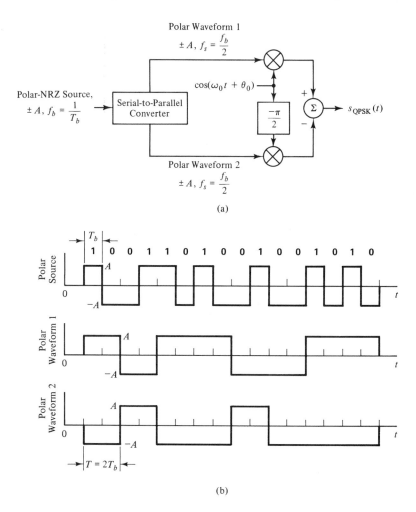

Figure 5.4-3. (a) A QPSK modulator for a single message source and (b) the applicable polar waveforms.

† A pair of bits is often referred to as a *dibit* [2].

The receiver is again two PSK channels in parallel that respond to their individual carrier component. Figure 5.4-4 illustrates the receiver. Processing channel 1 makes a decision (based on D_1) each symbol interval of duration T; these decisions correspond to the first bit of the 2-bit original message sequence during time T. Similarly, processing channel 2 is associated with the second bit. Since the two channels behave as independent PSK channels, their error probabilities are equal and give the bit error probability for the overall message data stream (at rate $1/T_b = 2/T$). From (5.3-5)

$$P_e = \frac{1}{2} \operatorname{erfc}\left(\sqrt{\frac{A^2 T}{2 \mathcal{N}_0}}\right) = \frac{1}{2} \operatorname{erfc}\left(\sqrt{\frac{A^2 T_b}{\mathcal{N}_0}}\right). \qquad (5.4\text{-}5)$$

Finally, the output data stream (bit rate of $1/T_b = f_b$) is generated in a parallel-to-serial converter using decisions made in the processor channels.

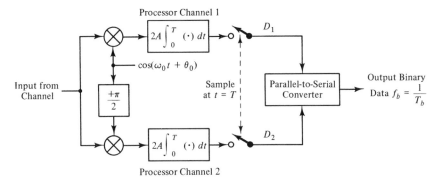

Figure 5.4-4. QPSK receiver functions for a single message source.

On casually comparing (5.4-5) with (5.3-5) it might appear that QPSK has lower P_e than PSK when both transmit the same message. An example will demonstrate that performances are actually the same.

Example 5.4-1
Let the message bit rate $(1/T_b)$ and channel noise levels $(\mathcal{N}_0/2)$ be the same when a fair comparison of bit error probability is to be made between PSK and QPSK systems. From (5.3-5) and (5.4-5) the two values of P_e will be the same if $A^2/2$ in the PSK system equals A^2 in the QPSK system. That this is indeed true can be seen from (5.3-1) and (5.4-2). The peak amplitude of the PSK *channel* pulse is A, so peak pulse power is $A^2/2$. The *channel* pulse in QPSK is $\sqrt{2}A$, so peak power is A^2. Clearly, a fair comparison of systems gives no power advantage to either. The equal-power condition, therefore, corresponds to equal values of P_e.

From the preceding example it is clear that the transmitted average energy per bit is $(\sqrt{2}\,A)^2 T_b/2 = A^2 T_b$, so (5.4-5) can be written as

$$P_e = \frac{1}{2} \operatorname{erfc}(\sqrt{\varepsilon}) \qquad (5.4\text{-}6)$$

where ε, as usual, is the ratio of average transmitted energy per bit to $\mathcal{N}_0$:

$$\varepsilon = \frac{A^2 T_b}{\mathcal{N}_0}. \tag{5.4-7}$$

The behavior of (5.4-6) is shown in Fig. 5.12-1.

Other Quadrature Modulations: QAM and OQPSK

In the QPSK systems of Figures 5.4-2 and 5.4-3(a), modulated carriers in phase quadrature were combined to form the output waveform. This process is called *quadrature modulation* (sometimes also called *quadrature multiplexing*). In QPSK the amplitudes of the modulating polar waveforms and modulator gains are made as nearly equal as possible. The resulting signal constellation has points located at the corners of a square (Fig. 5.4-1).

A generalization of quadrature modulation allows message polar waveforms to have unequal amplitudes. The signal constellation points now fall on the corners of a *rectangle*. The corresponding waveform is called *quadrature amplitude modulation* (QAM). Sometimes it is also known as *unbalanced* QPSK. Although there are some differences in where filters are placed in a practical system to establish spectral shaping [3], QAM and QPSK are quite similar. Indeed, the names are often used interchangeably in practice (equal-amplitude case).

The amplitude of a QPSK signal is ideally constant, a desirable property in one of its most important applications, a satellite link. Here the signal is bandlimited by a bandpass filter to conform to out-of-band spectral emission standards. The filtering degrades the constant-amplitude property of the QPSK signal. At the satellite repeater amplitude limiting restores this property but increases the out-of-band spectral levels again [4]. A modified form of QPSK, called *offset* QPSK (OQPSK), has been devised that is less susceptible to these effects.

The system of Fig. 5.4-3(a) becomes an OQPSK modulator if a delay of 1 bit (T_b) is added prior to the modulator in the line carrying polar waveform 2. The effect is to move the polar waveform 2 of (b) right by 1-bit duration T_b. The reader can no doubt recognize that the largest phase change is $\pm\pi/2$ per bit interval T_b, whereas the largest was $\pm\pi$ per *symbol* interval (2 bits) in QPSK. Thus phase changes are smaller and more frequent in OQPSK, which leads to smaller amplitude fluctuations following band-limiting. Even after amplitude limiting to remove fluctuations, tests [4, 5] indicate that the OQPSK spectrum remains nearly unchanged so that out-of-band spectral levels remain near their desired levels.

The receiver for OQPSK is a modified version of the QPSK receiver of Fig. 5.4-4. The lower integrator still integrates over each symbol interval T, but its intervals are delayed by T_b relative to the upper processor channel. Decision values D_1 then occur at times $T_b + T$, $T_b + 2T$, $T_b + 3T$, and so on, whereas D_2 occurs at times $2T$, $3T$, $4T$, and so on. Addition of a delay T_b will synchronize values of D_1 with values of D_2 for use in the converter. Since staggering bit streams does not change the properties of the quadrature modulation process, OQPSK has the same average probability of bit error as QPSK. Even in the presence of reference carriers with phase jitter, OQPSK seems to perform better than QPSK [4, 6]. Furthermore, the offset of T_b in the two bit streams in OQPSK has been shown to be optimum in terms of phase jitter immunity in the presence of additive Gaussian noise [4, 7].

Carrier Recovery in QAM and QPSK

Since QAM and QPSK are coherent systems, their receivers require locally generated coherent carriers. One method of generating these carriers uses the squaring loop shown in Fig. 5.4-5(a). It applies mainly to QAM, where it is used when more than 73% of transmitted power is due to one

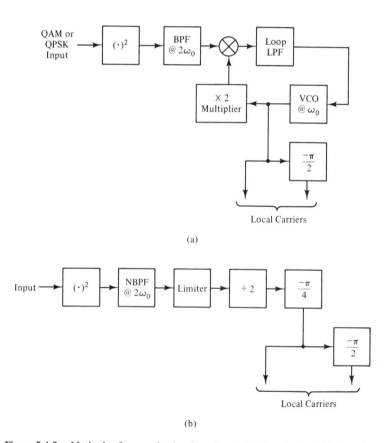

(a)

(b)

Figure 5.4-5. Methods of generating local carriers in QAM or QPSK. (a) Squaring loop and (b) an equivalent network not requiring a phase-locked loop.

of the quadrature components [8]. The phase-locked loop acts as a combined frequency divider and narrowband filter at frequency $2\omega_0$. In some applications the acquisition (lock-up) time of the loop is too long and the equivalent circuit of (b) is useful. It uses a passive narrowband filter followed by a divider.

When less than 73% of the total power is in the largest-power quadrature component in QAM, or when carrier recovery in QPSK is required, modified forms of the networks of Fig. 5.4-5 should be used [8]. Modifications involve replacing squares by fourth-power devices, $\div 2$ and $\times 2$ networks by $\div 4$ and $\times 4$ devices, and centering filters at $4\omega_0$. (See Prob. 5-26.) For constant-envelope infinite bandwidth input, a pure reference carrier can be recovered. However, in practice some bandlimiting occurs and the carrier contains some modulation-dependent interference called *pattern noise* [9]. These fourth-power nonlinear networks contain phase ambiguities at multiples of $\pi/2$ and some provision must be made to correct for their effects.

Another method of carrier recovery in QAM and QPSK uses a modified Costas loop, as illustrated in Fig. 5.4-6 [3, 10–12]. The network has phase ambiguities at multiples of $\pi/2$. (See Prob. 5-27.) Performance of Costas loops in noise can be optimized [13–14] by special selection of the filters at the carrier product device outputs or by using integrate-and-dump filters. The disadvantage of the latter approach is the need for symbol timing of the integrators.

Other methods of carrier recovery exist. We mention only one, called a *remodulator* [9, 15]. The received signal is first demodulated using quadrature versions of the local carrier, much as is done in the first stage of the receiver. The two message outputs are each lowpass filtered, symmetrically limited, and used to multiply (remodulate) quadrature versions of the input signal (which is delayed slightly at these multipliers to compensate for network delays). A summation of the product outputs (with appropriate sign) is bandpass filtered and limited to remove amplitude fluctuations. The final signal is the local carrier that was used in the start of the description.

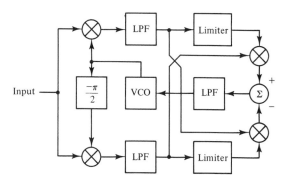

Figure 5.4-6.   Costas loop for local carrier recovery in QAM or QPSK systems.

When used with an interval of unmodulated carrier (called a preamble) for initial alignment, the remodulator does not contain a phase ambiguity [9, p. 140].

The reader is also referred to a special IEEE publication on synchronization topics [16].

5.5 COHERENT FREQUENCY SHIFT KEYING

In binary *frequency shift keying* (FSK), the frequency of the transmitted signal is switched between two values

$$\omega_1 = \omega_0 - \Delta\omega \tag{5.5-1}$$

$$\omega_2 = \omega_0 + \Delta\omega \tag{5.5-2}$$

according to messages m_1 and m_2 (for binary **0** and **1**), respectively. Here ω_0 represents a nominal carrier frequency, and $\Delta\omega$ represents a frequency deviation due to message modulation. Two principal versions of FSK can be defined based on how the frequency variations are imparted into the transmitted waveform. We consider both versions.

FSK Using Independent Oscillators

In one simple system FSK is generated by switching between two independent oscillators according to which message occurs, as shown in Fig. 5.5-1. In general, this form of FSK involves phase discontinuities at switching times and is not often implemented in practice. It is of historical interest, however, and does form a basis of comparison for other FSK systems. Thus although we consider this form of FSK, often called *discontinuous FSK*, our treatment is brief.

Transmitted waveforms involved are

$$s_{\text{FSK}}(t) = \begin{cases} s_1(t) = A\cos(\omega_1 t + \theta_1), & 0 \le t \le T_b, & m_1 \text{ sent} \\ s_2(t) = A\cos(\omega_2 t + \theta_2), & 0 \le t \le T_b, & m_2 \text{ sent.} \end{cases} \tag{5.5-3}$$

Here θ_1 and θ_2 are arbitrary phase angles and the transmitted energy per bit is

$$E_1 = E_2 = \frac{A^2 T_b}{2} \tag{5.5-4}$$

so

$$\varepsilon = \frac{E_1}{\mathcal{N}_0} = \frac{E_2}{\mathcal{N}_0} = \frac{A^2 T_b}{2\mathcal{N}_0}. \tag{5.5-5}$$

Receiver structure, received signal and noise levels, error probability, and threshold all derive from the work in Sec. 5.1. For simplicity, we assume equally probable messages [$P(m_1) = \frac{1}{2}$ and $P(m_2) = \frac{1}{2}$], even though

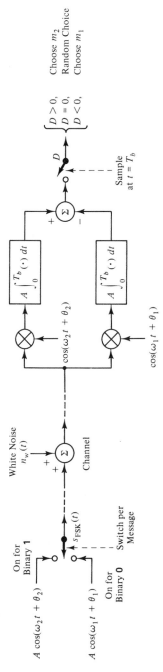

Figure 5.5-1. Block diagram for FSK system using independently switched oscillators.

the more general case could be stated. For convenience, we also assume ω_1 and ω_2 are positive integer multiples of the bit rate $\omega_b = 2\pi/T_b$. This condition guarantees that the forms of $s_1(t)$ and $s_2(t)$ will not change as a function of the bit interval in which they are to be transmitted. With these assumptions, γ is readily computed from (5.1-7) as follows:

$$
\begin{aligned}
\gamma &= \frac{1}{\sqrt{E_1 E_2}} \int_0^{T_b} s_1(t) s_2(t)\, dt \\
&= \frac{2}{T_b} \int_0^{T_b} \cos(\omega_1 t + \theta_1)\cos(\omega_2 t + \theta_2)\, dt \\
&= \mathrm{Sa}(\Delta\omega T_b)\cos(\Delta\omega T_b + \theta_2 - \theta_1) \\
&\quad + \mathrm{Sa}(\omega_0 T_b)\cos(\omega_0 T_b + \theta_2 + \theta_1).
\end{aligned}
\tag{5.5-6}
$$

Since $\omega_2 = M_2\omega_b$ and $\omega_1 = M_1\omega_b$, with M_1 and $M_2 > M_1$ both integers, we find from (5.5-1) and (5.5-2) that both $\Delta\omega T_b$ and $\omega_0 T_b$ are integral multiples of π. Hence

$$
\gamma = 0. \tag{5.5-7}
$$

This FSK system is called *orthogonal* because $\gamma = 0$.

After having established γ, other quantities readily follow. Receiver output signal levels and noise power, from (5.1-6) and (5.1-8), are

$$
s_o(T_b) = \begin{cases} V_1 = -A^2 T_b/2, & m_1 \text{ sent} \\ V_2 = A^2 T_b/2, & m_2 \text{ sent} \end{cases} \tag{5.5-8}
$$

$$
\sigma_o^2 = A^2 T_b \frac{\mathcal{N}_0}{2}. \tag{5.5-9}
$$

Equation (5.1-3) gives the optimum receiver threshold

$$
V_T = 0. \tag{5.5-10}
$$

The resulting receiver is shown in its correlation form in Fig. 5.5-1. An equivalent form using passive filters is also possible.

Finally, average bit error probability is

$$
P_e = \frac{1}{2}\operatorname{erfc}\left[\sqrt{\frac{A^2 T_b}{4\mathcal{N}_0}}\right] = \frac{1}{2}\operatorname{erfc}\left[\sqrt{\frac{\varepsilon}{2}}\right] \tag{5.5-11}
$$

from (5.1-10). This function is plotted in Fig. 5.12-1. In terms of peak signal power to average noise power (at the sampler) at the sample time, we have

$$
\hat{S}_i = V_1^2 = V_2^2 = \left(\frac{A^2 T_b}{2}\right)^2 \tag{5.5-12}
$$

$$
N_i = \sigma_o^2 = A^2 T_b \frac{\mathcal{N}_0}{2} \tag{5.5-13}
$$

$$\left(\frac{\hat{S}_i}{N_i}\right)_{\text{FSK}} = \frac{A^2 T_b}{2 \mathcal{N}_0} = \varepsilon \qquad (5.5\text{-}14)$$

$$P_e = \frac{1}{2} \operatorname{erfc}\left[\sqrt{\frac{1}{2}\left(\frac{\hat{S}_i}{N_i}\right)_{\text{FSK}}}\right]. \qquad (5.5\text{-}15)$$

Example 5.5-1

For comparison to the ASK system of Example 5.2-1, let an FSK system use the same values of A, T_b, and $\mathcal{N}_0$. Thus from (5.2-10), the previous example, and (5.5-14),

$$\left(\frac{\hat{S}_i}{N_i}\right)_{\text{ASK}} = 16.82 = \frac{1}{2}\left(\frac{\hat{S}_i}{N_i}\right)_{\text{FSK}}$$

From (5.5-15), $P_e = 0.5 \operatorname{erfc}[\sqrt{16.82}] \approx 3.41(10^{-9})$, which is smaller than P_e in the ASK system by a factor of 6026.4.

Example 5.5-1 demonstrates the superiority of FSK over ASK with equal peak signal amplitudes. If equal average energy becomes the basis of comparison, the two systems perform equally with noise.

The power density spectrum of the discontinuous FSK waveform is known [17, eq. (18)] but has a rather complicated form that depends intimately on phases θ_1 and θ_2. Generally, the power spectrum falls off as the reciprocal of the square of frequency for frequencies far from the carrier. It has two components, a continuous part arising from switching and a discrete component (with impulses) at the frequencies $\omega_0 \pm \Delta\omega$ and $-\omega_0 \pm \Delta\omega$. Half the total power is in the continuous part and half is in the discrete components. For the special case where θ_1 and θ_2 are assumed to be zero-mean independent random variables, uniform on $(0, 2\pi)$, and both ω_1 and ω_2 are integral multiples of the bit rate ω_b, the spectrum reduces to

$$\mathscr{S}_{\text{FSK}}(\omega) = \mathscr{S}(\omega) + \mathscr{S}(-\omega). \qquad (5.5\text{-}16)$$

where we define

$$\begin{aligned}
\mathscr{S}(\omega) = \frac{A^2 \pi}{8} & [\delta(\omega - \omega_0 - \Delta\omega) + \delta(\omega - \omega_0 + \Delta\omega)] \\
& + \frac{A^2 T_b}{16}\left\{ \operatorname{Sa}^2\left[\frac{(\omega - \omega_0 - \Delta\omega)T_b}{2}\right] \right. \\
& \left. + \operatorname{Sa}^2\left[\frac{(\omega - \omega_0 + \Delta\omega)T_b}{2}\right]\right\}.
\end{aligned} \qquad (5.5\text{-}17)$$

Figure 5.5-2 illustrates the behavior of $\mathscr{S}_{\text{FSK}}(\omega)$ for frequencies near ω_0 when (5.5-17) applies and $\Delta\omega = \omega_b/2$. Bandwidth between first nulls on either side of the carrier is $3\omega_b$.

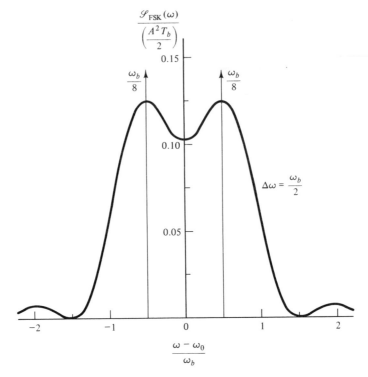

Figure 5.5-2. Power spectrum for discontinuous phase FSK, where θ_1 and θ_2 are independent uniformly distributed random variables on $(0, 2\pi)$.

Continuous Phase FSK

Another method of generating FSK is to frequency modulate a single oscillator by the message waveform. The resulting waveform is called *continuous phase* FSK (CPFSK). The applicable waveform is now

$$s_{\text{FSK}}(t) = \begin{cases} s_1(t) = A\cos(\omega_0 t - \Delta\omega t + \theta_0), & 0 \leq t \leq T_b, & m_1 \text{ sent}, \\ s_2(t) = A\cos(\omega_0 t + \Delta\omega t + \theta_0), & 0 \leq t \leq T_b, & m_2 \text{ sent}. \end{cases}$$

$$(5.5\text{-}18)$$

Again, if equal message probabilities are assumed, threshold, integrator output signal levels, output noise power, and bit error probabilities result from developments of Sec. 5.1:

$$V_T = 0 \qquad (5.5\text{-}19)$$

$$s_o(T_b) = \begin{cases} V_1 = -\dfrac{(1-\gamma)A^2 T_b}{2}, & m_1 \text{ sent} \\[2mm] V_2 = \dfrac{(1-\gamma)A^2 T_b}{2}, & m_2 \text{ sent} \end{cases}$$

$$(5.5\text{-}20)$$

$$\sigma_o^2 = \frac{(1 - \gamma)\,\mathcal{N}_0 A^2 T_b}{2} \tag{5.5-21}$$

$$P_e = \frac{1}{2}\operatorname{erfc}\left[\sqrt{\frac{(1 - \gamma)A^2 T_b}{4\,\mathcal{N}_0}}\,\right]. \tag{5.5-22}$$

The applicable optimum receiver is identical to that shown in Fig. 5.5-1. Of course, the transmitter is now just an FM oscillator having a polar modulating waveform.

On use of the definition of γ, (5.1-7), we find that

$$\gamma = \frac{2}{T_b} \int_0^{T_b} \cos(\omega_0 t - \Delta\omega t + \theta_0)\cos(\omega_0 t + \Delta\omega t + \theta_0)\,dt \tag{5.5-23}$$
$$= \operatorname{Sa}(2\Delta\omega T_b) + \operatorname{Sa}(\omega_0 T_b)\cos(\omega_0 T_b + 2\theta_0).$$

The second term in (5.5-23) is zero if $\omega_0 T_b$ is an integral multiple of π; alternatively, if $\omega_0 T_b \gg 1$ the term is negligible, a condition often true in practice. We assume one of these two conditions is true, so that

$$\gamma = \operatorname{Sa}(2\Delta\omega T_b) \tag{5.5-24}$$

in the remainder of this section. We shall consider two choices of γ.

First, suppose we make the FSK system orthogonal, as resulted with independent oscillators. Orthogonality requires $\gamma = 0$ which, in turn, requires $2\Delta\omega T_b$ to be an integral multiple of π in (5.5-24):

$$\Delta\omega = \frac{n\pi}{2T_b} = \frac{n\omega_b}{4}, \qquad n = 1, 2, \ldots. \tag{5.5-25}$$

Since we are dealing with frequency modulation of peak frequency deviation $\Delta\omega$, we expect the bandwidth of the FSK waveform to be smallest when $\Delta\omega$ is smallest. In other words, a spectrally efficient waveform follows the choice $n = 1$ in (5.5-25). The FSK system with $n = 1$, where $\Delta\omega = \omega_b/4$, has minimum separation between frequencies for orthogonality and is called *minimum shift keying* (MSK). We examine MSK in greater detail in the next section.

As a second choice, we find the value of γ that is most negative because it leads to the smallest value of P_e from (5.5-22). The largest negative value of γ occurs where $2\Delta\omega T_b \approx 3\pi/2$ in (5.5-24). It is $\gamma \approx -0.21$, which gives

$$P_e = \frac{1}{2}\operatorname{erfc}\left(\sqrt{\frac{1.21 A^2 T_b}{4\,\mathcal{N}_0}}\right) \tag{5.5-26}$$
$$= \frac{1}{2}\operatorname{erfc}\left(\sqrt{\frac{1.21\varepsilon}{2}}\right)$$

where

$$\varepsilon = \frac{A^2 T_b}{2\,\mathcal{N}_0}. \tag{5.5-27}$$

The factor 1.21 represents a performance improvement over orthogonal FSK of 0.83 dB.

A popular application of coherent FSK is in keyboard-type computer terminal modems that connect the terminal to a main computer, often by dial-up telephone lines. After the telephone connection is made by dialing the main computer, some modems will allow simultaneous two-way data communication, called *full-duplex* operation. To accomplish full-duplex, separate transmit and receive carrier frequencies are used. A *half-duplex* terminal can either transmit (talk) or receive (listen) at any one time, but not both. Some modems operate in a *simplex mode* where they *only* talk or listen. An example of a typical modem will illustrate some of the ideas previously discussed.

Example 5.5-2
The AT&T model 103 FSK modem can operate full-duplex using transmit and receive carrier frequencies of 1170 Hz and 2125 Hz, respectively.† The bit rate is 300 bits/s and $\Delta\omega/2\pi = 100$ Hz. For this modem $\omega_0 T_b = 2\pi(1170)/300 = 7.8\pi \gg 1$, so (5.5-24) applies. We calculate

$$\gamma = \text{Sa}(2\Delta\omega T_b) = \text{Sa}\left[\frac{4\pi(100)}{300}\right] = \text{Sa}\left(\frac{4\pi}{3}\right) = -0.207.$$

This value of γ is near the optimum value. We find P_e from (5.5-22) assuming $A = 0.5$ V and $\mathcal{N}_0 = 2.012(10^{-5})$ W/Hz:

$$P_e = \frac{1}{2}\text{erfc}\left[\sqrt{\frac{1.207(0.25)}{300(4)2.012(10^{-5})}}\right]$$

$$= \frac{1}{2}\text{erfc}[\sqrt{12.498}] \approx 2.98(10^{-7}).$$

Note that this value of P_e corresponds to a peak signal-to-noise power ratio at the sampler at the sample time of

$$\left(\frac{\hat{S}_i}{N_i}\right)_{\text{FSK}} = \frac{V_1^2}{\sigma_o^2} = \frac{V_2^2}{\sigma_o^2} = \frac{(1-\gamma)A^2 T_b}{2\mathcal{N}_0} = \frac{1.207(0.25)}{300(2)2.012(10^{-5})} = 25.0,$$

or 13.98 dB.

Power Spectrum of CPFSK

Anderson and Salz [18] and Bennett and Rice [17] have found the power density spectrum of the CPFSK waveform of (5.5-18). Because the analysis is quite involved, we only summarize the principal findings. The power spectrum $\mathcal{S}_{\text{CPFSK}}(\omega)$, valid for $-\infty < \omega < \infty$, can be written [17, eq. (48)] as

$$\mathcal{S}_{\text{CPFSK}}(\omega) = \mathcal{S}(\omega) + \mathcal{S}(-\omega) \tag{5.5-28}$$

† Frequencies apply to one of a pair of matched modems; the second modem's frequencies are 2125 Hz for transmitting and 1170 Hz for receiving.

where

$$\mathscr{S}(\omega) = \frac{A^2 T_b}{4} \cdot \frac{(\Delta\omega T_b)^2 \, \mathrm{Sa}^2[\gamma_1(\omega)] \, \mathrm{Sa}^2[\gamma_2(\omega)]}{\{1 + C_a^2 - 2 \, C_a \cos[(\omega - \omega_0)T_b]\}} \tag{5.5-29}$$

$$\gamma_1(\omega) = (\omega - \omega_0 + \Delta\omega)\frac{T_b}{2} \tag{5.5-30}$$

$$\gamma_2(\omega) = (\omega - \omega_0 - \Delta\omega)\frac{T_b}{2} \tag{5.5-31}$$

$$C_a = \cos(\Delta\omega T_b). \tag{5.5-32}$$

The spectrum of (5.5-28) applies rather generally. However, there are certain special cases that must be separately considered; these are subsequently discussed. Examples of power spectrum behavior are shown in Fig. 5.5-3. The curves were computed from (5.5-28), using (5.5-29).† For $\Delta\omega/\omega_b$ not greater than about 0.3, the spectrum is smooth and single-peaked at the carrier. As $\Delta\omega/\omega_b$ increases in the range $0.3 < \Delta\omega/\omega_b < 0.5$, the curves develop two increasingly sharp and large peaks at frequencies of magnitudes $|\omega - \omega_0| < \omega_b/2$. For $\Delta\omega/\omega_b > 0.5$ these peaks move outside frequencies $\omega_0 \pm (\omega_b/2)$ and decrease in sharpness and amplitude as $\Delta\omega/\omega_b$ increases. These spectrums are all continuous and have no discrete (impulse) components.

Bennett and Rice [17] have shown that three special cases arise in determining CPFSK power spectrum. These are summarized in Table 5.5-1. In cases I and II, the power spectrum is given by

$$\mathscr{S}_{\mathrm{CPFSK}}(\omega) = \frac{A^2 \pi}{8} [\delta(\omega - \omega_0 - \Delta\omega) + \delta(\omega - \omega_0 + \Delta\omega)$$

$$+ \delta(\omega + \omega_0 - \Delta\omega) + \delta(\omega + \omega_0 + \Delta\omega)] + \frac{1}{2}w_v(f), \tag{5.5-33}$$

where $w_v(f)$ is determined by equations of [17] as referenced in the table.‡ In a similar manner, for case III we have

$$\mathscr{S}_{\mathrm{CPFSK}}(\omega) = \frac{1}{2} w_u(f) \tag{5.5-34}$$

where $w_u(f)$ is given by equation (67) in [17]. We note that cases I and II result in impulses in the power spectrum while the spectrum of cases III and IV are entirely continuous. In all cases the spectrum decreases as the inverse fourth power of frequency for frequencies remote from the carrier; this behavior is typical of coherent phase FSK waveforms using rectangular bit pulses.

---

† The curves were generated from the term in (5.5-28) involving $\mathscr{S}(\omega)$. The term involving $\mathscr{S}(-\omega)$ is significant mainly at frequencies near $-\omega_0$ when $\omega_0 \gg \omega_b$.
‡ All our power spectrums are two-sided; those in [17] are one-sided.

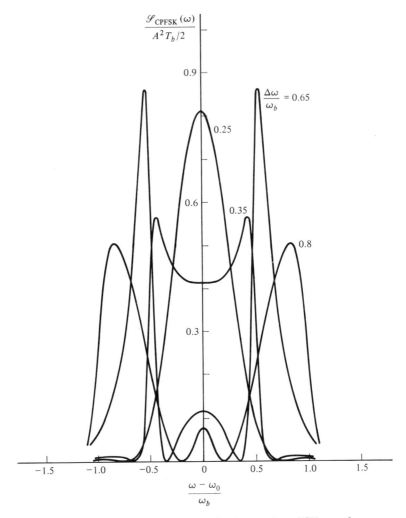

Figure 5.5-3. Power spectrums of coherent phase FSK waveforms.

TABLE 5.5-1. Special Cases in CPFSK Power Spectrums.

Case	Defining Conditions for Positive Integers r and l	Applicable Equations
I	$\Delta\omega/(\omega_b/4) =$ even integer $= 2r$ and $\omega_0/(\omega_b/4) \neq$ even integer $= 2l$	(53) or (54) in [17]
II	$\Delta\omega/(\omega_b/4) =$ even integer $= 2r$ and $\omega_0/(\omega_b/4) =$ even integer $= 2l$	(56) in [17]
III	$\Delta\omega/(\omega_b/4) =$ odd integer $= (2r - 1)$ and $\omega_0/(\omega_b/4) =$ odd integer $= (2l - 1)$	(67) in [17]
IV	All other continuous phase cases	(5.5-28), (5.5-29)

5.6 MINIMUM SHIFT KEYING

As noted earlier, minimum shift keying (MSK) is the name given to CPFSK when keying is between two frequencies separated by half the data bit rate. These frequencies are $\omega_1 = \omega_0 - \Delta\omega$ and $\omega_2 = \omega_0 + \Delta\omega$, where $\pm\Delta\omega$ are the frequency deviations caused by the digital data waveform (a polar NRZ format with amplitude ± 1). Thus MSK is defined by

$$\Delta\omega = \frac{\omega_b}{4} = \frac{\pi}{2T_b} \tag{5.6-1}$$

where T_b is the duration of a typical data bit and $\omega_b = 2\pi/T_b$.

One other characteristic of MSK serves to separate it from what we called CPFSK in the preceding section. The earlier continuous phase FSK system did not fully utilize all the phase information present in the received waveform, although it used enough to provide synchronization of the receiver to the transmitter through the local oscillator's phase. By making better use of phase information the noise performance of MSK can be better than the earlier CPFSK system.

MSK waveforms can be divided broadly into two categories of implementation, parallel and serial. Parallel MSK [4, 19–22] involves an implementation much like offset quadrature PSK (OQPSK). Serial MSK [23–26] has a somewhat simpler implementation than parallel and is sometimes preferred in high data rate systems. We subsequently discuss all these MSK implementations. However, it is first helpful to develop some properties of the general CPFSK waveform that help illustrate the ideas on which practical MSK systems are based.

CPFSK Signal Decomposition

Let $d(t)$ represent a polar-NRZ waveform representing a binary data source. Bits are assumed to occur independently each T_b seconds with equal probability. We let the sequence d_k represent the amplitudes, assumed to be $+1$ for a data **1** and -1 for a **0**, in the various bit intervals indexed by k, $k = \ldots, -1, 0, 1, 2, \ldots$. We assume the time origin occurs at the start of the zeroth interval ($k = 0$). These definitions are illustrated by Fig. 5.6-1(a). In interval k the frequency of the CPFSK waveform will be† $\omega_0 + \Delta\omega = \omega_0 + (\pi/2T_b)$ if $d_k = 1$ and $\omega_0 - (\pi/2T_b)$ if $d_k = -1$. Thus we have

$$s_{\text{CPFSK}}(t) = A \cos[\omega_0 t + \theta(t)]$$
$$= A \cos\left[\omega_0 t + \theta_k + \frac{d_k \pi t}{2T_b}\right], \qquad kT_b \le t \le (k + 1)T_b. \tag{5.6-2}$$

---

† We describe only the case where $\Delta\omega = \pi/2T_b$. Thus the CPFSK waveform described is the MSK waveform.

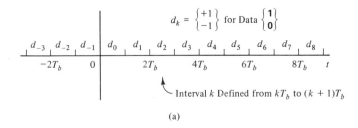

$$d_k = \begin{Bmatrix} +1 \\ -1 \end{Bmatrix} \text{ for Data } \begin{Bmatrix} 1 \\ 0 \end{Bmatrix}$$

Interval k Defined from kT_b to $(k+1)T_b$

(a)

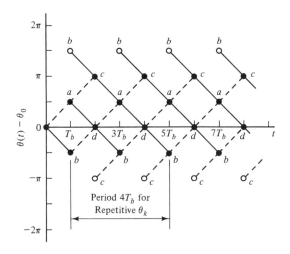

Node at Interval Start	d_k	$\theta_k - \theta_0$ for Value of k Shown								
		1	2	3	4	5	6	7	8	9
a	+1	0		π		0		π		0
	−1	π		0		π		0		π
b	+1	π		0		π		0		π
	−1	0		π		0		π		0
c	+1		0		π		0		π	
	−1		0		π		0		π	
d	+1		π		0		π		0	
	−1		π		0		π		0	

(b)

Figure 5.6-1. (a) Data interval definitions in CPFSK. (b) Trellis and table of $\theta_k - \theta_0$ to describe phase history of a CPFSK waveform.

If we think of $\theta(t)$ as the phase due to data modulation at the constant frequency $d_k\pi/2T_b$, then θ_k represents the phase at $t = 0$. Since d_k can change from interval to interval, θ_k can be different for each interval.

The phase history of $\theta(t)$ is best illustrated by a trellis diagram, as shown in Fig. 5.6-1(b) for $t \geq 0$. Plotted is $\theta(t) - \theta_0$ so that the function is zero at $t = 0$. In the zeroth interval, if $d_0 = 1$, frequency is larger than ω_0 by $\Delta\omega = 2\pi/4T_b$, so phase increases linearly (dashed line) by an amount $\Delta\omega T_b = \pi/2$ to arrive at the point marked a. If $d_0 = -1$ the solid line would lead to point b. From either of these points, the phase would again either increase or decrease linearly by another $\pm\pi/2$ according to d_1 being ± 1 in the interval. From a, the paths lead to points marked by heavy dots as c or d. From b, we either arrive at point d (for $d_1 = 1$) or c (the open dot). However, since phase is a modulo-2π function we have shown only four main phase corners (π, $\pi/2$, 0, $-\pi/2$) by heavy dots and auxiliary open points b or c to handle the extremes of the 2π interval. By proceeding in the same manner in other intervals, the phase history of $\theta(t) - \theta_0$ is seen to be a piecewise linear (continuous) function. The linear segment in interval k extrapolated back to $t = 0$ gives $\theta_k - \theta_0$. A table of a few values of $\theta_k - \theta_0$ is shown in the table of Fig. 5.6-1(b). We observe that $\theta_k - \theta_0$ can only have two values, 0 or π (modulo 2π), and its behavior is repetitive after four intervals.†

This phase history is useful in decomposing (5.6-2) into a form useful for generation of MSK signals. From trigonometric identities and the facts that $\theta_k - \theta_0 = 0$ or π and $d_k = \pm 1$, we have

$$s_{\text{CPFSK}}(t) = A \cos\left(\omega_0 t + \theta_0 + \theta_k - \theta_0 + d_k \frac{\pi t}{2T_b}\right)$$

$$= A \cos\left(\theta_k - \theta_0 + d_k \frac{\pi t}{2T_b}\right)\cos(\omega_0 t + \theta_0)$$

$$- A \sin\left(\theta_k - \theta_0 + d_k \frac{\pi t}{2T_b}\right)\sin(\omega_0 t + \theta_0)$$

$$= A \cos(\theta_k - \theta_0)\cos\left(\frac{\pi t}{2T_b}\right)\cos(\omega_0 t + \theta_0)$$

$$- A d_k \cos(\theta_k - \theta_0)\sin\left(\frac{\pi t}{2T_b}\right)\sin(\omega_0 t + \theta_0)$$

$$= s_I(t) A \cos(\omega_0 t + \theta_0) - s_Q(t) A \sin(\omega_0 t + \theta_0). \quad (5.6\text{-}3)$$

Here we define

$$s_I(t) \triangleq \cos(\theta_k - \theta_0)\cos\left(\frac{\pi t}{2T_b}\right) \quad (5.6\text{-}4)$$

$$s_Q(t) \triangleq d_k \cos(\theta_k - \theta_0)\sin\left(\frac{\pi t}{2T_b}\right). \quad (5.6\text{-}5)$$

Examination of the paths into and out of node c in the phase trellis of Fig. 5.6-1(b) shows that $\theta_k - \theta_0$ does not change at the node. A similar fact

---

† Since θ_0 is arbitrary, most authors simply assume $\theta_0 = 0$ so that the ordinate intercept is θ_k.

holds at node d. These facts mean that $\cos(\theta_k - \theta_0)$ does not change sign at nodes corresponding to kT_b, k even. However, at nodes a and b (at kT_b, k odd) changes in $\theta_k - \theta_0$ can occur. All these points show that the sign of $\cos(\theta_k - \theta_0)$ in (5.6-4) can change *only* at the zero crossings of $\cos(\pi t/2T_b)$.

Next, extend the examination to $d_k\cos(\theta_k - \theta_0)$. Since d_k can change at any interval we now find that the sign of $d_k\cos(\theta_k - \theta_0)$ *can* change at kT_b, k even, which is now at the zero crossings of $\sin(\pi t/2T_b)$ in (5.6-5). Again turning to nodes a and b, we find that $d_k\cos(\theta_k - \theta_0)$ does *not* change sign (the reader may wish to work through the logic by comparing input and output paths). These points show that $d_k\cos(\theta_k - \theta_0)$ in (5.6-5) changes sign only at the zero crossings of $\sin(\pi t/2T_b)$.

Our logic has shown that the factors $s_I(t)$ and $s_Q(t)$ multiplying the quadrature carriers in (5.6-3) are constant in sign over successive intervals of duration $2T_b$. The $2T_b$ intervals of $s_Q(t)$ are delayed one bit duration T_b relative to those of $s_I(t)$ because of the relative timing of $\cos(\pi t/2T_b)$ and $\sin(\pi t/2T_b)$. Such timing is exactly the type already discussed in offset QPSK. Since the *signs* of $s_I(t)$ and $s_Q(t)$ are related to the original data sequence d_k (as well as the alterating signs of the "carriers" at frequency $\Delta\omega = 2\pi/4T_b$), they can be taken as analogous to the modulating data waveforms in OQPSK. There is a difference, however, in that the symbol waveforms of duration $2T_b$ in OQPSK were rectangular, whereas in CPFSK the symbol waveforms are half-cycles of a sinusoid at frequency $\Delta\omega/2\pi = 1/4T_b$.

In summary, our developments have shown that a CPFSK waveform can be decomposed into the sum of two modulated carriers in quadrature according to (5.6-3). The characteristics of the data-related modulating waveforms $s_I(t)$ and $s_Q(t)$ were seen to be closely analogous to OQPSK. The analogy forms the basis on which most MSK systems are developed. We shall discuss several forms of MSK in following subsections, considering parallel implementations first because they most directly relate to the OQPSK analogy.

Parallel MSK Systems—Type I

A common MSK modulator that uses the analogy of (5.6-3) with OQPSK is shown in Fig. 5.6-2(a). The input data stream $d(t)$ with bit rate $1/T_b$ is separated into two data streams, $d_I(t)$ and $d_Q(t)$, each with symbol rate $1/2T_b$. The separation is exactly as in OQPSK where $d_I(t)$ is made up from odd-numbered bits in $d(t)$ and $d_Q(t)$ is developed from even-numbered bits in $d(t)$. Figure 5.6-3 illustrates the waveforms for a data stream having values $\{d_k\} = \{..., -1, -1, 1, 1, -1, 1, 1, -1, 1, 1, -1, ...\}$ for bit intervals $\{..., -2, -1, 0, ..., 8, ...\}$. In this modulator $d_I(t)$ and $d_Q(t)$ have replaced $\cos(\theta_k - \theta_0)$ and $d_k\cos(\theta_k - \theta_0)$, respectively, in (5.6-3) to obtain

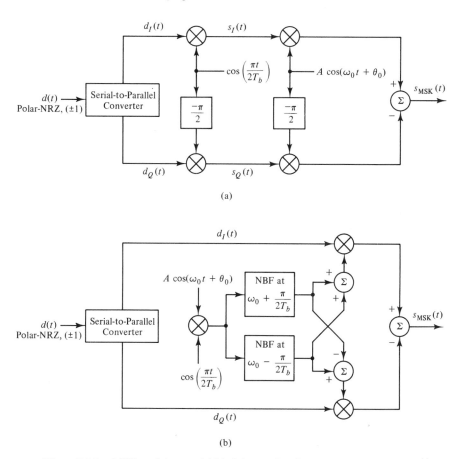

Figure 5.6-2. MSK modulators. (a) Modulator using direct carriers and (b) using equivalent carriers.

the MSK waveform. The modulator is called parallel because of the use of parallel data streams and two quadrature carriers. It is called type I following [9] to distinguish it from a second form (type II, discussed later) in which the half-cycles of $\cos(\pi t/2T_b)$ are rectified so that they do not alternate in sign. An equivalent modulator is shown in Fig. 5.6-2(b); it requires fewer product devices but requires two narrowband bandpass filters to select constant frequencies at the values indicated.

Except for the slight difference in the carriers involved, the type I system is just an OQPSK system. The optimum receiver is, therefore, quite similar to the OQPSK receiver. It is shown in Fig. 5.6-4. The integrators are timed with the symbols of duration $2T_b$ in each path. A sample is taken at the end of each interval to decide if the symbol was $+1$ or -1. The decision is in favor of $+1$ if $D_i > 0$, $i = 1$ or 2, and in favor of -1 otherwise.

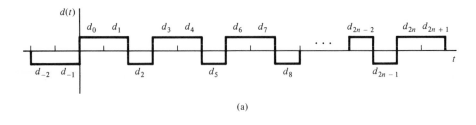

(a)

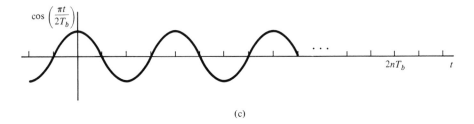

(b)

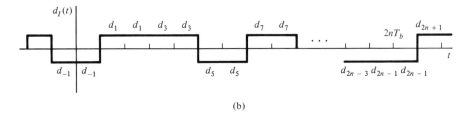

(c)

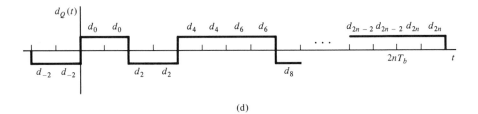

(d)

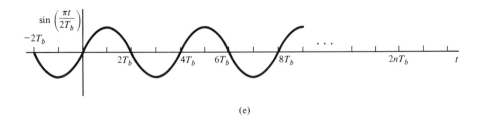

(e)

Figure 5.6-3. Timing diagrams for MSK modulators of Fig. 5.6-2.

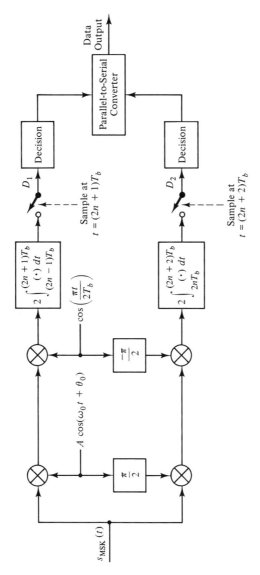

Figure 5.6-4. Receiver for MSK type I modulators of Fig. 5.6-2.

To determine the characteristics of the signal $s_{MSK}(t)$ generated as in Fig. 5.6-2, we consider the two time intervals $2n - 1$ and $2n$ centered at time $t = 2nT_b$, as shown in Fig. 5.6-3(a), (b), and (d). We readily write

$$s_{MSK}(t) = \begin{cases} Ad_{2n-1}\cos\left(\dfrac{\pi t}{2T_b}\right)\cos(\omega_0 t + \theta_0) \\[2mm] -Ad_{2n-2}\sin\left(\dfrac{\pi t}{2T_b}\right)\sin(\omega_0 t + \theta_0), & (2n-1)T_b \leqslant t \leqslant 2nT_b \\[2mm] Ad_{2n-1}\cos\left(\dfrac{\pi t}{2T_b}\right)\cos(\omega_0 t + \theta_0) \\[2mm] -Ad_{2n}\sin\left(\dfrac{\pi t}{2T_b}\right)\sin(\omega_0 t + \theta_0), & 2nT_b \leqslant t \leqslant (2n+1)T_b. \end{cases} \tag{5.6-6}$$

By use of standard trigonometric identities, this equation reduces to

$$s_{MSK}(t) = \begin{cases} A\left(\dfrac{d_{2n-1}-d_{2n-2}}{2}\right)\cos\left(\omega_0 t + \theta_0 - \dfrac{\pi t}{2T_b}\right) \\[2mm] \left. + A\left(\dfrac{d_{2n-1}+d_{2n-2}}{2}\right)\cos\left(\omega_0 t + \theta_0 + \dfrac{\pi t}{2T_b}\right) \right\} \begin{array}{l}(2n-1)T_b \leqslant t \leqslant \\ 2nT_b\end{array} \\[4mm] A\left(\dfrac{d_{2n-1}-d_{2n}}{2}\right)\cos\left(\omega_0 t + \theta_0 - \dfrac{\pi t}{2T_b}\right) \\[2mm] \left. + A\left(\dfrac{d_{2n-1}+d_{2n}}{2}\right)\cos\left(\omega_0 t + \theta_0 + \dfrac{\pi t}{2T_b}\right), \right. 2nT_b \leqslant t \leqslant (2n+1)T_b. \end{cases} \tag{5.6-7}$$

Next, we recognize that $(2n - 1)T_b \leqslant t \leqslant 2nT_b$ corresponds to data interval $k = 2n - 1$, k odd. Similarly, $2nT_b \leqslant t \leqslant (2n + 1)T_b$ corresponds to interval $k = 2n$, k even. By writing (5.6-7) in terms of k, we have

$$s_{MSK}(t) = \begin{cases} A\left(\dfrac{d_k-d_{k-1}}{2}\right)\cos\left(\omega_0 t + \theta_0 - \dfrac{\pi t}{2T_b}\right) \\[2mm] \left. + A\left(\dfrac{d_k+d_{k-1}}{2}\right)\cos\left(\omega_0 t + \theta_0 + \dfrac{\pi t}{2T_b}\right) \right\} \begin{array}{l}kT_b \leqslant t \leqslant (k+1)T_b \\ k \text{ odd}\end{array} \\[4mm] -A\left(\dfrac{d_k-d_{k-1}}{2}\right)\cos\left(\omega_0 t + \theta_0 - \dfrac{\pi t}{2T_b}\right) \\[2mm] \left. + A\left(\dfrac{d_k+d_{k-1}}{2}\right)\cos\left(\omega_0 t + \theta_0 + \dfrac{\pi t}{2T_b}\right) \right\} \begin{array}{l}kT_b \leqslant t \leqslant (k+1)T_b \\ k \text{ even}.\end{array} \end{cases} \tag{5.6-8}$$

Clearly, this result can be written for any interval k as:

$$s_{MSK}(t) = (-1)^{k+1}A\left(\frac{d_k - d_{k-1}}{2}\right)\cos\left(\omega_0 t + \theta_0 - \frac{\pi t}{2T_b}\right)$$

$$+ A\left(\frac{d_k + d_{k-1}}{2}\right)\cos\left(\omega_0 t + \theta_0 + \frac{\pi t}{2T_b}\right). \tag{5.6-9}$$

Equation (5.6-9) indicates that frequency modulation occurs according to

$$\omega_0 + \Delta\omega = \omega_0 + \frac{\pi}{2T_b} \qquad \text{when } d_k = d_{k-1} \tag{5.6-10a}$$

$$\omega_0 - \Delta\omega = \omega_0 - \frac{\pi}{2T_b} \qquad \text{when } d_k \neq d_{k-1}. \tag{5.6-10b}$$

Although keying is between the desired frequencies, there is not a one-to-one relationship between d_k and frequency deviations $\pm\Delta\omega$. Deviation in interval k depends not only on d_k but on d_{k-1} as well. Table 5.6-1 illustrates the waveforms that result from various combinations of d_k and d_{k-1}. By a slight modification the one-to-one relationship can be established. We next discuss this modification.

TABLE 5.6-1. Waveforms Possible with Type I MSK.

d_k	d_{k-1}	Waveform During Interval k	
		k odd	k even
+1	+1	$A\cos(\omega_0 t + \theta_0 + \Delta\omega t + 0)$	$A\cos(\omega_0 t + \theta_0 + \Delta\omega t + 0)$
−1	+1	$A\cos(\omega_0 t + \theta_0 - \Delta\omega t + \pi)$	$A\cos(\omega_0 t + \theta_0 - \Delta\omega t + 0)$
+1	−1	$A\cos(\omega_0 t + \theta_0 - \Delta\omega t + 0)$	$A\cos(\omega_0 t + \theta_0 - \Delta\omega t + \pi)$
−1	−1	$A\cos(\omega_0 t + \theta_0 + \Delta\omega t + \pi)$	$A\cos(\omega_0 t + \theta_0 + \Delta\omega t + \pi)$

Fast Frequency Shift Keying

By differentially encoding the data waveform $d(t)$ before it is applied to the type I MSK modulators of Fig. 5.6-2, the resulting MSK waveform will have a one-to-one correspondence between data waveform and transmitted frequency. The resulting system is often called *fast frequency shift keying* (FFSK) [9, 21], although we may still consider it a form of MSK.

Differential encoding can be done on the data source's binary digits using the encoder of Fig. 3.6-3(a) prior to polar-NRZ formatting. Decoding on the detected binary digits uses the decoder of (c). Alternatively, the encoder and decoder of Fig. 5.6-5 can be used directly with the polar-

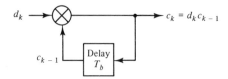

Truth Table for c_k

(a)

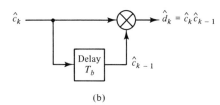

(b)

Figure 5.6-5. (a) Differential encoder and (b) decoder for a polar-NRZ data sequence.

formatted waveforms.† Here the encoded waveform levels are $c_k = \pm 1$, where

$$c_k = d_k \cdot c_{k-1} \qquad (5.6\text{-}11)$$

and the product is analog according to the truth table shown. On substitution of (5.6-11) into (5.6-9), we have

$$s_{\text{MSK}}(t) = (-1)^{k+1} A c_{k-1} \left(\frac{d_k - 1}{2} \right) \cos(\omega_0 t + \theta_0 - \Delta\omega t)$$
$$+ A c_{k-1} \left(\frac{d_k + 1}{2} \right) \cos(\omega_0 t + \theta_0 + \Delta\omega t) \qquad (5.6\text{-}12)$$

for any k. Transmitted frequency is now $\omega_0 + d_k \Delta\omega$, which has the desired dependence on d_k. There remains a dependence on the prior coded bit level c_{k-1} in that it affects the phase of the transmitted signal. Table 5.6-2 shows the possible waveforms, which may be compared directly to those in Table 5.6-1.

† If one (of either) type encoder is used in the transmitter while the other type of decoder is used in the receiver, the recovered binary digits must be complemented to obtain the correct data stream.

TABLE 5.6-2.   Waveforms Possible with Type I MSK Having Differentially Encoded Data (FFSK).

		Waveform During Interval k	
d_k	c_{k-1}	k odd	k even
$+1$	$+1$	$A \cos(\omega_0 t + \theta_0 + \Delta\omega t + 0)$	$A \cos(\omega_0 t + \theta_0 + \Delta\omega t + 0)$
-1	$+1$	$A \cos(\omega_0 t + \theta_0 - \Delta\omega t + \pi)$	$A \cos(\omega_0 t + \theta_0 - \Delta\omega t + 0)$
$+1$	-1	$A \cos(\omega_0 t + \theta_0 + \Delta\omega t + \pi)$	$A \cos(\omega_0 t + \theta_0 + \Delta\omega t + \pi)$
-1	-1	$A \cos(\omega_0 t + \theta_0 - \Delta\omega t + 0)$	$A \cos(\omega_0 t + \theta_0 - \Delta\omega t + \pi)$

Parallel MSK—Type II

In type I MSK the symbol waveforms in the quadrature channels were shaped by half-cycles of either $\cos(\pi t/2T_b)$ or $\sin(\pi t/2T_b)$ that alternated in sign. In another form of MSK, called type II [9], the half-cycle waveforms are always positive. The transmitted waveform can be generated as in Fig. 5.6-2(a) if full-wave rectifiers (having linear output-input characteristics) are placed prior to the multipliers in the lines carrying $\cos(\pi t/2T_b)$ and $\sin(\pi t/2T_b)$. An alternative modulator is shown in Fig. 5.6-6(a) [22], where a suitable pulse-shaping network is employed.† An appropriate receiver is shown in (b).

By using the analysis procedure developed above for the type I system it can be shown that the type II MSK waveform is

$$s_{\text{MSK}}(t) = \begin{cases} (-1)^{(k+1)/2}A\left[\left(\dfrac{d_k + d_{k-1}}{2}\right)\cos(\omega_0 t + \theta_0 - \Delta\omega t) \\ + \left(\dfrac{d_k - d_{k-1}}{2}\right)\cos(\omega_0 t + \theta_0 + \Delta\omega t)\right], \qquad k \text{ odd} \\ (-1)^{k/2}A\left[-\left(\dfrac{d_k - d_{k-1}}{2}\right)\cos(\omega_0 t + \theta_0 - \Delta\omega t) \\ + \left(\dfrac{d_k + d_{k-1}}{2}\right)\cos(\omega_0 t + \theta_0 + \Delta\omega t)\right], \qquad k \text{ even,} \end{cases} \quad (5.6\text{-}13)$$

in data bit interval k. This expression shows, as in type I MSK, that type II MSK does not have a one-to-one frequency/data correspondence. However, unlike type I, it can be shown that a one-to-one correspondence is not restored by differentially encoding the data waveform.

---

† This network is best implemented as an active switching network. A passive filter of the required transfer function is unrealizable (see Prob. 5-33).

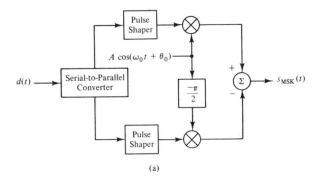

(a)

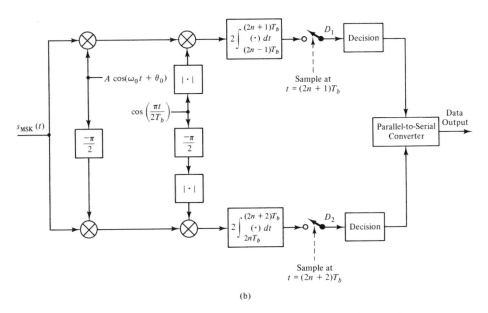

(b)

Figure 5.6-6. (a) Modulator and (b) demodulator for parallel MSK type II system.

Serial MSK

Most recent emphasis in MSK research has involved a serial implementation that is said to perform better in high data rate systems than the
parallel forms [23–26]. One form of serial MSK system is illustrated in
Fig. 5.6-7. In the transmitter a polar-NRZ data waveform $d(t)$, with levels
± 1 and bit duration T_b, multiplies a "carrier" of frequency $\omega_2 = \omega_0 +$
$\Delta\omega$, $\Delta\omega = \pi/2T_b$, and arbitrary phase θ_0. The product waveform, $s_p(t)$,

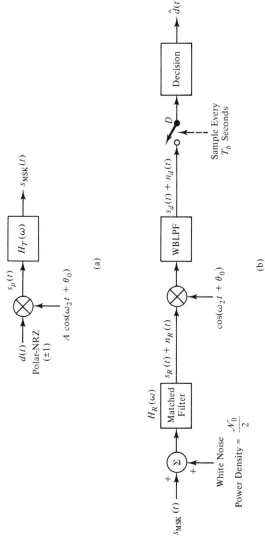

Figure 5.6-7. Block diagrams applicable to serial MSK. (a) Transmitter modulator and (b) receiver demodulator.

passes through a bandpass filter with transfer function

$$H_T(\omega) = \frac{\pi}{2j}\left\{e^{-j(\omega - \omega_1)T_b/2}\mathrm{Sa}\left[(\omega - \omega_1)\frac{T_b}{2}\right]\right.$$
$$\left. - e^{-j(\omega + \omega_1)T_b/2}\mathrm{Sa}\left[(\omega + \omega_1)\frac{T_b}{2}\right]\right\},$$

(5.6-14)

where $\omega_1 = \omega_0 - \Delta\omega$. The filter's impulse response is

$$h_T(t) = \frac{\pi}{T_b}\mathrm{rect}\left[\frac{t - (T_b/2)}{T_b}\right]\sin(\omega_1 t).$$

(5.6-15)

The final transmitted waveform is MSK, as we shall subsequently show. The receiver demodulator consists of a matched filter (for white Gaussian noise) that is proportional to the square root of the MSK signal's power spectrum [26]. Its transfer function can be taken as†

$$H_R(\omega) = \frac{AT_b\,\pi}{4}\left\{\frac{\cos[(\omega - \omega_0)T_b]}{(\pi/4)^2 - [(\omega - \omega_0)T_b/2]^2}\right.$$
$$\left. + \frac{\cos[(\omega + \omega_0)T_b]}{(\pi/4)^2 - [(\omega + \omega_0)T_b/2]^2}\right\}.$$

(5.6-16)

The reader may wish to demonstrate, as an exercise, that an equivalent form for $H_R(\omega)$ is (Prob. 5-35)

$$H_R(\omega) = \frac{AT_b\,\pi}{2}\left\{\mathrm{Sa}\left[(\omega - \omega_1)\frac{T_b}{2}\right]\mathrm{Sa}\left[(\omega - \omega_2)\frac{T_b}{2}\right]\right.$$
$$\left. + \mathrm{Sa}\left[(\omega + \omega_1)\frac{T_b}{2}\right]\mathrm{Sa}\left[(\omega + \omega_2)\frac{T_b}{2}\right]\right\}.$$

(5.6-17)

Following the matched filter is a coherent demodulator with a local oscillator coherent with the transmitted carrier. The wideband lowpass filter serves only to remove spectral terms near $2\omega_0$ and has no effect on the desired baseband waveform shapes (we presume ω_0 large relative to $1/T_b$). Finally, samples taken each *bit* interval are used to decide if the level of $d(t)$ was $+1$ or -1. If $D > 0$ the decision is $d(t) = -1$ in the interval, and $d(t) = +1$ if $D < 0$. If the original data waveform is desired, a reconstructed version $\hat{d}(t)$ is generated from the sequence of bit decisions.

To demonstrate that the output waveform of Fig. 5.6-7(a) is truly MSK, begin by defining bit interval k, as before, by $kT_b \leqslant t \leqslant (k + 1)T_b$. The component of $s_{\mathrm{MSK}}(t)$, due to data of interval k, denoted by $s_k(t)$, is

---

† The power spectrum of MSK is found in subsequent work. The final result is given by (5.6-32) to follow.

$$s_k(t) = \frac{d_k A\pi}{T_b} \int_{-\infty}^{\infty} \text{rect}\left[\frac{x - (k + 0.5)T_b}{T_b}\right] \cos(\omega_2 x + \theta_0)$$

$$\cdot \text{rect}\left[\frac{t - x - (T_b/2)}{T_b}\right] \sin[\omega_1(t - x)] \, dx$$

$$= \begin{cases} \dfrac{d_k A\pi}{T_b} \displaystyle\int_{kT_b}^{t} \cos(\omega_2 x + \theta_0)\sin[\omega_1(t - x)] \, dx, \\[2mm] \qquad kT_b \leq t \leq (k + 1)T_b \\[3mm] \dfrac{d_k A\pi}{T_b} \displaystyle\int_{t-T_b}^{(k+1)T_b} \cos(\omega_2 x + \theta_0)\sin[\omega_1(t - x)] \, dx, \\[2mm] \qquad (k + 1)T_b \leq t \leq (k + 2)T_b \\[3mm] 0, \qquad \text{elsewhere.} \end{cases} \qquad (5.6\text{-}18)$$

Equation (5.6-18) shows that a data bit in any interval causes a response in its interval as well as the interval following it. The full response in interval k therefore consists of the upper right-side term in (5.6-18) plus the middle right-side term with k replaced by $k - 1$. After algebraic manipulation we have

$$s_{\text{MSK}}(t) = \frac{d_k A\pi}{T_b} \int_{kT_b}^{t} \cos(\omega_2 x + \theta_0)\sin[\omega_1(t - x)] \, dx$$

$$+ \frac{d_{k-1}A\pi}{T_b} \int_{t-T_b}^{kT_b} \cos(\omega_2 x + \theta_0)\sin[\omega_1(t - x)] \, dx$$

$$= A\left(\frac{d_k + d_{k-1}}{2}\right)\cos(\omega_2 t + \theta_0)$$

$$+ (-1)^{k+1} A\left(\frac{d_k - d_{k-1}}{2}\right)\cos(\omega_1 t + \theta_0). \qquad (5.6\text{-}19)$$

Since $\omega_1 = \omega_0 - \Delta\omega = \omega_0 - (\pi/2T_b)$ and $\omega_2 = \omega_0 + \Delta\omega = \omega_0 + (\pi/2T_b)$, we see that (5.6-19) is exactly the same as (5.6-9). Thus serial MSK is exactly the same as type I MSK.

To show that the entire serial MSK system properly demodulates the data stream, consider its response to a typical bit in interval k, defined by

$$d_k(t) = d_k \, \text{rect}\left[\frac{t - (k + 0.5)T_b}{T_b}\right], \qquad (5.6\text{-}20)$$

having the spectrum

$$D_k(\omega) = d_k T_b \, e^{-j\omega(2k+1)T_b/2} \, \text{Sa}\left(\frac{\omega T_b}{2}\right). \qquad (5.6\text{-}21)$$

If $S_d(\omega)$, $S_R(\omega)$, and $S_p(\omega)$ denote the Fourier transforms of $s_d(t)$, $s_R(t)$, and $s_p(t)$ of Fig. 5.6-7, then clearly

$$S_d(\omega) = L_p\left[\frac{e^{j\theta_0}}{2} S_R(\omega - \omega_2) + \frac{e^{-j\theta_0}}{2} S_R(\omega + \omega_2)\right] \qquad (5.6\text{-}22)$$

$$S_R(\omega) = H_T(\omega)H_R(\omega)\, S_p(\omega) \qquad (5.6\text{-}23)$$

$$S_p(\omega) = \frac{A}{2}\, e^{j\theta_0}\, D_k(\omega - \omega_2) + \frac{A}{2}\, e^{-j\theta_0}\, D_k(\omega + \omega_2) \qquad (5.6\text{-}24)$$

where $L_p[\cdot]$ represents taking only the lowpass part of the quantity in brackets. By combining (5.6-21) through (5.6-24) and neglecting small terms permitted by our assumption that $\omega_0 \gg 2\pi/T_b$ we can obtain

$$S_d(\omega) = -\left(\frac{AT_b\pi}{4}\right)^2 d_k\, e^{-j\omega(k+1)T_b}\, \mathrm{Sa}^2\!\left(\frac{\omega T_b}{2}\right)H_c(\omega), \qquad (5.6\text{-}25)$$

where we define

$$H_c(\omega) \triangleq \mathrm{Sa}^2\!\left[\frac{(\omega + 2\Delta\omega)T_b}{2}\right] + \mathrm{Sa}^2\!\left[\frac{(\omega - 2\Delta\omega)T_b}{2}\right]. \qquad (5.6\text{-}26)$$

The exponential factor in (5.6-25) represents only a delay of $(k + 1)T_b$, whereas the product $\mathrm{Sa}^2(\omega T_b/2)H_c(\omega)$ represents convolution in the time domain of the functions that are the inverse Fourier transforms of $\mathrm{Sa}^2(\omega T_b/2)$ and $H_c(\omega)$. From the known pair

$$\frac{1}{T_b}\,\mathrm{tri}\!\left(\frac{t}{T_b}\right) \leftrightarrow \mathrm{Sa}^2\!\left(\frac{\omega T_b}{2}\right) \qquad (5.6\text{-}27)$$

we have

$$h_c(t) = \frac{2}{T_b}\mathrm{tri}\!\left(\frac{t}{T_b}\right)\cos(2\Delta\omega t) \leftrightarrow H_c(\omega). \qquad (5.6\text{-}28)$$

The inverse Fourier transform of (5.6-25) becomes

$$s_d(t) = \frac{-A^2 T_b\, d_k}{2}\, g[t - (k + 1)T_b], \qquad (5.6\text{-}29)$$

where

$$g(t) = \frac{\pi^2}{4T_b}\int_{-\infty}^{\infty} \mathrm{tri}\!\left(\frac{x}{T_b}\right)\cos\!\left(\frac{\pi x}{T_b}\right)\mathrm{tri}\!\left[\frac{(t - x)}{T_b}\right] dx$$

$$= \frac{1}{2}\left\{\left(1 - \frac{|t|}{2T_b}\right)\left[1 + \cos\!\left(\frac{\pi t}{T_b}\right)\right] + \frac{1}{\pi}\sin\!\left(\frac{\pi|t|}{T_b}\right)\right\}, \qquad -2T_b \le t \le 2T_b. \qquad (5.6\text{-}30)$$

To summarize, (5.6-29) with (5.6-30) defines the receiver's response to the original data waveform as it existed in interval k for $kT_b \le t \le (k + 1)T_b$. This function agrees with results given in [23]. Data waveform amplitude is $d_k = \pm 1$ according to the polar NRZ format.

Since the maximum of $g(t)$ occurs at $t = 0$ and is unity, the maximum magnitude of $s_d(t)$ is $A^2 T_b/2$; it occurs at $t = (k + 1)T_b$, which is the end of interval k, as one might expect. Behavior of $g(t)$ is shown in Fig. 5.6-8, as determined by (5.6-30). Even though $g(t)$ is nonzero over four bit intervals, it is zero at times $\pm T_b$ and $\pm 2T_b$ from its maximum. This means that samples of the complete waveform $s_d(t)$ of Fig. 5.6-7(b) taken at the ends of the bit intervals contain no intersymbol interference. The signal sample values are

$$s_d[(k + 1)T_b] = - \frac{d_k A^2 T_b}{2} \qquad (5.6\text{-}31)$$

which are similar to what one obtains with a polar waveform format. However, because of the negative sign, the decision logic is reversed. The receiver now decides $d_k = +1$ if a sample is negative and -1 if positive.

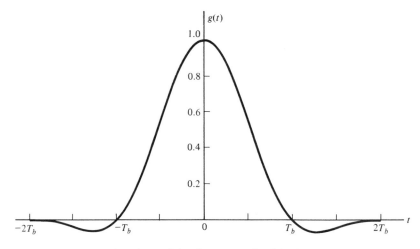

Figure 5.6-8. Typical shape of signal response of serial MSK receiver to a single transmitted data bit interval of duration T_b.

Power Spectrum of MSK

The MSK signal's power spectrum is derived from (5.5-29) with $\Delta\omega T_b = \pi/2$ from (5.5-25) when $n = 1$. After straightforward simplification we get

$$\mathcal{S}_{\text{MSK}}(\omega) = \left(\frac{\pi A}{4}\right)^2 \frac{T_b}{4} \left(\frac{\cos^2[(\omega - \omega_0)T_b]}{\{[\pi/4]^2 - [(\omega - \omega_0)T_b/2]^2\}^2} \right.$$
$$\left. + \frac{\cos^2[(\omega + \omega_0)T_b]}{\{[\pi/4]^2 - [(\omega + \omega_0)T_b/2]^2\}^2}\right). \qquad (5.6\text{-}32)$$

Noise Performance of MSK

In earlier work we showed that PSK (using a polar-NRZ format) and QPSK produce the same average bit error probability P_e, when both systems use optimum receivers, have the same average energy transmitted per bit, and have the same white channel noise density $\mathcal{N}_0/2$. P_e was given in (5.3-8) and (5.4-6). Furthermore, it was noted that OQPSK gave the same P_e as QPSK under similar conditions. Because the parallel MSK and OQPSK systems are so similar, they also have the same values of P_e under similar conditions [4, 22]. Thus PSK, QPSK, OQPSK, and parallel MSK systems produce the same value of P_e given by†

$$P_e = \left(\frac{1}{2}\right) \text{erfc}[\sqrt{\varepsilon}]. \tag{5.6-33}$$

We shall next demonstrate that the noise performance of serial MSK is also given by (5.6-33). Error probability is obtained from probability density functions applicable to the possible wideband lowpass filter output waveforms of Fig. 5.6-7(b) at the sample times. Since the signal at the sample times, $s_d[(k + 1)T_b]$, is either a positive or equal-amplitude negative level, according to (5.6-31), added to zero-mean Gaussian noise, the densities of Fig. 5.6-9 apply. Here $p_{n_d}(n_d)$ represents the density of the demodulator noise $n_d[(k + 1)T_b]$. Average bit error probability, for equally probable data bits, becomes

$$P_e = \frac{1}{2}\int_0^\infty p_{n_d}(x + V)\,dx + \frac{1}{2}\int_{-\infty}^0 p_{n_d}(x - V)\,dx$$

$$= \frac{1}{2}\text{erfc}\left[\sqrt{\frac{V^2}{2\sigma_o^2}}\right] = \frac{1}{2}\text{erfc}\left[\sqrt{\left(\frac{A^2 T_b}{2}\right)^2 \frac{1}{2\sigma_o^2}}\right], \tag{5.6-34}$$

where σ_o^2 is the power in the noise which remains to be found.

From Fig. 5.6-7(b) and Prob. B-54, the power spectrum of the noise is

$$\mathcal{S}_{n_d}(\omega) = L_p\left\{\frac{\mathcal{N}_0}{8}\left[|H_R(\omega - \omega_2)|^2 + |H_R(\omega + \omega_2)|^2\right]\right\}. \tag{5.6-35}$$

After substitution of (5.6-16) and neglect of small terms made possible by assumption that $\omega_0 \gg \Delta\omega$, (5.6-35) can be written as

$$\mathcal{S}_{n_d}(\omega) = \frac{\mathcal{N}_0}{8}\left(\frac{\pi A T_b}{4}\right)^2 \left(\frac{\cos^2[(\omega - \omega_0 + \omega_2)T_b]}{\{(\pi/4)^2 - [(\omega - \omega_0 + \omega_2)T_b/2]^2\}^2}\right.$$

$$\left. + \frac{\cos^2[(\omega + \omega_0 - \omega_2)T_b]}{\{(\pi/4)^2 - [(\omega + \omega_0 - \omega_2)T_b/2]^2\}^2}\right). \tag{5.6-36}$$

---

† As usual, ε is the ratio of average received (and transmitted for no channel loss) energy per bit to $\mathcal{N}_0$. P_e is plotted in Fig. 5.12-1.

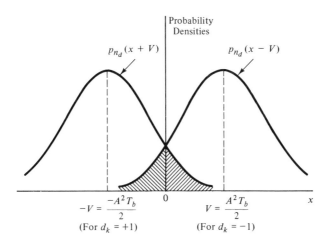

Figure 5.6-9. Probability density functions applicable to bit decisions in serial MSK.

Since each of these components is symmetric about one of the frequencies $\pm(\omega_2 - \omega_0) = \pm\Delta\omega = \pm\pi/2T_b$, it contains half the total power. By integration the total power is

$$\sigma_o^2 = \frac{\mathcal{N}_0}{8}\left(\frac{\pi A T_b}{4}\right)^2 \frac{1}{\pi}\int_{-\infty}^{\infty}\left\{\frac{\cos[(\omega - \Delta\omega)T_b]}{(\pi/4)^2 - [(\omega - \Delta\omega)T_b/2]^2}\right\}^2 d\omega. \quad (5.6\text{-}37)$$

Next, we use the frequency shifting property of Fourier transforms, along with the transform pair

$$\frac{4}{\pi T_b}\cos\left(\frac{\pi t}{2T_b}\right)\text{rect}\left(\frac{t}{2T_b}\right) \leftrightarrow \frac{\cos(\omega T_b)}{(\pi/4)^2 - (\omega T_b/2)^2} \quad (5.6\text{-}38)$$

and Parseval's theorem, to evaluate (5.6-37):

$$\sigma_o^2 = \frac{\mathcal{N}_0}{8}\left(\frac{\pi A T_b}{4}\right)^2 2\int_{-\infty}^{\infty}\left|\frac{4}{\pi T_b}\cos\left(\frac{\pi t}{2T_b}\right)\text{rect}\left(\frac{t}{2T_b}\right)e^{j\Delta\omega t}\right|^2 dt$$

$$= \frac{\mathcal{N}_0 A^2}{4}\int_{-T_b}^{T_b}\cos^2\left(\frac{\pi t}{2T_b}\right)dt = \frac{\mathcal{N}_0 A^2 T_b}{4}. \quad (5.6\text{-}39)$$

Finally, we substitute (5.6-39) into (5.6-34) to obtain

$$P_e = \frac{1}{2}\text{erfc}\left[\sqrt{\frac{A^2 T_b}{2\mathcal{N}_0}}\right] = \frac{1}{2}\text{erfc}[\sqrt{\varepsilon}]. \quad (5.6\text{-}40)$$

Here

$$\varepsilon = \frac{A^2 T_b}{2\mathcal{N}_0} \quad (5.6\text{-}41)$$

is the ratio of average energy transmitted per bit $(A^2 T_b/2)$ to twice the channel's noise density. Our final result (5.6-40) agrees with (5.6-33) as indicated at the start of the development.

Other MSK Implementations and Comments

Serial MSK can be implemented as shown in Fig. 5.6-7, except using a transmitter carrier frequency of ω_1 instead of ω_2 if the filter defined by (5.6-15) has ω_1 replaced by ω_2. The MSK signal is again given by (5.6-19) except ω_1 and ω_2 are respectively replaced by ω_2 and ω_1.

Although the sidelobes of the serial MSK signal's power spectrum decrease rapidly [as $1/(\omega - \omega_0)^4$ for $|\omega - \omega_0|$ large], it has been found that sidelobes can be reduced considerably more by use of a different transmitter filter (Fig. 5.6-7) [27]. The resulting signal is not constant-envelope as in true MSK, so an amplitude limiter is added. It is claimed that loss in communication efficiency, as measured by bit error rate versus signal-to-noise ratio, was so small as to be unmeasurable [27].

One of the problems of serial MSK, which is attractive for high data rates, is the implementation of the required broadband bandpass filters. Past implementations have used surface acoustic wave (SAW) devices that are limited in bandwidth to about 10 to 30% of center frequency. To relieve such problems, implementations have been devised that use baseband equivalents to the bandpass filters [25, 26]. An experimental 550 Megabit/s serial implementation using integrated circuit and stripline technology is described in [26].

Parallel MSK, as we have defined it, uses half-cycle sinusoidally shaped symbol pulses (duration $2T_b$) in the two quadrature channels. A lot of recent research in MSK has centered on reduction of spectral sidelobes by choice of symbol pulse shape. A class of pulse shapes due to Amoroso [28] has resulted in what is known as *sinusoidal* FSK (SFSK); it has a power spectrum that decreases as $1/(\omega - \omega_0)^8$ for $|\omega - \omega_0|$ large. Simon [29] also introduced another class of symbol pulse shapes and defined some conditions on pulse shapes that give desirable results. Rabzel and Pasupathy [30] have derived necessary and sufficient conditions on symbol pulse shapes for an MSK-type signal to have an asymptotic spectral rolloff behavior of $1/(\omega - \omega_0)^{2N+2}$, $N = 0, 1, \ldots$. They defined a class of pulse shapes for which spectral rolloff was $1/(\omega - \omega_0)^{4M+4}$, $M = 0, 1, \ldots$. The special cases $M = 0$ and 1 correspond to MSK and SFSK, respectively. Bazin [31] has given a class of pulse shapes for which the waveforms of [30] are a subset; he develops in particular a shape that defines *double sine* FSK (DSFSK), where spectral rolloff is $1/(\omega - \omega_0)^{12}$ for $|\omega - \omega_0| > 4.75 \, \omega_b$.

Though considerable efforts have been made to reduce MSK spectral sidelobes, less has been done to control the main lobe's width. It is known, however, that an MSK-type signal generated by using a monotonic symmetrical pulse will always have a main lobe wider than that of a conventional

PSK signal generated by using a polar-NRZ format with the same bit duration [32]. Some work on optimum pulse shaping has also been done by Deshpande and Wittke [33].

Because noise performance of MSK is better than that of orthogonal CPFSK (even for $\Delta\omega = \omega_b/4$, the MSK case),† it is instructive to inquire why this fact is true. The principal reason stems from observation times. The CPFSK receiver integrates over an interval T_b, whereas the MSK systems integrate over an interval $2T_b$. It is reasonable to wonder if systems using longer periods of observation can yield further improvements. Study has shown [34, 35] improvements compared to MSK are possible and are on the order of 1 dB ($3T_b$ interval) and 1.2 dB ($5T_b$ interval) for an optimum value of *modulation index* of 0.715 (ratio $2\Delta\omega/\omega_b$). Improvements are in the ratio of average energy per bit to $\mathcal{N}_0$ needed to achieve a specified P_e. For observation intervals longer than $5T_b$, improvements are minor [34].

Finally, we note that other pulse shapes [36] and correlative coding suggested in [4] (see also [37]) have been used in MSK. These methods improve spectral efficiency but recent work [38, 39] indicates a performance loss relative to MSK.

5.7 OPTIMUM NONCOHERENT BANDPASS SYSTEMS

By accepting a small loss in noise performance, it is possible to implement amplitude and frequency shift keying in a way that does not depend on carrier phase. Such implementations are called *noncoherent*, and they typically are simpler than their coherent counterparts. In this section we shall develop the forms of optimum ASK and FSK systems under the specification that they are to be noncoherent. Since the desired information is conveyed in the phase of a phase shift keyed waveform, no truly noncoherent version of PSK is possible.

Recall that an optimization procedure was developed in Chap. 4 (Secs. 4.1–4.3) for baseband binary systems. The same procedures were also applied to coherent bandpass binary systems in Sec. 5.1. With a small modification those procedures may be applied to the binary noncoherent systems. We shall begin by giving a short review of the basic developments to demonstrate where the modification takes place.

System, Noise, and Signal Definitions

The system is modeled as shown in Fig. 5.7-1. As usual, the modulator generates, during each bit interval of duration T_b, one of two possible waveforms $s_1(t)$ or $s_2(t)$, according to which of two binary messages m_1 or m_2 is generated by a source. Signals $s_1(t)$ and $s_2(t)$ are assumed nonzero

---

† Compare (5.5-22) with $\gamma = 0$ with (5.6-33) or (5.6-40).

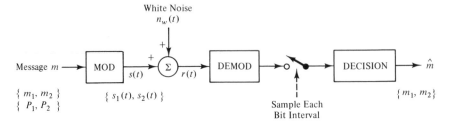

Figure 5.7-1. Model of optimum noncoherent binary system.

only in the bit interval in which they are generated. The receiver receives the signal with added white Gaussian, zero-mean, noise and processes the sum through a demodulator. It is the demodulator's structure that is found in the optimization procedure. At the end of each bit interval, the demodulation output is observed (sampled) and a decision is made as to which binary message is being received. Following the decisions the system can either output the recovered message sequence (m_1 or m_2, as shown) or some circuitry can be added (not shown) to reconstruct the original binary data sequence.

White noise is modeled as the limit of ideally bandlimited white noise $n_b(t)$ as the bandwidth W_N becomes infinite. As noted in Sec. 4.1, this model allows samples of the bandlimited noise that are statistically independent to be taken every ΔT s apart. ΔT and W_N are related by

$$\Delta T = \frac{\pi}{W_N}, \tag{5.7-1}$$

so $\Delta T \to 0$ as $W_N \to \infty$, which corresponds to white noise. The advantage of treating the white noise in this manner is that the receiver's observation of the received waveform can be modeled as a large number of samples ΔT apart spanning any given bit interval. As the limit is taken noise becomes white, summations become integrals, and the procedures become equivalent to continuous-time processing. Thus we presume K samples taken at times

$$t_k = k\,\Delta T, \qquad k = 1, 2, \ldots, K. \tag{5.7-2}$$

The joint probability density of the K noise samples is given by (4.1-7), where all samples have the same variance σ^2 related to ΔT and channel white noise density $\mathcal{N}_0/2$ through

$$\sigma^2 \Delta T = \frac{\mathcal{N}_0}{2}. \tag{5.7-3}$$

The two possible transmitted signals are assumed to have the forms

$$s_1(t) = g_1(t)\cos(\omega_1 t + \theta_1), \qquad 0 \leqslant t \leqslant T_b \tag{5.7-4}$$

$$s_2(t) = g_2(t)\cos(\omega_2 t + \theta_2), \qquad 0 \leqslant t \leqslant T_b. \tag{5.7-5}$$

Here $g_1(t)$ and $g_2(t)$ are arbitrary but nonzero only for $0 \leqslant t \leqslant T_b$. These functions, ω_1 and ω_2, and θ_1 and θ_2 depend on the modulation. For example, in ASK (OOK variety) $g_1(t) = 0$, $g_2(t) = A$ the carrier's peak amplitude, $\omega_2 = \omega_0$ the carrier's frequency, and $\theta_2 = \theta_0$ the carrier's phase. For discontinuous FSK, $g_1(t) = g_2(t) = A$, $\omega_1 = \omega_0 - \Delta\omega$, $\omega_2 = \omega_0 + \Delta\omega$, and θ_1 and θ_2 are arbitrary phases. In CPFSK $g_1(t) = g_2(t) = A$, $\omega_1 = \omega_0 - \Delta\omega$, $\omega_2 = \omega_0 + \Delta\omega$ and $\theta_1 = \theta_2 = \theta_0$, an arbitrary carrier phase. In all these cases, however, the receiver does *not* know, and *cannot obtain* phases (by specification), so we treat θ_1 and θ_2 as unknown, random phases. There is no reason to presume any phase more favorable than any other so the phases are assumed uniform on $(0, 2\pi)$. The assumption of random phases marks the distinction between coherent and noncoherent systems.

Optimum Receiver Decision Rule

The presence of the random phases alters the optimization procedure. In the previous coherent system optimization the receiver observed K values of the received waveform $r(t)$ at times t_k; these samples were denoted by ρ_k. The optimum system was that which implemented the rule of (4.1-17), which is:

$$\text{If } \frac{p_r(\rho_1, \rho_2, ..., \rho_K|m_2)}{p_r(\rho_1, \rho_2, ..., \rho_K|m_1)} > \frac{P_1}{P_2}, \qquad \text{choose } m_2$$

$$(5.7\text{-}6)$$

and choose m_1 otherwise.

Here $p_r(., \ldots, .|m_i)$ is the joint density of the observations of $r(t)$ given they were generated by transmitting $s_i(t)$ due to message m_i, $i = 1$ or 2. P_1 and P_2 are the probabilities that m_1 and m_2 were transmitted, respectively. In the present noncoherent system optimization, a retracing of the procedures of Chap. 4 that lead to (5.7-6) will again lead to (5.7-6) except that the conditional densities now depend on θ_1 or θ_2; the final optimization rule follows the averaging of these densities over the random values of θ_1 and θ_2. Thus we say the noncoherent system is optimum if it satisfies the rule:

$$\text{If } \frac{E_{\theta_2}[p_r(\rho_1, ..., \rho_K|m_2, \theta_2)]}{E_{\theta_1}[p_r(\rho_1, ..., \rho_K|m_1, \theta_1)]} > \frac{P_1}{P_2}, \qquad \text{choose } m_2$$

$$(5.7\text{-}7)$$

and choose m_1 otherwise.

The notation $E_{\theta_i}[\cdot]$ refers to taking the expected value of the quantity in brackets with respect to θ_i, $i = 1$ or 2.

Because noises are Gaussian and independent, the joint densities in (5.7-7) are given by

$$p_r(\rho_1, ..., \rho_K|m_i, \theta_i) = \prod_{k=1}^{K} p_{n_k}(\rho_k - s_{ik})$$

$$= (2\pi\sigma^2)^{-K/2} \exp\left\{ -\sum_{k=1}^{K} \frac{(\rho_k - s_{ik})^2}{(2\sigma^2)} \right\}, \qquad i = 1, 2.$$

$$(5.7\text{-}8)$$

Here $p_{n_k}(n_k)$ is the density function of noise sample k, defined by

$$n_k = n_b(t_k), \tag{5.7-9}$$

and s_{ik} is the signal sample given by

$$s_{ik} = s_i(t_k), \qquad i = 1, 2. \tag{5.7-10}$$

These samples form the observed values of $r(t)$

$$\rho_k = r(t_k) = s_{ik} + n_k, \qquad i = 1, 2. \tag{5.7-11}$$

Next, we expand (5.7-8), substitute for ρ_k and s_{ik} using (5.7-11) and (5.7-10), use (5.7-2) and substitute for σ^2 from (5.7-3). Then we allow noise to become white by letting $W_N \to \infty$. From (5.7-1) this means $\Delta T \to 0$ and $K \to \infty$ such that $K \Delta T = T_b$ holds true. All these operations allow sums to become integrals and $k \Delta T \to t$ while $\Delta T \to dt$. Thus (5.7-8) can be written as

$$p_r(\rho_1, \ldots, \rho_K | m_i, \theta_i) = K_r \exp \left\{ \frac{-1}{\mathcal{N}_0} \int_0^{T_b} r^2(t)\, dt - \frac{1}{\mathcal{N}_0} \int_0^{T_b} s_i^2(t)\, dt \right.$$
$$\left. + \frac{2}{\mathcal{N}_0} \int_0^{T_b} r(t) s_i(t)\, dt \right\}, \qquad i = 1, 2, \tag{5.7-12}$$

where

$$K_r \triangleq \lim_{\substack{\Delta T \to 0 \\ K \to \infty}} \left(\frac{\Delta T}{\pi \mathcal{N}_0} \right)^{K/2}. \tag{5.7-13}$$

To reduce (5.7-12) further we substitute (5.7-4) and (5.7-5). The required integrals are

$$\int_0^{T_b} s_i^2(t)\, dt = \int_0^{T_b} g_i^2(t) \cos^2(\omega_i t + \theta_i)\, dt$$
$$= \frac{1}{2} \int_0^{T_b} g_i^2(t)\, dt + \frac{1}{2} \int_0^{T_b} g_i^2(t) \cos(2\omega_i t + 2\theta_i)\, dt \tag{5.7-14}$$
$$\approx E_i, \qquad i = 1, 2$$

where

$$E_i \triangleq \frac{1}{2} \int_0^{T_b} g_i^2(t)\, dt, \qquad i = 1, 2, \tag{5.7-15}$$

and

$$\int_0^{T_b} r(t) s_i(t)\, dt = x_i \cos(\theta_i) - y_i \sin(\theta_i)$$
$$= r_{oi} \cos(\theta_i + \psi_i), \tag{5.7-16}$$

where

$$x_i \triangleq \int_0^{T_b} r(t) g_i(t) \cos(\omega_i t)\, dt, \qquad i = 1, 2 \tag{5.7-17}$$

$$y_i \triangleq \int_0^{T_b} r(t)g_i(t)\sin(\omega_i t)\, dt, \qquad i = 1, 2 \tag{5.7-18}$$

$$r_{oi} \triangleq [x_i^2 + y_i^2]^{1/2}, \qquad i = 1, 2 \tag{5.7-19}$$

$$\psi_i \triangleq \tan^{-1}\left(\frac{y_i}{x_i}\right), \qquad i = 1, 2. \tag{5.7-20}$$

In obtaining (5.7-14) the assumption that $2\omega_i$ is large relative to the highest frequency in $g_i^2(t)$ has been made. Finally, we substitute the required integrals into (5.7-12) and take the expected value

$$
\begin{aligned}
E_{\theta_i}[p_r(\rho_1, \ldots, \rho_K | m_i, \theta_i)] &= K_r \exp\left\{ -\frac{1}{\mathcal{N}_0} \int_0^{T_b} r^2(t)\, dt - \frac{E_i}{\mathcal{N}_0} \right\} \\
&\quad \cdot \frac{1}{2\pi} \int_0^{2\pi} e^{(2/\mathcal{N}_0)r_{oi}\,\cos(\theta_i + \psi_i)}\, d\theta_i \\
&= K_r \exp\left\{ -\frac{1}{\mathcal{N}_0} \int_0^{T_b} r^2(t)\, dt \right. \\
&\quad \left. - \frac{E_i}{\mathcal{N}_0} \right\} I_0\!\left(\frac{2r_{oi}}{\mathcal{N}_0}\right), \qquad i = 1, 2.
\end{aligned}
\tag{5.7-21}
$$

$I_0(\alpha)$ is the modified Bessel function of order zero; it is a monotonically increasing function of its argument for $\alpha \geq 0$ with a minimum of 1.0 at $\alpha = 0$.

The optimum system's decision rule (5.7-7) can now be written in final form by use of (5.7-21). After cancellation of the common factors we have:

$$\text{If } P_2 e^{-E_2/\mathcal{N}_0} I_0\!\left(\frac{2r_{o2}}{\mathcal{N}_0}\right) > P_1 e^{-E_1/\mathcal{N}_0} I_0\!\left(\frac{2r_{o1}}{\mathcal{N}_0}\right), \qquad \text{choose } m_2 \tag{5.7-22}$$

and choose m_1 otherwise.

Correlation Receiver Implementation

The demodulator shown in Fig. 5.7-2 is a direct implementation of the decision rule of (5.7-22) using (5.7-17)–(5.7-19). The decision to be made is now:

$$
\begin{aligned}
&\text{Choose } m_2 \text{ if } D > 0, \\
&\text{Choose } m_1 \text{ if } D < 0, \\
&\text{Random choice if } D = 0.
\end{aligned}
\tag{5.7-23}
$$

We observe that this receiver is nonlinear.

In the special case where

$$P_2 e^{-E_2/\mathcal{N}_0} = P_1 e^{-E_1/\mathcal{N}_0} \tag{5.7-24}$$

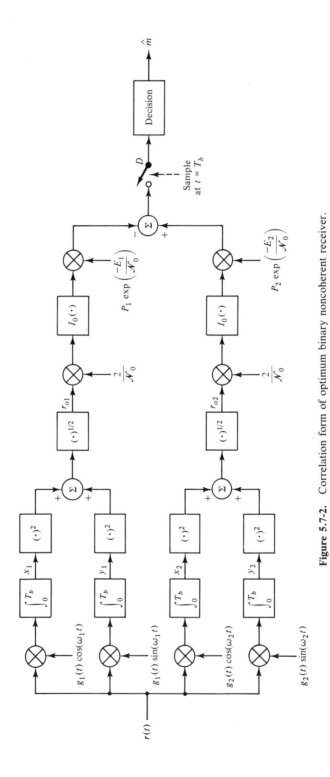

Figure 5.7-2. Correlation form of optimum binary noncoherent receiver.

296

the decision rule of (5.7-22) reduces to a comparison of modified Bessel functions. However, because they are the same functions and are monotonic in their arguments, the rule reduces to a simple comparison of arguments. The simplified receiver of Fig. 5.7-3(a) applies in this case. The condition (5.7-24) is satisfied if $P_1 = P_2$ and $E_1 = E_2$, as may be the case in FSK, but does not require these specific equalities in general.

Matched Filter Implementation

Consider the response, denoted by $R_{oi}(t)$, of a filter having impulse response

$$h_i(t) = \begin{cases} g_i(T_b - t)\cos(\omega_i t), & 0 \le t \le T_b \\ 0, & \text{elsewhere} \end{cases} \qquad (5.7\text{-}25)$$

to an input $r(t)$. It is

$$
\begin{aligned}
R_{oi}(t) &= \int_{-\infty}^{\infty} r(\alpha)h_i(t - \alpha)\, d\alpha \\
&= \int_{t-T_b}^{t} r(\alpha)\, g_i(T_b - t + \alpha)\cos[\omega_i(t - \alpha)]\, d\alpha \\
&= \cos(\omega_i t) \int_{t-T_b}^{t} r(\alpha)\, g_i(T_b - t + \alpha)\cos(\omega_i \alpha)\, d\alpha \\
&\quad + \sin(\omega_i t) \int_{t-T_b}^{t} r(\alpha)\, g_i(T_b - t + \alpha)\sin(\omega_i \alpha)\, d\alpha.
\end{aligned}
\qquad (5.7\text{-}26)
$$

By defining

$$x_i(t) \triangleq \int_{t-T_b}^{t} r(\alpha)g_i(T_b - t + \alpha)\cos(\omega_i \alpha)\, d\alpha \qquad (5.7\text{-}27)$$

$$y_i(t) \triangleq \int_{t-T_b}^{t} r(\alpha)g_i(T_b - t + \alpha)\sin(\omega_i \alpha)\, d\alpha, \qquad (5.7\text{-}28)$$

we can write

$$R_{oi}(t) = r_{oi}(t)\cos[\omega_i t - \psi_i(t)] \qquad (5.7\text{-}29)$$

where

$$r_{oi}(t) = \{x_i^2(t) + y_i^2(t)\}^{1/2} \qquad (5.7\text{-}30)$$

$$\psi_i(t) = \tan^{-1}\left[\frac{y_i(t)}{x_i(t)}\right]. \qquad (5.7\text{-}31)$$

Clearly, $r_{oi}(t)$ is the envelope of $R_{oi}(t)$. If the filter is followed by an envelope detector, its output is simply $r_{oi}(t)$. At time $t = T_b$ this output is

$$
\begin{aligned}
r_{oi} \triangleq r_{oi}(T_b) &= \{x_i^2(T_b) + y_i^2(T_b)\}^{1/2} \\
&= \{x_i^2 + y_i^2\}^{1/2},
\end{aligned}
\qquad (5.7\text{-}32)
$$

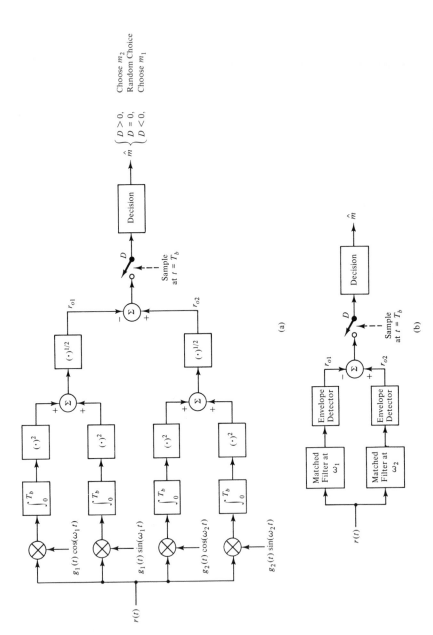

(a)

(b)

Figure 5.7.3. Optimum binary noncoherent receivers that apply when $P_1 \exp(-E_1/\mathcal{N}_0) = P_2 \exp(-E_2/\mathcal{N}_0)$. (a) Correlator form and (b) matched filter form.

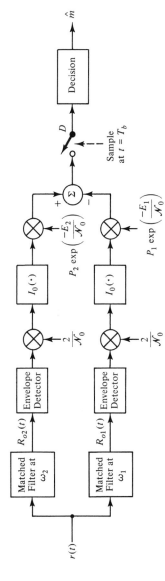

Figure 5.7-4. Matched filter receiver equivalent to that of Fig. 5.7-2.

where $x_i \triangleq x_i(T_b)$ and $y_i \triangleq y_i(T_b)$ are found from (5.7-27) and (5.7-28) to be exactly the same as (5.7-17) and (5.7-18). Thus we conclude that r_{oi}, $i = 1, 2$, in Fig. 5.7-2 can be generated equally well by filters defined by (5.7-25) followed by envelope detectors. The filters are actually matched filters (in white noise) for the two transmitted waveforms. Thus we may construct a matched filter receiver, as shown in Fig. 5.7-4, that is equivalent to the correlation system of Fig. 5.7-2.

If $G_i(\omega)$ is defined as the Fourier transform of $g_i(t)$, it can be shown that the transfer functions of the matched filters, denoted by $H_i(\omega)$, $i = 1, 2$, are

$$H_i(\omega) = \frac{1}{2} G_i^*(\omega - \omega_i)e^{-j(\omega - \omega_i)T_b}$$

$$+ \frac{1}{2} G_i^*(\omega + \omega_i)e^{-j(\omega + \omega_i)T_b}.$$

(5.7-33)

For the special case where (5.7-24) applies, the system of Fig. 5.7-4 will reduce to that shown in Fig. 5.7-3(b).

5.8 NONCOHERENT ASK

For ASK (OOK variety), $g_1(t) = 0$, $g_2(t) = A$, $0 \le t \le T_b$, and $\omega_2 = \omega_0$, so

$$s_2(t) = \begin{cases} A \cos(\omega_0 t + \theta_0), & 0 \le t \le T_b \\ 0, & \text{elsewhere} \end{cases}$$

(5.8-1)

$$s_1(t) = 0.$$

(5.8-2)

From (5.7-15) the energy in $s_2(t)$ is

$$E_2 = \frac{A^2 T_b}{2}.$$

(5.8-3)

Average energy per bit is $P_2 E_2$, so

$$\varepsilon = \frac{P_2 A^2 T_b}{2 \mathcal{N}_0}.$$

(5.8-4)

We next apply the preceding optimum receiver results to ASK.

Noncoherent System and Threshold

We shall consider only the matched filter receiver, since it is the form most often implemented in practice. Figure 5.7-4 applies. However, with $s_1(t) = 0$ an equivalent receiver is shown in Fig. 5.8-1(a). The decision rule of this receiver is

$$\text{If } I_0 \left(\frac{2r_{o2}}{\mathcal{N}_0} \right) > \frac{P_1}{P_2} e^{E_2/\mathcal{N}_0}, \qquad \text{choose } m_2$$

(5.8-5)

and choose m_1 otherwise.

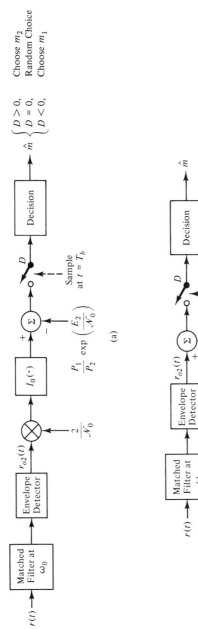

Figure 5.8-1. (a) Optimum noncoherent ASK receiver and (b) a simplified form for $P_1 = P_2 = \frac{1}{2}$.

Although this receiver structure is valid, it would be implemented in a more practical form. Because $r_{o2} \geq 0$ and $I_0(\cdot)$ is a monotonic function, the test of (5.8-5) is entirely equivalent to comparing r_{o2} to a properly chosen threshold, denoted by V_T. The value of V_T can be determined by noting that when $r_{o2} = V_T$, the output level is at the dividing point of the two regions of decision. This fact means that V_T is the solution of (5.8-5) with an equality substituted:

$$I_0\left(\frac{2V_T}{\mathcal{N}_0}\right) = \frac{P_1}{P_2} e^{E_2/\mathcal{N}_0}. \qquad (5.8\text{-}6)$$

Consider only the case where $P_1 = P_2 = \tfrac{1}{2}$ and write (5.8-6) as

$$I_0\left(V_T\sqrt{\frac{2}{E_2\mathcal{N}_0}}\sqrt{\frac{2E_2}{\mathcal{N}_0}}\right) = e^{E_2/\mathcal{N}_0}. \qquad (5.8\text{-}7)$$

Functions of this form—namely, $I_0(b_0\sqrt{2\gamma}) = \exp(\gamma)$—are known [40, p. 291] to have an excellent approximate solution $b_0 = [2 + (\gamma/2)]^{1/2}$. The solution of (5.8-7) becomes

$$V_T = \sqrt{\frac{E_2\mathcal{N}_0}{2}}\sqrt{2 + \frac{E_2}{2\mathcal{N}_0}}. \qquad (5.8\text{-}8)$$

Now it can be shown (Prob. 5-39 and Example 5.8-1) that the signal envelope amplitude, denoted by V_{mf}, and noise power, denoted by N_i, at the input to the envelope detector are

$$V_{mf} = E_2 \qquad (5.8\text{-}9)$$

$$N_i = \frac{\mathcal{N}_0 E_2}{2} = \frac{\mathcal{N}_0 V_{mf}}{2}. \qquad (5.8\text{-}10)$$

On substitution into (5.8-8), the optimum threshold becomes

$$V_T = \frac{V_{mf}}{2}\sqrt{1 + \frac{8N_i}{V_{mf}^2}}. \qquad (5.8\text{-}11)$$

V_T is, therefore, approximately half the signal's amplitude at the input to the envelope detector when signal-to-noise ratio is large. Otherwise V_T depends on signal-to-noise ratio. The optimum receiver, using the equivalent threshold, is shown in Fig. 5.8-1(b).

Example 5.8-1

As an example, we prove (5.8-10). For ASK

$$g_2(t) = A \operatorname{rect}\left[\frac{t - (T_b/2)}{T_b}\right]$$

so

$$G_2(\omega) = AT_b\, e^{-j\omega T_b/2}\operatorname{Sa}\left(\frac{\omega T_b}{2}\right).$$

From (5.7-33) the filter's transfer function is

$$H_2(\omega) = \frac{AT_b}{2}\left\{ \mathrm{Sa}\left[\frac{(\omega - \omega_0)T_b}{2}\right]\exp\left[\frac{-j(\omega - \omega_0)T_b}{2}\right]\right.$$

$$\left. + \mathrm{Sa}\left[\frac{(\omega + \omega_0)T_b}{2}\right]\exp\left[\frac{-j(\omega + \omega_0)T_b}{2}\right]\right\}.$$

After forming $|H_2(\omega)|^2$ and neglecting cross terms as small, because $\omega_0 \gg 2\pi/T_b$ in most systems, we compute envelope detector input noise power to be

$$N_i = \frac{1}{2\pi}\int_{-\infty}^{\infty}\frac{\mathcal{N}_0}{2}|H_2(\omega)|^2\,d\omega$$

$$= \frac{A^2T_b^2\mathcal{N}_0}{16\pi}\left\{\int_{-\infty}^{\infty}\mathrm{Sa}^2\left[\frac{(\omega - \omega_0)T_b}{2}\right]d\omega + \int_{-\infty}^{\infty}\mathrm{Sa}^2\left[\frac{(\omega + \omega_0)T_b}{2}\right]d\omega\right\}$$

$$= A^2T_b\frac{\mathcal{N}_0}{4} = E_2\frac{\mathcal{N}_0}{2}.$$

Bit Error Probability

Denote by r_{o2} the amplitude of the envelope detector's output in Fig. 5.8-1(b) at the sample time. The probability density of r_{o2} when $s_1(t)$ is transmitted is due to noise only, as shown in Fig. 5.8-2 as $p_0(r_{o2})$. It has the Rayleigh distribution

$$p_0(r_{o2}) = \frac{r_{o2}}{N_i}e^{-r_{o2}^2/2N_i}u(r_{o2}). \tag{5.8-12}$$

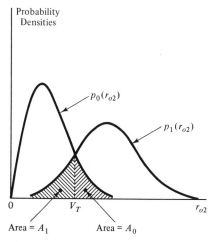

Probability
Densities

$p_0(r_{o2})$

$p_1(r_{o2})$

0   V_T r_{o2}

Area = A_1 Area = A_0

Figure 5.8-2. Probability density functions applicable to noncoherent ASK reception.

When $s_2(t)$ is transmitted, the density of r_{o2} can be shown to be the *Rice* function

$$p_1(r_{o2}) = \frac{r_{o2}}{N_i} \exp\left[-\left(\frac{r_{o2}^2 + V_{mf}^2}{2N_i}\right)\right] I_0\left(\frac{r_{o2} V_{mf}}{N_i}\right) u(r_{o2}), \qquad (5.8\text{-}13)$$

as illustrated in Fig. 5.8-2. Here V_{mf} and N_i are given by (5.8-9) and (5.8-10), respectively.

Average probability of bit error derives from the areas shown in Fig. 5.8-2. Since $P_1 = P_2 = \frac{1}{2}$ is assumed, we get

$$P_e = \frac{1}{2} A_1 + \frac{1}{2} A_0 = \frac{1}{2} \int_0^{V_T} p_1(r_{o2}) \, dr_{o2} + \frac{1}{2} \int_{V_T}^{\infty} p_0(r_{o2}) \, dr_{o2}. \qquad (5.8\text{-}14)$$

V_T is the threshold developed in (5.8-11); it can be shown to occur at the intersection of the curves $p_0(r_{o2})$ and $p_1(r_{o2})$. After using (5.8-12) and (5.8-13) in (5.8-14), assuming a large signal-to-noise ratio so $V_T \approx V_{mf}/2$ and so the approximations

$$I_0(x) \approx \frac{e^x}{\sqrt{2\pi x}}, \qquad x \gg 1 \qquad (5.8\text{-}15)$$

$$\mathrm{erfc}(x) \approx \frac{e^{-x^2}}{\sqrt{\pi}\, x}, \qquad x \gg 1, \qquad (5.8\text{-}16)$$

hold, we obtain

$$P_e \approx \frac{1}{2}\left(1 + \sqrt{\frac{1}{2\pi\varepsilon}}\right) e^{-\varepsilon/2}, \qquad (5.8\text{-}17)$$

where ε is given by (5.8-4). P_e is plotted in Fig. 5.12-1.

5.9 NONCOHERENT FSK

The FSK waveforms involve $g_1(t) = A$, $g_2(t) = A$, both for $0 \le t \le T_b$. Thus

$$s_2(t) = \begin{cases} A \cos(\omega_2 t + \theta_2), & 0 \le t \le T_b \\ 0, & \text{elsewhere} \end{cases} \qquad (5.9\text{-}1)$$

$$s_1(t) = \begin{cases} A \cos(\omega_1 t + \theta_1), & 0 \le t \le T_b \\ 0, & \text{elsewhere.} \end{cases} \qquad (5.9\text{-}2)$$

Energies in $s_2(t)$ and $s_1(t)$ are equal:

$$E_2 = E_1 = \frac{A^2 T_b}{2}. \qquad (5.9\text{-}3)$$

Average energy per bit divided by $\mathcal{N}_0$ is, therefore,

$$\varepsilon = \frac{A^2 T_b}{2\,\mathcal{N}_0}. \qquad (5.9\text{-}4)$$

System Block Diagram

The optimum noncoherent FSK receiver is given by either Fig. 5.7-2 or Fig. 5.7-4, the latter being the more practical implementation. In the important case where $P_1 = P_2 = \frac{1}{2}$, the receiver reduces to the simple form of Fig. 5.9-1. The optimum decision rule becomes a comparison of the output of the two envelope detectors at the sample time. If $r_{o2} > r_{o1}$ the decision is that the message was m_2, while $r_{o2} < r_{o1}$ results in a decision in favor of m_1. If $r_{o2} = r_{o1}$ an arbitrary choice is made. We observe that, in effect, the output of the receiver channel at frequency ω_1 becomes the reference for the other channel.

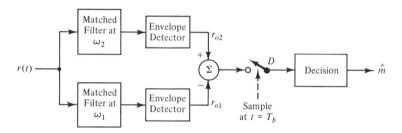

Figure 5.9-1.   Optimum noncoherent FSK receiver when $P_1 = P_2 = \frac{1}{2}$.

Bit Error Probability

From the previous discussion, we know that a decision error will occur when $s_2(t)$ is transmitted if $r_{o2} < r_{o1}$. Call the probability of this error $P(e|m_2)$. When $s_1(t)$ is transmitted an error occurs with probability $P(e|m_1)$ if $r_{o2} > r_{o1}$. Average bit error probability for $P_1 = P_2 = \frac{1}{2}$, as assumed, becomes

$$
\begin{aligned}
P_e &= P(e|m_1)P_1 + P(e|m_2)P_2 \\
&= \frac{1}{2}P(e|m_1) + \frac{1}{2}P(e|m_2).
\end{aligned}
\tag{5.9-5}
$$

Consider $s_2(t)$ transmitted. The matched filter at frequency ω_2 will have a signal response, and the probability density of the envelope detector's output, r_{o2}, at the sample time is Rician according to (5.8-13). The other detector contains no signal so its density is Rayleigh according to (5.8-12) but in terms of r_{o1}. For our assumed white Gaussian channel noise and the FSK signals of (5.9-1) and (5.9-2), the noises out of the matched filters are statistically independent if both $2\omega_0 T_b$ and $2\Delta\omega T_b$ are integral multiples of π. (See Prob. 5-43.) We assume these conditions are true. The joint probability density of r_{o1} and r_{o2} becomes $p_1(r_{o2})p_0(r_{o1})$, so

$$
P(e|m_2) = \int_{r_{o2}=0}^{\infty} p_1(r_{o2}) \int_{r_{o1}=r_{o2}}^{\infty} p_0(r_{o1}) \, dr_{o1} \, dr_{o2}.
\tag{5.9-6}
$$

In a similar development, except with $s_1(t)$ transmitted, we find

$$P(e|m_1) = \int_{r_{o1}=0}^{\infty} p_1(r_{o1}) \int_{r_{o2}=r_{o1}}^{\infty} p_0(r_{o2}) \, dr_{o2} \, dr_{o1} = P(e|m_2). \quad (5.9\text{-}7)$$

On substituting (5.9-6) and (5.9-7) into (5.9-5) and evaluating the integrals [40, p. 298], we obtain

$$P_e = \frac{1}{2} \exp\left(-\frac{\varepsilon}{2}\right) \quad (5.9\text{-}8)$$

where ε is given by (5.9-4). Equation (5.9-8) is plotted in Fig. 5.12-1.

Example 5.9-1

In a coherent orthogonal FSK system, $P_e = 1.105(10^{-5})$ when $\varepsilon = 18.0$. We find the necessary value of ε in a noncoherent system that will produce the same value of P_e. From (5.9-8),

$$\varepsilon = -2 \ln(2P_e) = -2 \ln[2.21(10^{-5})] = 21.44,$$

which is 0.76 dB larger than in the coherent system.

Example 5.9-1 illustrates that only a relatively small penalty must be paid in transmitted signal power to obtain the benefits of the noncoherent FSK system. The largest obvious advantage is that no locally generated phase-coherent carrier is required. However, bit synchronization *is* required in both noncoherent and coherent systems so that samples can be taken at the proper time in each bit interval.

5.10 DIFFERENTIAL PSK

As we have seen, all the coherent keyed carrier systems require a locally generated coherent carrier in the receiver to accomplish demodulation. The need for this local carrier could be overcome in some cases, such as in ASK and FSK systems, by implementing noncoherent versions of the receiver. While the noncoherent systems are usually preferred for their simplicity, there is no truly noncoherent method of demodulating PSK because the information is carried in the carrier's phase. There is, however, a sort of semicoherent method of demodulating PSK that requires no locally generated carrier; it is called *differential* PSK (DPSK).

DPSK System

DPSK is an ingenious technique whereby the carrier reference for a given bit interval is derived from the received waveform in the *preceding* bit interval by use of a 1-bit delay. In essence, the received waveform, delayed by 1 bit (duration T_b), serves as its own reference. A DPSK system based on these ideas is diagramed in Fig. 5.10-1.

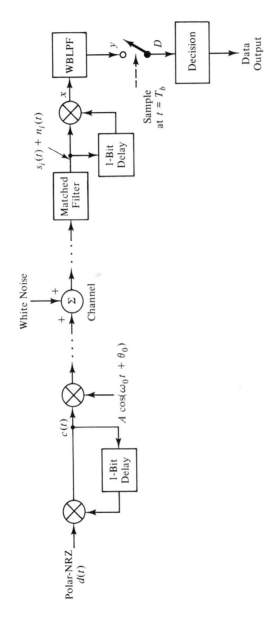

Figure 5.10-1. Block diagram of a binary differential phase-shift keyed system.

307

In the receiver, a sample taken at the end of any given bit interval will depend on the data in the interval of interest and the preceding interval. The dependence on the preceding interval can be removed by differentially encoding the data stream back at the transmitter prior to PSK modulation— thus the name *differential*.

To gain a general idea of system behavior, we shall neglect noise and let the matched filter be replaced by a direct connection. Later, a more careful analysis of the full system is given to support the error probability analysis. At the transmitter of Fig. 5.10-1 a polar-NRZ data waveform $d(t)$ is differentially encoded to form a new data waveform $c(t)$ using the product device with the 1-bit feedback path. The encoding function is described in Fig. 5.6-5, and example waveforms are illustrated in Fig. 5.10-2(a). In the receiver the key to operation is the synchronous detector (product and wideband lowpass filter). Let $s_i(t)$ be the input signal and $s_d(t)$ be the detector output signal. The action of the detector is such that $s_d(t)$ is a positive voltage if the phases of $s_i(t)$ and $s_i(t - T_b)$ are both the same. On the other hand, if the two phases differ by π radians, $s_d(t)$ is a negative voltage. Thus the sign of $s_d(t)$ at the sample time depends on the phase relationship between $s_i(t)$ and its delayed replica. Figure 5.10-2(b) illustrates applicable phase relationships. It is apparent that the sign of $s_d(t)$ is of the same form as $d(t)$. Consequently, sampling to determine the sign of the detector output can be used to reconstruct a replica of the original data signal.

Next, we return to the system of Fig. 5.10-1 and establish signal and noise relationships that will then be used to determine bit error probability.

Signals and Noises in DPSK

At the transmitter we let d_k (having values ± 1) represent $d(t)$ in interval k defined by $kT_b \leq t \leq (k + 1)T_b$. Similarly, c_k represents $c(t)$ in interval k. From Fig. 5.6-5 we have

$$c_k = d_k c_{k-1}, \qquad (5.10\text{-}1)$$

or, since d_k can only have values $+1$ or -1,

$$c_k d_k = c_{k-1}. \qquad (5.10\text{-}2)$$

The transmitted waveform, with peak amplitude A, frequency ω_0, and arbitrary phase θ_0, can be written as the sum of waveforms in all bit intervals. For rectangular pulses we have

$$s_{\text{DPSK}}(t) = \left\{ \sum_{k=-\infty}^{\infty} c_k \, \text{rect}\left[\frac{t - (k + 0.5)T_b}{T_b} \right] \right\} A \cos(\omega_0 t + \theta_0). \quad (5.10\text{-}3)$$

The channel adds white Gaussian noise of power density $\mathcal{N}_0/2$ (for $-\infty < \omega < \infty$) to the DPSK signal. The sum is processed by a receiver filter matched to the *envelope* of a single bit. That is, its impulse response and transfer function are taken as

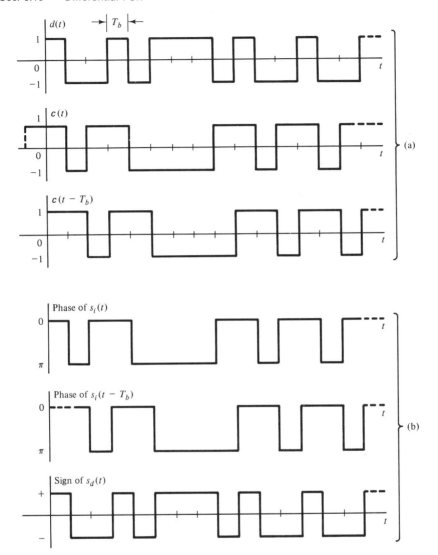

Figure 5.10-2. (a) Waveforms applicable to DPSK signal generation and (b) waveform phases applicable to DPSK signal recovery [1].

$$h_R(t) = \left(\frac{1}{T_b}\right)\text{rect}\left[\frac{(t - 0.5T_b)}{T_b}\right]\cos(\omega_0 t) \qquad (5.10\text{-}4)$$

$$
\begin{aligned}
H_R(\omega) = &\frac{1}{2}\,\text{Sa}\left[\frac{(\omega - \omega_0)T_b}{2}\right]\exp\left[\frac{-j(\omega - \omega_0)T_b}{2}\right] \\
&+ \frac{1}{2}\,\text{Sa}\left[\frac{(\omega + \omega_0)T_b}{2}\right]\exp\left[\frac{-j(\omega + \omega_0)T_b}{2}\right],
\end{aligned}
\qquad (5.10\text{-}5)
$$

respectively. After assuming $\omega_0 \gg 2\pi/T_b$, the filter's output signal, which is the input to the synchronous detector, can be found to be

$$s_i(t) = \left\{ \frac{1}{2} \sum_{k=-\infty}^{\infty} c_k \operatorname{tri}\left[\frac{t - (k + 1)T_b}{T_b} \right] \right\} A \cos(\omega_0 t + \theta_0). \quad (5.10\text{-}6)$$

The noise, denoted by $n_i(t)$, at the same point is readily found to have average power

$$N_i = E[n_i^2(t)] = \frac{1}{2\pi} \int_{-\infty}^{\infty} \frac{\mathcal{N}_0}{2} |H_R(\omega)|^2 \, d\omega = \frac{\mathcal{N}_0}{4T_b}, \quad (5.10\text{-}7)$$

when (5.10-5) is used. Noise autocorrelation function is the inverse transform of the power spectrum, which is $(\mathcal{N}_0/2)|H_R(\omega)|^2$. Calculation reveals

$$R_{n_i}(\tau) = E[n_i(t)n_i(t + \tau)]$$
$$= \left(\frac{\mathcal{N}_0}{4T_b} \right) \operatorname{tri}(\tau/T_b) \cos(\omega_0 \tau). \quad (5.10\text{-}8)$$

This expression is important because it shows that $R_{n_i}(T_b) = 0$, which means that samples of $n_i(t)$ and those of $n_i(t)$ delayed by T_b are uncorrelated. They are also statistically independent because $n_i(t)$ is Gaussian.

Since $n_i(t)$ is bandpass, it can be written in the form

$$n_i(t) = n_1(t)\cos(\omega_0 t + \theta_0) - n_2(t)\sin(\omega_0 t + \theta_0), \quad (5.10\text{-}9)$$

where $n_1(t)$ and $n_2(t)$ are quadrature lowpass noises that each have the same power as $n_i(t)$. (See Sec. B.8 of Appendix B.) That is,

$$N_i = E[n_i^2(t)] = E[n_1^2(t)] = E[n_2^2(t)]. \quad (5.10\text{-}10)$$

We now assume that ω_0 is a multiple of $2\pi/T_b$ so that the delayed noise becomes

$$n_i(t - T_b) = n_1(t - T_b)\cos(\omega_0 t - \omega_0 T_b + \theta_0)$$
$$- n_2(t - T_b)\sin(\omega_0 t - \omega_0 T_b + \theta_0) \quad (5.10\text{-}11)$$
$$= n_3(t)\cos(\omega_0 t + \theta_0) - n_4(t)\sin(\omega_0 t + \theta_0).$$

Here

$$n_3(t) \triangleq n_1(t - T_b) \quad (5.10\text{-}12)$$
$$n_4(t) \triangleq n_2(t - T_b). \quad (5.10\text{-}13)$$

Because $|H_R(\omega)|$ is symmetric about the carrier frequency and because $R_{n_i}(T_b) = 0$, we have the facts that all four noises $n_1(t)$, $n_2(t)$, $n_3(t)$, and $n_4(t)$ are statistically independent (at the same time of interest) and have the same power N_i.

From (5.10-6) the reader can verify that $s_i(t)$ and $s_i(t - T_b)$ comprise a carrier $\cos(\omega_0 t + \theta_0)$ with amplitudes that depend on c_k, c_{k-1}, and c_{k-2} in bit interval k. (See Prob. 5-48.) However, for times near the interval's end (at sample time), we have

$$s_i(t) = \frac{c_k A}{2} \cos(\omega_0 t + \theta_0), \qquad t \text{ near } (k + 1)T_b \qquad (5.10\text{-}14)$$

$$s_i(t - T_b) = \frac{c_{k-1} A}{2} \cos(\omega_0 t + \theta_0), \qquad t \text{ near } (k + 1)T_b. \qquad (5.10\text{-}15)$$

If we let w_1 and w_2 represent the direct and delayed inputs to the product device, respectively, *for times near the end of interval* k,† we have

$$w_1 = \left[\frac{c_k A}{2} + n_1(t) \right] \cos(\omega_0 t + \theta_0) - n_2(t)\sin(\omega_0 t + \theta_0) \qquad (5.10\text{-}16)$$

$$w_2 = \left[\frac{c_{k-1} A}{2} + n_3(t) \right] \cos(\omega_0 t + \theta_0) - n_4(t)\sin(\omega_0 t + \theta_0). \qquad (5.10\text{-}17)$$

The product $w_1 w_2$ is filtered by the wideband lowpass filter (WBLPF), which removes components near $2\omega_0$ in frequency but does not alter the shapes of the baseband components.‡ The signal component of the filtered waveform when sampled at $t = (k + 1)T_b$ is

$$s_i[(k + 1)T_b]s_i[kT_b] \text{ (filtered)} = c_k c_{k-1} A^2/8. \qquad (5.10\text{-}18)$$

There are two cases of interest here. First is when $c_k = c_{k-1}$, and the second is when $c_k \neq c_{k-1}$. In the first case the signal component is positive; it is negative in the second (with the same magnitude, however). When noise is added to the signal component at the sample time, an error occurs in case one when $y \leq 0$ in Fig. 5.10-1, whereas $y > 0$ corresponds to an error in case two. The average probability of these two errors is P_e, the average bit error probability found later. By use of (5.10-2) in (5.10-18), we readily demonstrate that the two cases correspond to the original data level (d_k) being $+1$ or -1, respectively, because

$$s_i[(k + 1)T_b]s_i(kT_b) \text{ (filtered)} = \frac{c_k^2 d_k A^2}{8} = \frac{d_k A^2}{8}. \qquad (5.10\text{-}19)$$

Thus the decoding logic in DPSK is simply to observe the polarity of the sample taken at the end of a bit interval: If positive, we decide $d_k = +1$; if negative, we choose $d_k = -1$.

Bit Error Probability

One can reason that the two error cases described earlier have equal probabilities of occurrence and the probabilities of error are the same when the two values of d_k are equally probable. Thus we describe only case one, where $c_k = c_{k-1}$, since its error probability equals P_e. Similar to Carlson [2, p. 401] let us define quantities α_i, α_q, β_i, and β_q as follows:

---

† For $t = (k + 1)T_b$ these results are exact.

‡ Filter bandwidth can be nearly as large as $\omega_0/2$ in a practical system.

$$\alpha_i = \frac{c_k A + n_1 + n_3}{2\sqrt{2}} \qquad (5.10\text{-}20a)$$

$$\alpha_q = \frac{n_2 + n_4}{2\sqrt{2}} \qquad (5.10\text{-}20b)$$

$$\beta_i = \frac{n_1 - n_3}{2\sqrt{2}} \qquad (5.10\text{-}20c)$$

$$\beta_q = \frac{n_2 - n_4}{2\sqrt{2}}. \qquad (5.10\text{-}20d)$$

Then (5.10-16) and (5.10-17) can be written as

$$w_1 = \sqrt{2}\{[\alpha_i\cos(\omega_0 t + \theta_0) - \alpha_q\sin(\omega_0 t + \theta_0)]$$
$$+ [\beta_i\cos(\omega_0 t + \theta_0) - \beta_q\sin(\omega_0 t + \theta_0)]\} \qquad (5.10\text{-}21)$$

$$w_2 = \sqrt{2}\{[\alpha_i\cos(\omega_0 t + \theta_0) - \alpha_q\sin(\omega_0 t + \theta_0)]$$
$$- [\beta_i\cos(\omega_0 t + \theta_0) - \beta_q\sin(\omega_0 t + \theta_0)]\}. \qquad (5.10\text{-}22)$$

Some thought will reveal that $\alpha_i\cos(\omega_0 t + \theta_0) - \alpha_q\sin(\omega_0 t + \theta_0)$ is the sum of a sinusoidal signal of peak amplitude $c_k A/2\sqrt{2}$ added to Gaussian noise with power $N_i/4$ (see Prob. 5-49). The envelope, denoted by r_1, of this sum has a Rice probability density function

$$p_1(r_1) = \frac{4r_1}{N_i}I_0\left(\frac{\sqrt{2}\,c_k A r_1}{N_i}\right)\exp\left\{-\frac{2}{N_i}\left(r_1^2 + \frac{A^2}{8}\right)\right\}u(r_1). \qquad (5.10\text{-}23)$$

Similarly, $\beta_i\cos(\omega_0 t + \theta_0) - \beta_q\sin(\omega_0 t + \theta_0)$ is seen to be a Gaussian bandpass noise with power $N_i/4$. Its envelope, denoted by r_0, is Rayleigh, given by

$$p_0(r_0) = \frac{4\,r_0}{N_i}\exp(-2r_0^2/N_i)u(r_0). \qquad (5.10\text{-}24)$$

The filtered product $w_1 w_2$ at the sampler at the sample time becomes

$$y = (\alpha_i^2 + \alpha_q^2) - (\beta_i^2 + \beta_q^2) = r_1^2 - r_0^2. \qquad (5.10\text{-}25)$$

Since an error occurs in case one when $y \le 0$, we have the probability

$$P_e = P\{r_1^2 < r_0^2\} = P\{r_1 < r_0\} = \int_0^\infty p_1(r_1)\int_{r_0=r_1}^\infty p_0(r_0)\,dr_0\,dr_1. \qquad (5.10\text{-}26)$$

When (5.10-23) and (5.10-24) are substituted into (5.10-26) and the easy integral over r_0 is computed, the remaining integral is complicated but known [41]. Final evaluation reveals the very simple expression

$$P_e = \frac{1}{2}\exp\left(\frac{-A^2 T_b}{2\,\mathcal{N}_0}\right) = \frac{1}{2}\exp(-\varepsilon) \qquad (5.10\text{-}27)$$

which is plotted in Fig. 5.12-1; in developing (5.10-27) we have defined

$$\varepsilon = \frac{A^2 T_b}{2 \mathcal{N}_0}. \tag{5.10-28}$$

When (5.10-27) is compared to P_e of other systems by determining ε (average energy per bit divided by $\mathcal{N}_0$) required to produce the same P_e, we find that DPSK is 3 dB superior to noncoherent FSK and noncoherent ASK. It is about 2 dB superior to coherent FSK and coherent ASK. However, compared to PSK, QPSK and MSK it is about 1 dB inferior. An example will illustrate this last point.

Example 5.10-1

We find ε in a DPSK system that gives the same P_e as a PSK system with $\varepsilon = $ 6.76 (or 8.30 dB). In the PSK system

$$P_e = \tfrac{1}{2}\text{erfc}[\sqrt{6.76}] = \tfrac{1}{2}\text{erfc}[2.6] = 1.18(10^{-4}).$$

For the DPSK system

$$\tfrac{1}{2}e^{-\varepsilon} = 1.18(10^{-4}),$$

which gives $\varepsilon = 8.352$, or 0.92 dB larger than required in the PSK system. For increasingly larger values of ε (for $P_e < 10^{-4}$), the difference in the two systems approaches zero.

Various implementations of DPSK are possible, including nonbinary, or M-ary, systems. Indeed, even in the binary case, various forms of optimum system structure are possible [14, 42], but these forms are equivalent as noted by Simon [43]. Error probability P_e for these optimum systems is given by (5.10-27), and since the system described above gave this P_e, we note that it is optimum (see also [44, 45]). Finally, a number of books have shown various versions of DPSK system; these are not optimum and their performances have been given by Park [42].

5.11 DIFFERENTIAL DETECTION OF FSK AND MSK SIGNALS

Detector Operation

Binary FM signals, such as FSK and MSK, can also be demodulated by a receiver having a differential detector, as shown in Fig. 5.11-1. To demonstrate this fact let the undelayed input to the detector be the *constant* frequency signal $A\cos[(\omega_0 \pm \Delta\omega)t + \theta_0]$. Here A is peak amplitude, θ_0 is an arbitrary phase angle, frequency is either $\omega_0 + \Delta\omega$ or $\omega_0 - \Delta\omega$, and $\Delta\omega$ is an offset from a carrier frequency ω_0. After delaying this waveform by τ, multiplying the delayed and undelayed waveforms, and eliminating

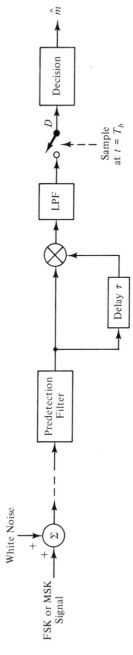

Figure 5.11-1. Binary FM receiver that uses a differential detector for demodulation.

terms at $2\omega_0$ due to the action of the lowpass filter (LPF), the detector's response is $(A^2/2)\cos(\omega_0\tau \pm \Delta\omega\tau)$. Now suppose we require

$$\omega_0\tau = -\frac{\pi}{2} \quad \text{(modulo } 2\pi). \qquad (5.11\text{-}1)$$

The response becomes $(A^2/2)\sin(\pm\Delta\omega\tau)$, which clearly has a maximum response $\pm(A^2/2)$ if τ is selected as

$$\tau = \frac{\pi}{2\Delta\omega}. \qquad (5.11\text{-}2)$$

Since $\pm\Delta\omega$ represents frequency modulations associated with data waveform levels ± 1, we see the detector output is maximum with a sign that correctly corresponds to data levels when (5.11-1) and (5.11-2) are true. In practical systems the smallest $\Delta\omega$ of interest is for MSK, where $\Delta\omega = \pi/2T_b$ (T_b being the bit duration), so the largest τ of interest is $\tau = T_b$.

The preceding discussion assumed a constant frequency (static) waveform. We shall next demonstrate, by means of an example, that the concepts hold true for the true binary FM (dynamic) waveform.

Example 5.11-1

We examine the demodulation process for the first 9 bits of a polar waveform sequence, as shown in Fig. 5.11-2(a) when the system is orthogonal FSK for which $\Delta\omega = \pi/T_b$. For this system the transmitted waveform will advance linearly in phase by $\Delta\omega T_b = \pi$ during each bit interval corresponding to a $+1$ data level and by $-\pi$ for a -1 level. The transmitted signal's phase history is shown as the solid line in (b) (we assume arbitrarily that phase is zero at $t = 0$ for illustration purposes). According to (5.11-2) $\tau = T_b/2$ in the receiver. The $\frac{1}{2}$-bit delayed signal's phase is, therefore, the dashed line in (b). The difference phase between direct and delayed signals is sketched in (c). Due to the requirement that $\omega_0\tau = -\pi/2$ from (5.11-1), the product device response will be $A^2/2$ times the sine of the difference phase, as shown in (d). Clearly, samples taken near the ends of the bit intervals are of the proper signs to recover the original data sequence of (a), and demodulation has occurred.

Noise Performance

Ekanayake [46] has examined the bit error probability of the receiver of Fig. 5.11-1 under the following assumptions: The predetection filter has a symmetrical transfer function about the carrier frequency and is wideband enough that it passes the FM signal with negligible distortion; τ is chosen according to (5.11-2); the carrier frequency satisfies (5.11-1). For an FM signal of peak amplitude A and noise power σ^2 at the predetection filter output, Ekanayake [46] finds

$$P_e = \frac{1}{2}[1 - Q(\sqrt{b}, \sqrt{a}) + Q(\sqrt{a}, \sqrt{b})], \qquad (5.11\text{-}3)$$

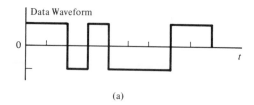

(a)

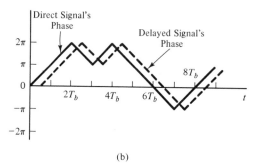

(b)

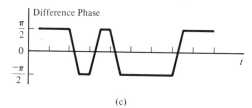

(c)

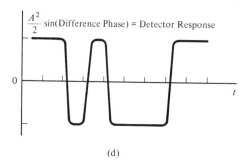

(d)

Figure 5.11-2. (a) Data waveform, (b) phases of direct and delayed FSK signals to product device, (c) difference in phase between direct and delayed waveforms, and (d) detector response of the differential detector.

where

$$a = \frac{A^2(1 - K)^2}{4\sigma^2(1 + |\rho|)} \tag{5.11-4}$$

$$b = \frac{A^2(1 + K)^2}{4\sigma^2(1 + |\rho|)} \tag{5.11-5}$$

$$K^2 = \frac{1 + |\rho|}{1 - |\rho|}. \tag{5.11-6}$$

Here ρ is the normalized autocorrelation function (autocorrelation function divided by σ^2) and $Q(.,.)$ is Marcum's Q function defined by [41]

$$Q(\alpha, \beta) = \int_{\beta}^{\infty} x I_0(\alpha x) \exp\left[-\frac{(\alpha^2 + x^2)}{2} \right] dx \qquad (5.11\text{-}7)$$

$$Q(\alpha, 0) = 1 \qquad (5.11\text{-}8)$$

$$Q(0, \beta) = \exp\left(-\frac{\beta^2}{2} \right). \qquad (5.11\text{-}9)$$

For sources of the tabulated Q function, the reader is referred to the references in [41].

If the shape of the predetection filter's transfer function is chosen such that $\rho = 0$ when evaluated at delay τ, (5.11-3) will reduce to

$$P_e = \left(\frac{1}{2}\right) \exp\left(-\frac{A^2}{2\sigma^2} \right). \qquad (5.11\text{-}10)$$

Because the form of (5.11-10) is the same as (5.10-27) for P_e in DPSK, the potential exists for differential FM detection to perform as well as DPSK. Actual performance depends greatly on the selection of the predetection filter.

If the predetection filter's 3-dB bandwidth, denoted by B (Hz), is large, noise power σ^2 is excessive and P_e degrades. If B is too small, the signal waveform is distorted and intersymbol interference increases, which tends to degrade P_e. Anderson, Bennett, Davy, and Salz, [47] studied an orthogonal FSK system ($\Delta\omega = \pi/T_b$) that used a $\frac{1}{2}$-bit delay and a cosine-shaped filter of bandwidth $B = 1/T_b$; performance was degraded 4.8 dB relative to DPSK (which is about 5.7 dB relative to coherent MSK for P_e near 10^{-4}).† Only 1.8 dB of this degradation was attributed to the demodulation; 3 dB was considered to be the result of power in discrete spectral components that represent wasted power. Suzuki [48] studied an MSK system (1-bit delay) with a Gaussian shaped filter; he found that the optimum bandwidth $B = 1.21/T_b$ gave a degradation minimum of 4.02 dB relative to coherent MSK. Crozier, Mazur, and Matyas [49] studied six filters: Gaussian, two- and four-pole Butterworth (both phase equalized and nonequalized), and ideal rectangular. They found degradation and optimum B in each case. Degradations ranged from 2.9 to 3.4 dB relative to coherent MSK; optimum B ranged from $1.0/T_b$ to $1.2/T_b$. The four-pole equalized filter gives the smallest degradation (2.9 dB) when its bandwidth is $1.1/T_b$ [49]. It was also found in [49] that degradation is minimum for large bandwidth lowpass filters following the detector. These studies tend to show that practical systems tend to give about 3 dB degradation relative to coherent MSK and the best predetection filter bandwidth is approximately $1.1/T_b$.

---

† *Degradation* is defined as the increase in signal-to-noise ratio required, relative to the reference system, to produce the same values of P_e in the two systems.

5.12 NOISE PERFORMANCE COMPARISONS OF SYSTEMS

To compare performance of various systems and to serve as a convenient reference, the various equations that give average bit error probability are summarized in Table 5.12-1. Entries are in order of decreasing performance (higher P_e for ε above about 5 dB). Equations are shown in terms of ε, the average transmitted energy per bit divided by $\mathcal{N}_0$, because this parameter is a fair basis on which to compare systems. A few baseband system entries are also given for completeness. Plots of these expressions are shown in Fig. 5.12-1.

TABLE 5.12-1.   Average Bit Error Probabilities, P_e, of Various Systems.

Equation	Applicable Systems
$P_e = \dfrac{1}{2} \operatorname{erfc}[\sqrt{\varepsilon}\,]$	PSK, QPSK, OQPSK, MSK, Polar-NRZ baseband, Manchester baseband
$P_e = \dfrac{1}{2} \exp(-\varepsilon)$	DPSK
$P_e \approx \dfrac{3}{4} \operatorname{erfc}\left[\dfrac{\pi}{4}\sqrt{\varepsilon}\right]$	Duobinary baseband, Modified duobinary baseband
$P_e = \dfrac{1}{2} \operatorname{erfc}[\sqrt{\varepsilon/2}\,]$	ASK (coherent), FSK (coherent), CPFSK (orthogonal), Unipolar-NRZ baseband
$P_e = \dfrac{1}{2} \exp(-\varepsilon/2)$	FSK (noncoherent)
$P_e \approx \dfrac{1}{2}[1 + (2\pi\varepsilon)^{-1/2}]\exp(-\varepsilon/2)$	ASK (noncoherent)

Of the baseband systems, polar-NRZ and Manchester formats have a clear performance advantage over the duobinary formats, although these latter systems have a small advantage over the unipolar formatted system.

Of the carrier modulated systems, the noncoherent forms (ASK, FSK) are only slightly poorer (by less than 1 dB) than their coherent versions when $P_e < 10^{-4}$. Correspondingly, the DPSK system is less than 1 dB poorer than PSK under similar conditions. Such small losses in performance are often a minor price to pay for the implementation simplicity of the noncoherent systems, which explains, in part, why they enjoy such wide use. Of the noncoherent systems, the DPSK system clearly has superior performance compared to the noncoherent ASK and FSK systems; the performance advantage is 3 dB in ε for a fixed value of P_e.

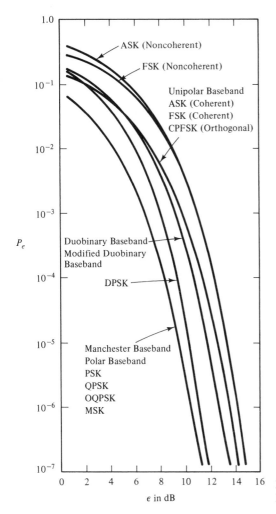

Figure 5.12-1. Plots of P_e versus $10 \log_{10}(\varepsilon)$ for various bandpass systems.

5.13 SUMMARY AND DISCUSSION

When some characteristic of a carrier signal, such as its amplitude, phase, or frequency, is modulated by a digital message, the result is a bandpass waveform. This chapter examines a number of binary digital systems that transmit bandpass waveforms through the channel. These systems can be broadly classified as coherent, noncoherent, and semicoherent.

A coherent system presumes knowledge of the transmitted carrier's phase, timing of bit (symbol) intervals, and the forms (shapes) of the two possible transmitted waveforms. Waveform shape is presumed to be a priori knowledge. The first two quantities are usually determined by special circuitry designed to observe the received waveform to obtain estimates of carrier phase and bit timing. These estimates essentially correspond to *synchronization* of the receiver with the transmitter. Synchronization is not a principal consideration in this chapter, and is assumed, for the most part, to exist without error. Within this constraint, the coherent binary system is found that maximizes the probability that a given symbol is correctly demodulated when the channel has white Gaussian noise and infinite bandwidth. This is defined as the optimum system and its structure is given in Fig. 5.1-1 for both correlator and matched filter forms. The average probability P_e of a symbol error is given by (5.1-9).

Special cases of the general coherent system are also developed. The amplitude shift keyed (ASK) system has the structure of Fig. 5.2-1 and its error probability is given by (5.2-5) when the two transmitted waveforms are equally probable. A system using phase shift keying (PSK) has the block diagram of Fig. 5.3-1 for equally probable messages; its error probability P_e is given by (5.3-5). Quadrature PSK (or QPSK) is a variation of PSK that allows twice the data rate of PSK while requiring no additional channel bandwidth or average energy per bit. QPSK also gives the same bit error probability as PSK. Its modulator and receiver are given in Figs. 5.4-3 and 5.4-4, respectively. Offset QPSK (or OQPSK) is a minor variation of QPSK that is slightly better than QPSK in some practical applications.

Another special case of the general coherent binary system is frequency shift keying (FSK), where carrier frequency is modulated. Two versions of FSK are described for the case of equally probable messages. If keying is between independent oscillator frequencies at the modulator, the system of Fig. 5.5-1 applies; if the frequencies are chosen so that the transmitted waveforms are orthogonal ($\gamma = 0$), P_e is given by (5.5-11). A second version of FSK follows the use of a voltage-controlled oscillator modulated by the binary message; this form of modulation is called continuous phase FSK (CPFSK). The optimum receiver is again given in Fig. 5.5-1. Bit error probability in CPFSK is given by (5.5-22). The smallest frequency deviation $\Delta\omega$ that makes the correlation between the transmitted waveforms in CPFSK zero ($\gamma = 0$) defines minimum shift keying (MSK).

Other possible implementations of MSK are next described. Some implementations used parallel forms for the modulator (Fig. 5.6-2) and receiver (Fig. 5.6-4) and are called type I. When the data stream at the modulator is precoded, the type I MSK system is called fast FSK (FFSK). A slight variation of the type I system leads to what is known as type II MSK (Fig. 5.6-6). Still another form of MSK is called serial (Fig. 5.6-7). Serial MSK has a simpler form than the parallel types and is preferred in some high data-rate systems. Average bit error probability, as given by (5.6-40) for equally probable messages, is the same for both the parallel and serial types of MSK.

The chapter next turns to noncoherent bandpass systems where the receiver has no knowledge of carrier phase. Based on maximizing average probability of making correct bit decisions, the general optimum receiver structure is found to be nonlinear. Both correlator (Fig. 5.7-2) and matched filter forms (Fig. 5.7-4) are possible. Special cases considered are noncoherent ASK (Fig. 5.8-1) and noncoherent FSK (Fig. 5.9-1). Average bit error probability is given by (5.8-17) for noncoherent ASK and by (5.9-8) for noncoherent FSK, both for equally probable messages.

No truly noncoherent form of PSK is possible. However, differential PSK (DPSK), a sort of semicoherent system, requires no carrier phase knowledge. The system (Fig. 5.10-1) uses the received waveform as its own reference. P_e for this system is given by (5.10-27).

Differential demodulation can also be applied to FSK and MSK waveforms, as discussed in Sec. 5.11. Performance (P_e) for this type of demodulator is difficult to determine because it is heavily dependent on the form of the predetection filter that precedes the differential detector.

The chapter ends by grouping the various equations defining P_e for both bandpass and baseband systems (Table 5.12-1). Curves (Fig. 5.12-1) generally show that coherent systems outperform their respective noncoherent (or semicoherent) counterparts, although by less than 1 dB for $P_e < 10^{-4}$. Of five baseband systems compared, those using polar or Manchester waveform formats perform the same and are the most superior (equal also to the best coherent bandpass systems). A system with the unipolar format performs the poorest of the five formats. Duobinary and modified duobinary systems perform equally well and fall in between the other baseband cases, being less than 1 dB superior to the unipolar system.

REFERENCES

[1] Peebles, Jr., P. Z., *Communication System Principles*, Addison-Wesley Publishing Co., Inc., Reading, Massachusetts, 1976. (Figures 5.2-2 and 5.10-2 have been adapted.)

[2] Carlson, A. B., *Communication Systems, An Introduction to Signals and Noise in Electrical Communication*, 2nd ed., McGraw-Hill Book Co., Inc., New York, 1975. (See also third edition, 1986.)

[3] Stremler, F. G., *Introduction to Communication Systems*, 2nd ed., Addison-Wesley Publishing Co., Reading, Massachusetts, 1982.

[4] Pasupathy, S., Minimum Shift Keying: A Spectrally Efficient Modulation, *IEEE Communications Magazine*, Vol. 17, No. 4, July 1979, pp. 14–22.

[5] Rhodes, S. A., Effects of Hardlimiting on Bandlimited Transmissions with Conventional and Offset QPSK Modulation, *Proc. National Telecommunications Conf.*, Houston, Texas, 1972, pp. 20F/1–20F/7.

[6] Rhodes, S. A., Effect of Noisy Phase Reference on Coherent Detection of

Offset QPSK Signals, *IEEE Trans. on Communications*, Vol. COM-22, No. 8, August 1974, pp. 1046–1055.

[7] Gitlin, R. D. and Ho, E. H., The Performance of Staggered Quadrature Amplitude Modulation in the Presence of Phase Jitter, *IEEE Trans. on Communications*, Vol. COM-23, No. 3, March 1975, pp. 348–352.

[8] Lesh, J. R., Costas Loop Tracking of Unbalanced QPSK Signals, *Conference Record 1978 International Conf. on Communications*, Vol. 1, Toronto, Canada, June 4–7, 1978, pp. 16.2.1–16.2.5.

[9] Bhargava, V. K., Haccoun, D., Matyas, R., and Nuspl, P. P., *Digital Communications by Satellite*, John Wiley & Sons, New York, 1981.

[10] Gagliardi, R. M., *Satellite Communications*, Lifetime Learning Publications, Inc., Belmont, California, 1984.

[11] Ziemer, R. E., and Tranter, W. H., *Principles of Communications*, Houghton Mifflin Co., Boston, Massachusetts, 1976.

[12] Spilker, Jr., J. J., *Digital Communications by Satellite*, Prentice-Hall, Englewood Cliffs, New Jersey, 1977.

[13] Simon, M. K., and Lindsey, W. C., Optimum Performance of Suppressed Carrier Receivers with Costas Loop Tracking, *IEEE Trans. on Communications*, Vol. COM-25, No. 2, February 1977, pp. 215–227.

[14] Lindsey, W. C., and Simon, M. K., *Telecommunication Systems Engineering*, Prentice-Hall, Inc., Englewood Cliffs, New Jersey, 1973.

[15] Yamamoto, H., Hirade, K., and Watanabe, Y., Carrier Synchronizer for Coherent Detection of High-Speed Four-Phase-Shift-Keyed Signals, *IEEE Trans. on Communications*, Vol. COM-20, No. 4, August 1972, pp. 803–808.

[16] Gardner, F. M., and Lindsey, W. C. (Editors), Special Issue on Synchronization, *IEEE Trans. on Communications*, Part 1, Vol. COM-28, No. 8, August 1980.

[17] Bennett, W. R., and Rice, S. O., Spectral Density and Autocorrelation Functions Associated with Binary Frequency Shift Keying, *Bell System Technical Journal*, September 1963, pp. 2355–2385.

[18] Anderson, R. R., and Salz, J., Spectra of Digital FM, *Bell System Technical Journal*, July-August 1965, pp. 1165–1189.

[19] Doelz, M. L., and Heald, E. H., Minimum-Shift Data Communication System, U.S. Patent 2,977,417, March 28, 1961, assigned to Collins Radio Company.

[20] Sullivan, W. A., High-Capacity Microwave System for Digital Data Transmission, *IEEE Trans. on Communications*, Vol. COM-20, No. 3, June 1972, pp. 466–470.

[21] De Buda, R., Coherent Demodulation of Frequency-Shift Keying with Low Deviation Ratio, *IEEE Trans. on Communications*, Vol. COM-20, No. 3, June 1972, pp. 429–435.

[22] Gronemeyer, S. A., and McBride, A. L., MSK and Offset QPSK Modulation, *IEEE Trans. on Communications*, Vol. COM-24, No. 8, August 1976, pp. 809–820.

[23] Amoroso, F., and Kivett, J. A., Simplified MSK Signaling Technique, *IEEE Trans. on Communications*, Vol. COM-25, No. 4, April 1977, pp. 433–441.

[24] Ryan, C. R., Hambley, A. R., and Voght, D. E., 760 Mbit/s Serial MSK

Microwave Modem, *IEEE Trans. on Communications*, Vol. COM-28, No. 5, May 1980, pp. 771–777.

[25] Ziemer, R. E., Ryan, C. R., and Stilwell, J. H., Conversion and Matched Filter Approximations for Serial Minimum-Shift Keyed Modulation, *IEEE Trans. on Communications*, Vol. COM-30, No. 3, March 1982, pp. 495–509.

[26] Ziemer, R. E., and Ryan, C. R., Minimum-Shift Keyed Modem Implementations for High Data Rates, *IEEE Communications Magazine*, October 1983, pp. 28–37.

[27] Amoroso, F., Experimental Results on Constant Envelope Signaling with Reduced Spectral Sidelobes, *IEEE Trans. on Communications*, Vol. COM-31, No. 1, January 1983, pp. 157–160.

[28] Amoroso, F., Pulse and Spectrum Manipulation in the Minimum (Frequency) Shift Keying (MSK) Format, *IEEE Trans. on Communications*, Vol. COM-24, No. 3, March 1976, pp. 381–384.

[29] Simon, M. K., A Generalization of Minimum-Shift-Keying (MSK)-Type Signaling Based Upon Input Data Symbol Pulse Shaping, *IEEE Trans. on Communications*, Vol. COM-24, No. 8, August 1976, pp. 845–856.

[30] Rabzel, M., and Pasupathy, S., Spectral Shaping in Minimum Shift Keying (MSK)-Type Signals, *IEEE Trans. on Communications*, Vol. COM-26, No. 1, January 1978, pp. 189–195.

[31] Bazin, B., A Class of MSK Baseband Pulse Formats with Sharp Spectral Roll-Off, *IEEE Trans. on Communications*, Vol. COM-27, No. 5, May 1979, pp. 826–829.

[32] Boutin, N., and Morissette, S., Do all MSK-Type Signaling Waveforms Have Wider Spectra than those for PSK?, *IEEE Trans. on Communications*, Vol. COM-29, No. 7, July 1981, pp. 1071–1072.

[33] Deshpande, G. S., and Wittke, P. H., Optimum Pulse Shaping in Digital Angle Modulation, *IEEE Trans. on Communications*, Vol. COM-29, No. 2, February 1981, pp. 162–168.

[34] Osborne, W. P., and Luntz, M. B., Coherent and Noncoherent Detection of CPFSK, *IEEE Trans. on Communications*, Vol. COM-22, No. 8, August 1974, pp. 1023–1036.

[35] De Buda, R., About Optimal Properties of Fast Frequency-Shift Keying, *IEEE Trans. on Communications*, Vol. COM-22, No. 10, October 1974, pp. 1726–1727.

[36] Murota, K., and Hirade, K., GMSK Modulation for Digital Mobile Radio Telephony, *IEEE Trans. on Communications*, Vol. COM-29, No. 7, July 1981, pp. 1044–1050.

[37] Pasupathy, S., Correlative Encoding: A Bandwidth-Efficient Signaling Scheme, *IEEE Communications Society Magazine*, Vol. 15, No. 4, July 1977, pp. 4–11.

[38] Svensson, A., and Sundberg, C.-E., Serial MSK-Type Detection of Partial Response Continuous Phase Modulation, *IEEE Trans. on Communications*, Vol. COM-33, No. 1, January 1985, pp. 44–52.

[39] Galko, P., and Pasupathy, S., Linear Receivers for Correlatively Coded MSK,

IEEE Trans. on Communications, Vol. COM-33, No. 4, April 1985, pp. 338–347.

[40] Schwartz, M., Bennett, W. R., and Stein, S., *Communication Systems and Techniques*, McGraw-Hill Book Co., New York, 1966.

[41] Stein, S., Unified Analysis of Certain Coherent and Noncoherent Binary Communication Systems, *IEEE Trans. on Information Theory*, Vol. IT-10, No. 1, January 1964, pp. 43–51.

[42] Park, Jr., J. H., On Binary DPSK Detection, *IEEE Trans. on Communications*, Vol. COM-26, No. 4, April 1978, pp. 484–486.

[43] Simon, M. K., Comments on "On Binary DPSK Detection," *IEEE Trans. on Communications*, Vol. COM-26, No. 10, October 1978, pp. 1477–1478.

[44] Wozencraft, J. M., and Jacobs, I. M., *Principles of Communication Engineering*, John Wiley & Sons, Inc., New York, 1965.

[45] Couch, II, L. W., *Digital and Analog Communication Systems*, Macmillan Publishing Co., Inc., New York, 1983.

[46] Ekanayake, N., On Differential Detection of Binary FM, *IEEE Trans. on Communications*, Vol. COM-32, No. 4, April 1984, pp. 469–470.

[47] Anderson, R. R., Bennett, W. R., Davey, J. R., and Salz, J., Differential Detection of Binary FM, *Bell System Technical Journal*, January 1965, pp. 111–159.

[48] Suzuki, H., Optimum Gaussian Filter for Differential Detection of MSK, *IEEE Trans. on Communications*, Vol. COM-29, No. 6, June 1981, pp. 916–918.

[49] Crozier, S., Mazur, B., and Matyas, R., Performance Evaluation of Differential Detection of MSK, *Conference Record*, Vol. 1, IEEE Global Telecommunications Conference, Miami, Florida, November 29–December 2, 1982, pp. 131–135.

PROBLEMS

5-1. The two waveforms transmitted in a binary digital system are

$$s_2(t) = A \sin\left(\frac{4\pi t}{T_b}\right) \text{rect}\left[\frac{t - (T_b/2)}{T_b}\right] \cos(\omega_0 t + \theta_2)$$

$$s_1(t) = A \sin\left(\frac{2\pi t}{T_b}\right) \text{rect}\left[\frac{t - (T_b/2)}{T_b}\right] \cos(\omega_0 t + \theta_1),$$

where θ_1 and θ_2 are arbitrary phase angles. (a) Find expressions for the energies in these signals. (b) For a specified value of T_b, what values of ω_0 are allowed if the two energies are to be independent of ω_0, θ_1, and θ_2? (c) If $\omega_0 T_b \gg 4\pi$, show that $E_1 = E_2 = A^2 T_b/4$.

5-2. In an optimum coherent binary digital system, it is desired that the optimum threshold be zero. Show that this condition is equivalent to energies E_1 and E_2, probabilities P_1 and P_2, and $\mathcal{N}_0$ satisfying

$$P_2 \exp\left[\frac{-E_2}{\mathcal{N}_0}\right] = P_1 \exp\left[\frac{-E_1}{\mathcal{N}_0}\right].$$

5-3. For T_b specified, determine what values of ω_0 will give zero normalized correlation between the waveforms of Prob. 5-1. (*Hint:* Calculate the integral required to find γ first.) If $\omega_0 \gg 3\pi/T_b$, what does γ become?

5-4. In an optimum coherent binary system transmitted waveform energies per bit are $E_1 = 5(10^{-4})$ J and $E_2 = 20(10^{-4})$ J while $\gamma = -0.3$. (a) If channel noise level is determined by $\mathcal{N}_0 = 1.6012(10^{-4})$ W/Hz, what is P_e for this system? (b) If $\mathcal{N}_0$ is reduced to 66.39% of its value in (a) by improved receiver design, what does P_e become? By what factor has P_e improved? Assume $P_1 = P_2 = \frac{1}{2}$ in every case.

5-5. If $E_1 = E_2$ and $\gamma = -1$ in an optimum coherent binary system what minimum value of the ratio of average energy transmitted per bit to $\mathcal{N}_0$ is required to give $P_e = 3.605(10^{-3})$?

5-6. In an ASK system (OOK variety), the probability of a pulse is P_2 and that of no pulse is $1 - P_2$. Plot curves for $\varepsilon = 2.5$, 5.0, and 10.0 to show how the optimum normalized threshold $V_T/\mathcal{N}_0$ of the receiver deviates, as a ratio, from its value when $P_2 = 1 - P_2$. That is, plot $[(V_T/\mathcal{N}_0) - \varepsilon]/\varepsilon$ for various P_2.

5-7. Use (5.2-4) to show that the responses of the receiver of Fig. 5.2-1(c) to the signals of (5.2-1) are identical to those of the receiver of (b) if all responses are evaluated at the sample time $t = T_b$.

5-8. An optimum ASK system, for which $P_1 = P_2$ and $P_e = 5.715(10^{-4})$, uses pulses with peak amplitude 0.2 V. On the channel $\mathcal{N}_0 = 10^{-5}$ W/Hz. (a) What peak signal-to-noise ratio occurs at the sampler at the sample time? (b) What is T_b?

5-9. An optimum coherent ASK system produces a peak signal-to-noise ratio of 12.5 (or 10.97 dB). What is P_e if $P_1 = P_2 = \frac{1}{2}$?

5-10. By integration of (5.2-11) find the power in (a) the discrete spectral components, and (b) the remaining terms. (c) What should the total average power be? Does the sum of powers in (a) and (b) equal what it should?

5-11. An ASK signal does not go to full off when a code **0** is transmitted. If the two peak carrier levels are A_1 and A_0, obtain an expression for the transmitted signal's power density spectrum. Assume $\omega_0 T_b$ is an integral multiple of π if needed.

5-12. In the ASK system of Prob. 5-11, assume message probabilities are $P_1 = P_2 = \frac{1}{2}$, $A_1 = 3$ V, $A_0 = 1$ V, $T_b = 1$ ms, and $\mathcal{N}_0 = 8.681(10^{-5})$ W/Hz. (a) Find the optimum threshold voltage, (b) Find bit error probability P_e.

5-13. A coherent ASK system (OOK variety) transmits with a peak voltage of 5 V over a channel having an unknown loss. If $\mathcal{N}_0 = 6.0(10^{-18})$ W/Hz, bit duration is 0.5 μs and the system performs with $P_e = 10^{-4}$, what is the loss? Assume message probabilities are $P_1 = P_2 = \frac{1}{2}$.

5-14. Work Prob. 5-13 except for a channel having $P_e = 2.035(10^{-4})$.

5-15. In a PSK system ω_0 is an integral multiple of $2\pi/T_b$. If $s_2(t) = A \sin(\pi t/T_b)\sin(\omega_0 t + \theta_0)$ for $0 \leqslant t \leqslant T_b$ and zero elsewhere and $s_1(t) = -s_2(t)$, sketch the block diagram of a correlation-type optimum receiver (for white noise). Find the possible signal components of output voltage at the sampler at the sample time. Assume $P_1 = P_2 = \frac{1}{2}$.

5-16. Show that the system of Fig. 5.3-1 remains optimum even if ω_0 is a multiple of half the bit rate ω_b if the local carrier is phase coherent with the transmitted carrier.

5-17. Find the transfer function of a matched filter that can replace the integrator in Fig. 5.3-1 without loss of performance. Prove that the filter produces the same signal responses and noise power at the sample time as the integrator.

5-18. PSK and coherent ASK systems use the same peak pulse amplitudes and have the same values of P_e and $\mathcal{N}_0$. How are the bit durations T_b related?

5-19. In a PSK system $P_e = 10^{-7}$, $A = 10$ V, $P_1 = P_2 = \frac{1}{2}$, $\mathcal{N}_0 = 3.69(10^{-7})$ W/Hz and the bit stream is the result of time multiplexing bits from N binary PCM encoded sources. What is N if the original analog messages (all similar) had spectral extent $W_f = \pi(10^4)$ rad/s and were sampled at the Nyquist rate with 5-bit PCM encoding?

5-20. Find P_e for a PSK system having $\varepsilon = 10$. If local carrier phase causes the signal levels at the sampler to decrease by 1.5 dB, to what value will P_e change?

5-21. For the PSK system of Prob. 5-19 what is the peak signal-to-noise power ratio that occurs at the sampler at the sample time?

★5-22. As indicated in the text, show that two coherent PSK systems that use the same carrier but in phase-quadrature—that is, one is $\cos(\omega_0 t + \theta_0)$ while the other is $\sin(\omega_0 t + \theta_0)$—operate independently if $\omega_0 T_b = m\pi$, m a positive integer. The proof must show (a) that the output signal of either system is not dependent on the other at the sample times and (b) that noises out of the two systems are independent at the sample times.

5-23. Two separate digital messages each with bit rate $f_b = 1/T_b = 10^4$ bits/s are conveyed to a receiver by QPSK over a channel for which $\mathcal{N}_0 = 9.645(10^{-5})$ W/Hz. If the composite QPSK signal's peak amplitude is $2.5\sqrt{2}$ V, what is the probability that either or both of the messages will have a bit error occur in a given bit interval?

5-24. (a) Sketch the signal constellation for a QAM waveform using polar waveforms of ± 2 V (waveform 1) and ± 3 V (waveform 2). (b) Demonstrate that the QAM signal has constant amplitude in every bit interval. (c) What are the four phases possible in the QAM signal? What is its amplitude?

5-25. (a) Sketch the three polar waveforms applicable to a OQPSK modulator for a source data sequence **0101110010010**. (b) Sketch the carrier phase values with time for this sequence.

5-26. Assume a QPSK signal and no noise are applied to the network of Fig. 5.4-5(b) modified to use a fourth-power nonlinearity, filter at $4\omega_0$ and $\div 4$ divider. Find the divider's output signal. What phase shifts are now needed?

★5-27. Represent a QPSK input signal to the Costas loop of Fig. 5.4-6 by $c_{1k}\cos(\omega_0 t + \theta_0) - c_{2k}\sin(\omega_0 t + \theta_0)$, where $c_{1k} = \mp A$ in symbol interval k according to odd-numbered bits in a data message stream and $c_{2k} = \mp A$ according to the even-numbered data bits. Let the VCO output be $\sin(\omega_0 t + \theta_0 + \phi_\varepsilon)$, where ϕ_ε is a *small* phase error. (a) Find the VCO's input control voltage and show that it tends to force ϕ_ε to zero. (b) Add a phase shift (ambiguity) $\theta_m = m\pi/2$, $m = 1, 2, 3, 4$, to the input QPSK signal and show that the same control voltage is obtained as in part (a). Assume no noise in both (a) and (b).

5-28. Work Prob. 5-13 for a coherent FSK system that uses independently switched oscillators.

5-29. A nonorthogonal CPFSK system has $\varepsilon = 9.68$, $\Delta \omega T_b = 2.0$ and transmits bits of 100 μs duration on a high-frequency carrier where $\omega_0/2\pi \gg 1/T_b$. In the receiver the average noise power at the sampler is known to be $\sigma_o^2 = 1.92(10^{-7})$ W. Find: (a) γ, (b) $\mathcal{N}_0$, (c) A, and (d) P_e for this system.

5-30. Find an equation for the power spectrum of the transmitted FSK waveform in Prob. 5-29. Plot the spectrum normalized to $(A^2 T_b/2)$ versus the variable $-0.7 \leq (\omega - \omega_0)/\omega_b \leq 0.7$. Compare your results to those in Fig. 5.5-3.

5-31. Amplitudes for the first 16 bit intervals of a data waveform used in CPFSK are defined by $\{1, -1, -1, 1, -1, 1, 1, 1, -1, 1, -1, 1, 1, -1, -1, -1\}$. If $\Delta \omega = \pi/4T_b$ for this system, construct a phase trellis showing the history of phase due to modulation. Assume phase is zero at $t = 0$ and note this is *not* an MSK system.

5-32. Find expressions for the carriers that form inputs to the products in Fig. 5.6-2(b) that have other inputs $d_I(t)$ and $d_Q(t)$. Does your work verify that the two networks of (a) and (b) are equivalent?

5-33. Assume a rectangular symbol pulse rect$[(t - T_b)/2T_b]$ is applied to a passive pulse shaping network as in Fig. 5.6-6(a). If the desired response must be $\cos[\pi(t - T_b)/2T_b]$rect$[(t - T_b)/2T_b]$, show that the required network is unrealizable.

5-34. Begin with (5.6-13) and use (5.6-11) to show that type II MSK with differential encoding of data does not give a one-to-one correspondence of FM with data. Give a table similar to Table 5.6-2 to support your conclusion.

5-35. Show that (5.6-17) and (5.6-16) are equivalent.

★5-36. By using the power spectrum of (5.6-32) show that the total power in an MSK signal is $A^2/2$. (*Hint:* Use Parseval's theorem.)

5-37. A serial MSK system transmits a signal with 1.2-V peak amplitude and $9(10^{-4})$ J/bit energy over a channel defined by $\mathcal{N}_0 = 1.8(10^{-4})$ W/Hz. Find: (a) the peak signal power to average noise power ratio at the sampler at the sample time, (b) average bit error probability P_e, and (c) the duration of a bit.

5-38. Show that (5.7-24) can be restated in the form

$$\frac{E_2 - E_1}{2} + \frac{\mathcal{N}_0}{2}\ln\left(\frac{P_1}{P_2}\right) = 0.$$

Note that in an optimum *coherent* system, this condition is equivalent to having a zero threshold V_T [see (5.1-3)].

5-39. Show that (5.8-9) is true. That is, show that the matched filter with transfer function $H_2(\omega)$ found in Example 5.8-1 produces a peak envelope $V_{mf} = E_2 = A^2 T_b/2$ at its output when its input is a typical ASK pulse.

★5-40. Figure 5.8-2 applies to noncoherent ASK system output voltages from the envelope detector at the sample times. Densities are given by (5.8-12) and (5.8-13) and V_{mf} is given by (5.8-9). (a) As a second derivation of ASK results, set up the error probability integrals assuming V_T is *arbitrary*. By differentiation show that the optimum V_T occurs at the crossover of the densities when data bits are equally probable. (b) From your result in (a) use (5.8-9) and (5.8-10) to show it to be equivalent to (5.8-6).

★5-41. Carry out the steps indicated in the text to obtain (5.8-17).

5-42. Use (5.8-16) to obtain an approximate expression for P_e in a coherent ASK system. Relate the coherent and noncoherent ASK systems to each other using (5.8-17) and the approximate expression. If $\varepsilon = 23.8$ in both systems, what are the two error probabilities? Check your values with Fig. 5.12-1.

5-43. Assume both $2\omega_0 T_b$ and $2\Delta\omega T_b$ are integral multiples of π and show that the noises at the matched filter outputs in Fig. 5.9-1 are statistically independent, if taken at the same time. Assume white, zero-mean Gaussian channel noise.

5-44. Consider P_e for noncoherent FSK as given by (5.9-8) and (5.5-11) for coherent FSK. (a) Use the approximation of (5.8-16) and derive the relationship between ε in coherent FSK (denote by ε_c) and ε in the noncoherent system (denote by ε_n) that must hold if both systems give the same values of P_e. (b) Show that $\varepsilon_n/\varepsilon_c \to 1$ as $\varepsilon_c \to \infty$—that is, both systems require the same ε and neither has any advantage over the other for large ε.

5-45. Work Prob. 5-13, except assume a noncoherent FSK system.

5-46. In a certain noncoherent FSK system $\varepsilon = 20.0$. If the channel suddenly increases 1 dB in attenuation, by what factor will P_e increase? Find the two values of P_e.

5-47. Show that the DPSK signal of (5.10-3) causes the response of (5.10-6) at the output of the receiver's matched filter defined by (5.10-4) and (5.10-5).

5-48. Use (5.10-6) to obtain expressions for the direct and 1-bit delayed signals at the DPSK receiver's product device in Fig. 5.10-1 that are applicable in bit interval k—that is, for $kT_b \le t \le (k + 1)T_b$. For t near $(k + 1) T_b$, do your results yield (5.10-14) and (5.10-15)?

★5-49. Justify that the powers in the noises β_i, β_q, and α_q and the noise part of α_i, as given by (5.10-20), are all the same and equal to $N_i/4$, where N_i is the DPSK system noise power at the matched filter's output.

5-50. A DPSK system uses the same energy per bit and the same channel (same $\mathcal{N}_0$) as a PSK system for which $\varepsilon = 5.1$. Find P_e in the two systems. Do your values agree with results from Fig. 5.12-1?

5-51. An FSK signal having $\Delta\omega = 2\pi/3T_b$ is to be differentially detected. (a) Find the required detector delay τ. (b) Sketch waveforms analogous to those in Fig. 5.11-2 if the data sequence is $\{-1, 1, -1, -1, 1, -1, 1, 1, 1, -1\}$ for the first 10 bit intervals past $t = 0$.

★Chapter 6

M-ary Digital Systems

6.0 INTRODUCTION

Except for a small number of digressions (*M*-ary PAM in Chap. 4; QAM and QPSK in Chap. 5), our preceding work on digital systems has concentrated on binary message sources. There are, however, a number of current-day applications that involve nonbinary systems. We call these *M*-ary systems, which refers to the fact that the transmission interval can now contain any one of *M* possible waveforms as opposed to only two in the binary system. The transmission interval is called a *symbol interval* (bit interval in the binary case), and its duration will be labeled T (instead of T_b in the binary system).

To define the *M*-ary system more carefully, we consider Fig. 6.0-1, which is an extension of the binary system of Fig. 4.1-1. The message source can now generate any one of *M* messages in the set $\{m_i\}$ during any symbol interval. To each message the modulator assigns one waveform of a set of *M* waveforms $\{s_i(t)\}$ for transmission over the channel. Thus each waveform is uniquely assigned to one message; its probability of being transmitted is equal to the message's probability of occurrence, denoted by

$$P_i = P(m_i), \qquad i = 1, 2, ..., M. \tag{6.0-1}$$

These signals are assumed to be nonzero only in the symbol intervals in which they are transmitted, so they have duration T.

The waveform at the receiver, denoted generally by $r(t)$, consists of the transmitted signal corrupted by broadband channel noise that we model

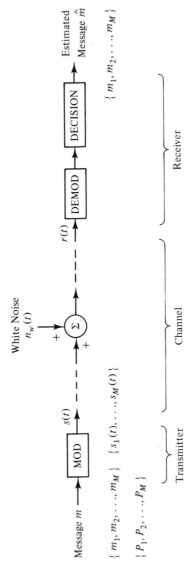

Figure 6.0-1. Basic *M*-ary digital system.

as Gaussian white noise $n_w(t)$. The power density of $n_w(t)$ is denoted as $\mathcal{N}_0/2$, $-\infty < \omega < \infty$. Hence

$$r(t) = r_i(t) = s_i(t) + n_w(t), \qquad i = 1, 2, ..., M. \qquad (6.0\text{-}2)$$

We seek the receiver that is optimum under certain constraints. For example, we assume that it observes $r(t)$ during each symbol interval and then decides which of the messages m_i, $i = 1, 2, ..., M$, was most probably sent. The system is constrained, therefore, to make symbol-by-symbol decisions. The overall system output, represented by $\hat{m}$, consists of one of the M possible messages in each interval.

The optimization criterion that we use is similar to those used for binary systems in Chaps. 4 and 5. That is, the receiver makes the optimum decision by choosing, in each interval, that message with the largest probability of having been sent, given that the observed waveform $r(t)$ occurred. The analysis procedure will again depend on representing the received signal and noise waveforms by a suitable set of random variables. In the binary analyses these random variables were generated by samples taken by the receiver over the bit interval. Although the sampling technique was simple and direct, we shall adopt a second, more elegant and powerful, approach in this chapter. The approach is based on representing both signals and noise by a vector (geometric) model. In the following sections we first establish these vector representations and then develop the optimum systems. Our work is based heavily on the pioneering works of Rice [1] on noise theory, and Kotel'nikov [2] on waveform and system modeling, as well as other literature giving refinements and extensions of the earlier theory, such as Golomb [3], Wozencraft and Jacobs [4], Viterbi [5], Sakrison [6], and Lindsey and Simon [7], to name just a few.

★6.1 VECTOR REPRESENTATION OF SIGNALS

Orthogonal Functions

Let $\{\phi_n(t)\}$ be a set of possibly complex functions of time with the property that

$$\int_{t_1}^{t_2} \phi_n(t)\phi_m^*(t)\, dt = \begin{cases} 0, & n \neq m \\ \lambda, & n = m, \end{cases} \qquad (6.1\text{-}1)$$

where $t_2 > t_1$ and λ is a real constant (the energy in $\phi_n(t)$ on the interval $[t_1, t_2]$). The functions are called *orthogonal* on the interval $[t_1, t_2]$ due to their behavior in (6.1-1). They are said to be *orthonormal* if $\lambda = 1$ (unit energy waveforms).

To illustrate that orthogonal functions are not new to the reader we develop a simple example.

Example 6.1-1

Consider the set of cosines $\{\cos(n2\pi t/T)\}$, $n = 1, 2, ...$, where T is the period of the lowest-frequency waveform. Let $t_1 = 0$ and $t_2 = T$ in (6.1-1). Then

$$\int_0^T \cos\left(\frac{n2\pi t}{T}\right)\cos\left(\frac{m2\pi t}{T}\right) dt = \frac{1}{2} \int_0^T \left\{ \cos\left[\frac{(n + m)2\pi t}{T}\right] \right.$$

$$\left. + \cos\left[\frac{(n - m)2\pi t}{T}\right] \right\} dt$$

$$= \frac{\sin[(n + m)2\pi]}{2(n + m)2\pi/T} + \frac{\sin[(n - m)2\pi]}{2(n - m)2\pi/T}$$

$$= \begin{cases} 0, & n \neq m \\ \dfrac{T}{2}, & n = m, \end{cases}$$

and we find that the cosines are orthogonal.† By dividing by $\sqrt{T/2}$, the set $\{\sqrt{2/T} \cos(n2\pi t/T)\}$ contains orthonormal cosines.

By repeating the above procedure the waveforms $\{\sqrt{2/T} \sin(n2\pi t/T)\}$, $n = 1, 2, ...$, are also found to be orthonormal on $[0, T]$.

Vector Signals and Signal Space

One of the most important uses of orthogonal functions is in representing waveforms. The reader is already familiar with two such representations. The sine and cosine functions of Example 6.1-1 may be used to define a periodic function through the Fourier series of (A.4-2). (Also see Prob. 6-2.) In a similar manner the orthonormal exponentials can be used in the complex Fourier series (A.4-6) to develop a second representation of an arbitrary periodic waveform (Prob. 6-3).

Of more interest to the *M*-ary system of Fig. 6.0-1 is an orthogonal function representation of the transmitted signals $s_i(t)$. We summarize the important ideas in a theorem and then illustrate the theorem's proof:

Theorem. Let $s_i(t)$, $i = 1, 2, ..., M$ be M (possibly complex) signals that are arbitrary except they are all nonzero only for $0 \leq t \leq T$ and have finite energy

$$E_i = \int_0^T |s_i(t)|^2 dt < \infty, \qquad i = 1, 2, ..., M. \qquad (6.1-2)$$

Then a set of $N \leq M$ orthonormal "basis" functions $\phi_n(t)$, $n = 1, 2, ...,$

---

† From L'Hospital's rule, $\lim_{x \to 0} x^{-1}\sin(x) = 1$, which is needed for the case $m = n$.

N, that are nonzero only for $0 \leqslant t \leqslant T$ can always be found such that

$$s_i(t) = \sum_{n=1}^{N} s_{in}\phi_n(t), \qquad i = 1, 2, \ldots, M \qquad (6.1\text{-}3)$$

where

$$s_{in} = \int_0^T s_i(t)\phi_n^*(t) \, dt, \qquad (6.1\text{-}4)$$

for $i = 1, 2, \ldots, M$ and $n = 1, 2, \ldots, N$.

Proof of the above theorem consists of giving a procedure, called the *Gram-Schmidt procedure*, that will always yield the set $\{\phi_n(t)\}$. First, however, we shall introduce the important concept of representing a waveform by a suitably defined *vector*.

Observe from (6.1-3) that once the functions $\phi_n(t)$ have been determined that apply to all M waveforms $s_i(t)$, each waveform is completely defined by its coefficients s_{in}. We visualize these coefficients as components of an N-dimensional geometric space we call the *signal space* that has N mutually orthogonal axes. Each axis, or coordinate, is associated with one of the orthonormal functions. From this point of view the M signals $s_i(t)$ are defined by M points in the N-dimensional signal space through their vectors $(s_{i1}, s_{i2}, \ldots, s_{iN})$, $i = 1, 2, \ldots, M$; vector component (amplitude) s_{i1} is associated with coordinate ϕ_1, component s_{i2} with coordinate ϕ_2, and so on, for signal i. An example will help make these concepts more concrete.

Example 6.1-2
Consider $M = 4$ signals, the first of which is $s_1(t) = A \cos(\omega_0 t + \theta_0) + A \sin(\omega_0 t + \theta_0)$, where ω_0 is a positive integer multiple of $2\pi/T$; A and θ_0 are constants. As in Example 6.1-1, these two sinusoids can be shown to be orthonormal if we define

$$\phi_1(t) = \sqrt{\frac{2}{T}} \cos(\omega_0 t + \theta_0), \qquad 0 \leqslant t \leqslant T$$

$$\phi_2(t) = \sqrt{\frac{2}{T}} \sin(\omega_0 t + \theta_0), \qquad 0 \leqslant t \leqslant T.$$

We use these results to rewrite $s_1(t)$ and define other waveforms as follows:

$$s_1(t) = \sqrt{\frac{E}{2}} \phi_1(t) + \sqrt{\frac{E}{2}} \phi_2(t)$$

$$s_2(t) = -A \cos(\omega_0 t + \theta_0) + A \sin(\omega_0 t + \theta_0)$$

$$= -\sqrt{\frac{E}{2}} \phi_1(t) + \sqrt{\frac{E}{2}} \phi_2(t)$$

$$s_3(t) = -A \cos(\omega_0 t + \theta_0) - A \sin(\omega_0 t + \theta_0)$$

$$= -\sqrt{\frac{E}{2}} \phi_1(t) - \sqrt{\frac{E}{2}} \phi_2(t)$$

$$s_4(t) = A \cos(\omega_0 t + \theta_0) - A \sin(\omega_0 t + \theta_0)$$

$$= \sqrt{\frac{E}{2}} \, \phi_1(t) - \sqrt{\frac{E}{2}} \, \phi_2(t)$$

all for $0 \le t \le T$, where $E = A^2 T$. Signal $s_1(t)$ is defined by its vector ($\sqrt{E/2}$, $\sqrt{E/2}$) in $N = 2$-dimensional signal space having coordinates ϕ_1 and ϕ_2. Others are similarly defined. These vectors are illustrated in Fig. 6.1-1. We call the array of signal vectors a *signal constellation*. All vectors have the same length (called the *norm*), which means all four signals $s_i(t)$ have the same energy per symbol.

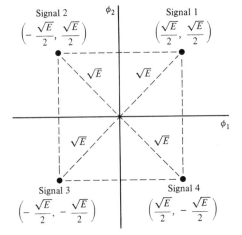

Signal 2
$$\left(-\frac{\sqrt{E}}{2}, \frac{\sqrt{E}}{2} \right)$$

Signal 1
$$\left(\frac{\sqrt{E}}{2}, \frac{\sqrt{E}}{2} \right)$$

Signal 3
$$\left(-\frac{\sqrt{E}}{2}, -\frac{\sqrt{E}}{2} \right)$$

Signal 4
$$\left(\frac{\sqrt{E}}{2}, -\frac{\sqrt{E}}{2} \right)$$

Figure 6.1-1. Signal constellation of four signals in two-dimensional signal space.

The signals above can also be written as

$$s_i(t) = \sqrt{2} \, A \cos \left[\omega_0 t + \theta_0 - \frac{(2i - 1)\pi}{4} \right], \qquad i = 1, 2, 3, 4,$$

which shows that they collectively form a 4-ary (4-phase) PSK signal. Since this same signal constellation was observed in QPSK (Fig. 5.4-1), we conclude that QPSK is the same as 4-ary PSK.

Gram-Schmidt Procedure

Proof of the theorem involving the signal representation of (6.1-3) consists of giving a procedure whereby the orthonormal functions $\phi_n(t)$ are found. The procedure is outlined in the following steps [4, 6, 8, 9] called the *Gram-Schmidt procedure*.

Step 1.   Order (number) signals so $s_1(t) \ne 0$.†
Find the energy in $s_1(t)$:

$$E_1 = \int_0^T |s_1(t)|^2 \, dt. \qquad (6.1\text{-}5)$$

† It is possible that a desired transmission could be no pulse.

Normalize $s_1(t)$ and let the result be the first orthonormal function $\phi_1(t)$:

$$\phi_1(t) = \left(\frac{1}{\sqrt{E_1}}\right)s_1(t). \tag{6.1-6}$$

Clearly,

$$s_1(t) = \sqrt{E_1}\,\phi_1(t) = s_{11}\phi_1(t), \tag{6.1-7}$$

so that

$$s_{11} = \sqrt{E_1}. \tag{6.1-8}$$

Step 2.   Find the projection of $s_2(t)$ into coordinate ϕ_1 by

$$s_{21} = \int_0^T s_2(t)\phi_1^*(t)\,dt. \tag{6.1-9}$$

Subtract the component of $s_2(t)$ that is in the ϕ_1 direction; call the remainder an auxiliary function $\theta_2(t)$ for this second step:

$$\theta_2(t) = s_2(t) - s_{21}\phi_1(t). \tag{6.1-10}$$

If $\theta_2(t) = 0$, disregard $s_2(t)$, renumber remaining signals and repeat this step for a new $s_2(t)$ so that $\theta_2(t) \neq 0$. Compute the energy in $\theta_2(t)$. The energy becomes the square of the component of $s_2(t)$ in direction ϕ_2. Thus

$$E_{\theta_2} = \int_0^T |\theta_2(t)|^2\,dt \triangleq s_{22}^2. \tag{6.1-11}$$

Next, set

$$\phi_2(t) = (1/\sqrt{E_{\theta_2}})\theta_2(t). \tag{6.1-12}$$

Hence

$$\begin{aligned} s_2(t) &= \theta_2(t) + s_{21}\phi_1(t) \\ &= \sqrt{E_{\theta_2}}\,\phi_2(t) + s_{21}\phi_1(t) \\ &= s_{22}\phi_2(t) + s_{21}\phi_1(t) \end{aligned} \tag{6.1-13}$$

from (6.1-10) and (6.1-12).

Step 3.   Find the projections of $s_3(t)$ in coordinates ϕ_1 and ϕ_2:

$$s_{31} = \int_0^T s_3(t)\phi_1^*(t)\,dt \tag{6.1-14}$$

$$s_{32} = \int_0^T s_3(t)\phi_2^*(t)\,dt. \tag{6.1-15}$$

Subtract projections from $s_3(t)$ to obtain the auxiliary function $\theta_3(t)$:

$$\theta_3(t) = s_3(t) - s_{32}\phi_2(t) - s_{31}\phi_1(t). \tag{6.1-16}$$

If $\theta_3(t) = 0$, disregard $s_3(t)$ and renumber remaining signals so $\theta_3(t) \neq 0$. Compute energy in $\theta_3(t)$ and normalize to obtain $\phi_3(t)$:

$$E_{\theta_3} = \int_0^T |\theta_3(t)|^2 \, dt \stackrel{\Delta}{=} s_{33}^2 \tag{6.1-17}$$

$$\phi_3(t) = (1/\sqrt{E_{\theta_3}})\theta_3(t). \tag{6.1-18}$$

Clearly,

$$s_3(t) = s_{33}\phi_3(t) + s_{32}\phi_2(t) + s_{31}\phi_1(t). \tag{6.1-19}$$

Step 4.   Continue by repeating the procedure of Step 3 until no auxiliary functions can be found that are nonzero. There will be $N \leq M$ functions $\phi_n(t)$. If the M signals $s_i(t)$ are not linearly independent, it will result that $N < M$. If all are linearly independent, then $N = M$. This is the end of the procedure.

We call N the *dimension* of the signal space. The set $\{\phi_n(t)\}$, $n = 1, 2, \ldots, N$ is said to be a *basis function set* that spans the N-dimensional signal space.

Signal Energy and Average Power

Energy, denoted by E_i, in signal $s_i(t)$ is related to its vector representation (6.1-3) as follows

$$\begin{aligned}
E_i &= \int_0^T |s_i(t)|^2 \, dt = \int_0^T \sum_{n=1}^N s_{in}\phi_n(t) \sum_{m=1}^N s_{im}^*\phi_m^*(t) \, dt \\
&= \sum_{n=1}^N \sum_{m=1}^N s_{in}s_{im}^* \int_0^T \phi_n(t)\phi_m^*(t) \, dt = \sum_{n=1}^N |s_{in}|^2.
\end{aligned} \tag{6.1-20}$$

Average power $\overline{P}_i$ on $[0, T]$ is energy divided by T, or

$$\overline{P}_i = \frac{1}{T} \sum_{n=1}^N |s_{in}|^2, \qquad i = 1, 2, \ldots, M. \tag{6.1-21}$$

★6.2 VECTOR REPRESENTATION OF NOISE

After having introduced the vector representation of signals, we next extend the concepts to include noise so that they may be used in the analysis of the system of Fig. 6.0-1.

Basic Theorem.   Let $X(t)$ be a zero-mean stationary, possibly complex, random process† with a continuous autocorrelation function $R_X(\cdot)$, and let

---

† The developments can also be extended to include nonstationary processes [10].

$\{\psi_n(t)\}$ be a set of functions orthonormal on the interval $t_1 \leqslant t \leqslant t_2$ that satisfy the integral equation

$$\int_{t_1}^{t_2} R_X(t - \xi)\psi_n(\xi)\, d\xi = |\sigma_n|^2\psi_n(t), \qquad t_1 \leqslant t \leqslant t_2 \qquad (6.2\text{-}1)$$

where σ_n is a constant, then $X(t)$ has the representation (called the *Karhunen-Loeve expansion*)

$$X(t) = \sum_{n=1}^{\infty} X_n\psi_n(t) \qquad (6.2\text{-}2)$$

where the coefficients X_n are random variables given by

$$X_n = \int_{t_1}^{t_2} X(t)\psi_n^*(t)\, dt. \qquad (6.2\text{-}3)$$

The random variables X_n have the property

$$E[X_n^* X_m] = \begin{cases} 0, & n \neq m \\ |\sigma_n|^2, & n = m, \end{cases} \qquad (6.2\text{-}4)$$

which means they are uncorrelated. The equality in (6.2-2) is taken in the sense of zero mean-squared error $\overline{\varepsilon^2}$†

$$\overline{\varepsilon^2} = \lim_{N\to\infty} E\left[\left| X(t) - \sum_{n=1}^{N} X_n\psi_n(t) \right|^2\right] = 0. \qquad (6.2\text{-}5)$$

We shall not give a proof of this theorem. A proof is available in [10] (see also Probs. 6-13–6-15). Rather, we shall apply the theorem to our specific problem.

White Noise Case

For a white noise random process $N(t)$ of power density $\mathcal{N}_0/2$ we have‡

$$R_N(t - \xi) = \left(\frac{\mathcal{N}_0}{2}\right)\delta(t - \xi), \qquad (6.2\text{-}6)$$

so (6.2-1) with $t_1 = 0$ and $t_2 = T$ becomes

$$\int_0^T R_N(t - \xi)\psi_n(\xi)\, d\xi = \int_0^T \left(\frac{\mathcal{N}_0}{2}\right)\delta(t - \xi)\psi_n(\xi)\, d\xi$$

$$= \left(\frac{\mathcal{N}_0}{2}\right)\psi_n(t) = |\sigma_n|^2\psi_n(t). \qquad (6.2\text{-}7)$$

---

† The set $\{\psi_n(t)\}$ is said to be *complete* if $\overline{\varepsilon^2} \to 0$ as $N \to \infty$.

‡ We justify the need for a continuous noise autocorrelation function by treating white noise as the limit as bandwidth becomes infinite of bandlimited noise of constant power density $\mathcal{N}_0/2$ as shown in Fig. 4.1-2.

Thus we find that *any* complete orthonormal set will satisfy (6.2-7) and

$$|\sigma_n|^2 = \frac{\mathcal{N}_0}{2}. \tag{6.2-8}$$

When these results are combined with (6.2-2)–(6.2-4), we have a representation of white noise $N(t)$

$$N(t) = \sum_{n=1}^{\infty} N_n \psi_n(t) \tag{6.2-9}$$

$$N_n = \int_0^T N(t) \psi_n^*(t) \, dt \tag{6.2-10}$$

valid on $[0, T]$, where

$$E[N_n^* N_m] = \begin{cases} 0, & n \neq m \\ \dfrac{\mathcal{N}_0}{2}, & n = m. \end{cases} \tag{6.2-11}$$

★6.3 OPTIMIZATION OF THE *M*-ARY DIGITAL SYSTEM

We are now in a position to develop the optimum *M*-ary system of Fig. 6.0-1 based on the preceding vector modeling of signals and noise.

Signal and Noise Vectors

The transmitted (received) signals $\{s_i(t)\}$ have been represented by the orthonormal representation of (6.1-3), where the set of orthonormal functions $\{\phi_n(t)\}$ was found by the Gram-Schmidt procedure. White noise has the representation (6.2-9), where the functions $\psi_n(t)$, $n = 1, 2, \ldots$, can be any complete set of orthonormal functions. For analysis purposes we desire both signals and noise to use the same orthonormal functions. The conversion is readily accomplished by first arranging signals and noise functions $\psi_n(t)$ in a sequence [6, p. 236]:

$$s_1(t), s_2(t), \ldots, s_M(t), \psi_1(t), \psi_2(t), \ldots . \tag{6.3-1}$$

We then use the Gram-Schmidt procedure to form a new set of orthonormal functions based on the sequence. Clearly, the first N functions will be those found before for the signals. They were labeled $\phi_n(t)$, $n = 1, 2, \ldots, N$. The set now extends beyond N due to the signals $\psi_n(t)$.

Henceforth in this chapter we shall assume all signals and noises are real functions. From our procedure above, the signals become

$$s_i(t) = \sum_{j=1}^{N} s_{ij} \phi_j(t), \qquad i = 1, 2, \ldots, M, \tag{6.3-2}$$

where

$$s_{ij} = \int_0^T s_i(t)\phi_j(t)\,dt, \qquad i = 1, 2, \ldots, M. \qquad (6.3\text{-}3)$$

White Gaussian noise is represented by

$$n_w(t) = \sum_{j=1}^{\infty} n_j\phi_j(t), \qquad (6.3\text{-}4)$$

where the n_j are zero-mean statistically independent Gaussian† random variables:

$$n_j = \int_0^T n_w(t)\phi_j(t)\,dt \qquad (6.3\text{-}5)$$

$$E[n_j n_m] = \begin{cases} 0, & j \neq m \\ \dfrac{\mathcal{N}_0}{2}, & j = m. \end{cases} \qquad (6.3\text{-}6)$$

Noise terms in (6.3-4) for $j > N$ exist in coordinates orthogonal to all coordinates for $j \leq N$. These terms therefore have no affect on the coordinates in which the signals are defined, and it is known [4] that the terms are irrelevant to the optimum receiver. On dropping the irrelevant noise, we denote the relevant noise by $n(t)$:

$$n(t) = \sum_{j=1}^{N} n_j\phi_j(t). \qquad (6.3\text{-}7)$$

In general, the statistically optimum receiver can observe only the received waveform in deciding which message was transmitted in a symbol interval. It is, however, reasonable to assume it has symbol synchronization as well as knowledge of the *shapes* of the possible transmitted waveforms and of message probabilities $P(m_i) = P_i$. Since $r(t)$ can be represented by its vector random variable $(r_1, r_2, \ldots, r_N)$ according to

$$r(t) = \sum_{j=1}^{N} r_j\phi_j(t) \qquad (6.3\text{-}8)$$

$$r_j = \int_0^T r(t)\phi_j(t)\,dt, \qquad j = 1, 2, \ldots, N, \qquad (6.3\text{-}9)$$

its statistical properties are determined by this vector. Now let $\rho(t)$ be a particular value of $r(t)$. Since $r(t) = s_i(t) + n(t)$ in general, we have, in

---

† Since $n_w(t)$ is assumed Gaussian, the coefficients n_j must be Gaussian. Because the n_j are uncorrelated and Gaussian, they must also be statistically independent.

particular†

$$\rho(t) = s_i(t) + n(t) = \sum_{j=1}^{N} \rho_j \phi_j(t), \qquad (6.3\text{-}10)$$

where

$$\rho_j = \int_0^T \rho(t)\phi_j(t)\, dt = s_{ij} + n_j. \qquad (6.3\text{-}11)$$

Decision Rule

We call the receiver optimum if it chooses as the transmitted message that which has the largest probability of being sent, given that $\rho(t)$ is observed in the symbol interval. Thus the optimum receiver sets the message estimate to $\hat{m} = m_k$ if, for $i = 1, 2, \ldots, M$,

$$P(m_k \,|\, r_1 = \rho_1, \ldots, r_N = \rho_N) > P(m_i \,|\, r_1 = \rho_1, \ldots, r_N = \rho_N), \qquad i \neq k. \tag{6.3-12}$$

Equation (6.3-12) defines the optimum receiver's decision rule. The probabilities indicated are called *a posteriori* because they relate to message probabilities after the fact, that is, after a specific waveform has been received. The receiver based on (6.3-12) is, therefore, called a *maximum a posteriori* (MAP) receiver.

By use of Bayes' theorem (Prob. 6-16), it can be shown that

$$P(m_i \,|\, r_1 = \rho_1, \ldots, r_N = \rho_N) = \frac{p_r(r_1 = \rho_1, \ldots, r_N = \rho_N \,|\, m_i)P_i}{p_r(r_1 = \rho_1, \ldots, r_N = \rho_N)}, \tag{6.3-13}$$

where $p_r(r_1, \ldots, r_N \,|\, m_i)$ is the conditional joint probability density of the random variables defining $r(t)$, and $p_r(r_1, \ldots, r_N)$ is the density without condition. On substituting (6.3-13) into (6.3-12), the MAP decision rule becomes

$$p_r(r_1 = \rho_1, \ldots, r_N = \rho_N \,|\, m_k)P_k > p_r(r_1 = \rho_1, \ldots, r_N = \rho_N \,|\, m_i)P_i, \qquad i \neq k. \tag{6.3-14}$$

A receiver that uses (6.3-14) for a decision rule, except without regard for probabilities P_i, is called a *maximum likelihood* receiver [4]. When all messages are equally probable, the maximum likelihood and MAP receivers are the same.

In our system noises n_j are Gaussian and statistically independent. Thus using (6.3-6) and (6.3-11), we have

$$p_r(\rho_1, \ldots, \rho_N \,|\, m_i) = (\pi \mathcal{N}_0)^{-N/2} \exp\left\{ -\frac{1}{\mathcal{N}_0} \sum_{j=1}^{N} (\rho_j - s_{ij})^2 \right\}. \tag{6.3-15}$$

† We think of ρ_j as *specific values* of the random variables r_j in (6.3-8).

The decision rule then reduces to choosing $\hat{m} = m_k$ if

$$\ln(P_k) - \frac{1}{\mathcal{N}_0} \sum_{j=1}^{N} (\rho_j - s_{kj})^2 > \ln(P_i) - \frac{1}{\mathcal{N}_0} \sum_{j=1}^{N} (\rho_j - s_{ij})^2, \qquad i \neq k,$$

(6.3-16)

where $\ln(\cdot)$ represents the natural logarithm.†

Decision Regions

Equation (6.3-16) is helpful in defining those specific values ρ_j that correspond to a correct message decision. First, we observe that $\sum_{j=1}^{N}$ $(\rho_j - s_{ij})^2$ is the square of the Euclidean distance (norm) between the received waveform vector $(\rho_1, ..., \rho_N)$ and the vector $(s_{i1}, ..., s_{iN})$ of signal $s_i(t)$. For equally probable messages (6.3-16) is equivalent to deciding in favor of the message having the least distance between the received waveform vector and the message's waveform vector. The region of all such vector points $(\rho_1, ..., \rho_N)$ in N-dimensional signal space that are closer to point $(s_{i1}, ..., s_{iN})$ than any other point is called the *decision region* for message i and is denoted by I_i. Boundaries of I_i for equally probable messages become perpendicular bisectors of lines drawn between signal points, as illustrated in Fig. 6.3-1 for $M = 4$ messages and $N = 2$ dimensions. When messages have unequal probabilities, boundaries are shifted according to

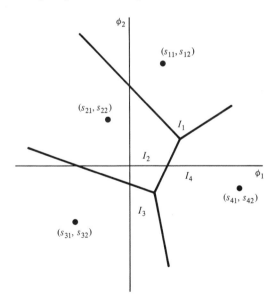

Figure 6.3-1.   Decision regions for four signals in two-dimensional signal space.

† In reducing (6.3-14) the natural logarithm has been used for convenience; this is an allowable operation because the natural logarithm is a monotonic function of its argument.

(6.3-16) toward the signal vector having the smaller probability P_i of each vector pair.

Generally, the determination of decision regions is quite difficult. Only in some simple (but useful) situations can the regions be evaluated. For example, with equally probable messages and signal vectors on a rectangular lattice, decision boundaries are rectangular [4, p. 253].

Error Probability

The probability of symbol error, denoted by P_w, is 1.0 minus the probability that a symbol is correct. The probability of a correct symbol is obtained by averaging all of the conditional probabilities of a correct symbol, denoted by $P(C \mid m_i)$.

Thus

$$P_w = 1 - \sum_{i=1}^{M} P(C \mid m_i)P_i. \qquad (6.3\text{-}17)$$

Since $P(C \mid m_i)$ is the result of points $(\rho_1, \ldots, \rho_N)$ falling in decision region I_i, we have

$$P_w = 1 - \sum_{i=1}^{M} P_i \int_{I_i} \cdots \int p_r(\rho_1, \ldots, \rho_N \mid m_i)\, d\rho_1 \cdots d\rho_N. \qquad (6.3\text{-}18)$$

In certain special cases where regions I_i can be specified, (6.3-18) can be evaluated using (6.3-15). We subsequently consider some of these cases.

Some Signal Constellation Properties

We state some useful properties of signal vectors and decision regions when messages are equally probable, are transmitted on a white Gaussian noise channel, and $M > 2$. For additional discussion and proofs the reader is referred to Wozencraft and Jacobs [4].

1. The probability that a received signal vector $(\rho_1, \rho_2, \ldots, \rho_N)$ falls in decision region I_i corresponding to signal vector $(s_{i1}, s_{i2}, \ldots, s_{iN})$ is unaffected if the signal vector and I_i are translated together through signal space.
2. The probability in property 1 above is also unaffected by a rotation of I_i about the signal vector of $s_i(t)$.
3. The origin of signal space can be shifted to a new location that minimizes the average energy in the signals $s_i(t)$ without affecting the probability of symbol error. The new origin is the centroid of the signal constellation.
4. Systems with the same signal constellations but different orthonormal functions will give the same minimum symbol error probability.

★6.4 OPTIMUM RECEIVERS

Even though decision regions and error probability may be difficult to calculate, the structure of the optimum receiver can readily be found. By expanding the squares in (6.3-16) and using (6.1-20), the decision rule can be written as: choose $\hat{m} = m_k$ if

$$D_k > D_i, \qquad \text{all } i \neq k, \tag{6.4-1}$$

where we define

$$D_i = R_i - B_i, \tag{6.4-2}$$

$$B_i = \left(\frac{E_i}{2}\right) - \left(\frac{\mathcal{N}_0}{2}\right)\ln(P_i), \tag{6.4-3}$$

and

$$R_i = \sum_{j=1}^{N} \rho_j s_{ij}. \tag{6.4-4}$$

We refer to R_i, B_i, and D_i as response, bias, and decision variables, respectively. The optimum receiver structure depends on how R_i is represented.

Correlation Receiver Structures

First, let s_{ij} in (6.4-4) be replaced according to (6.3-3) and then use (6.3-10) to get

$$R_i = \int_0^T \rho(t)s_i(t)\,dt, \qquad i = 1, 2, \ldots, M. \tag{6.4-5}$$

The optimum receiver of Fig. 6.4-1(a) implements the decision rule (6.4-1), when R_i is represented as in (6.4-5). Because $\rho(t)$ is just a specific value of $r(t)$, we have chosen to replace it by $r(t)$ in the figure to represent the fact that the receiver operates on $r(t)$ as shown, whatever its particular value.

Alternatively, the response variables of (6.4-4) can be written differently by replacing ρ_j according to (6.3-11) to obtain

$$R_i = \sum_{j=1}^{N} s_{ij} \int_0^T \rho(t)\phi_j(t)\,dt. \tag{6.4-6}$$

The optimum receiver can now be put in the form shown in Fig. 6.4-1(b). This form of receiver may be more attractive than that in (a) when many signals with few dimensions are to be used, that is, when $N \ll M$.

The difference in the two optimum correlation receivers lies only in the manner in which the response variables R_i are generated. The receiver of Fig. 6.4-1(a) uses the fact that it knows the signals $s_i(t)$ available for transmission. The receiver of Fig. 6.4-1(b) uses the set of orthonormal

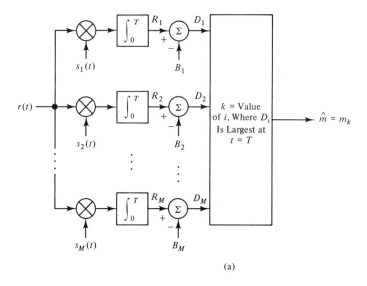

(a)

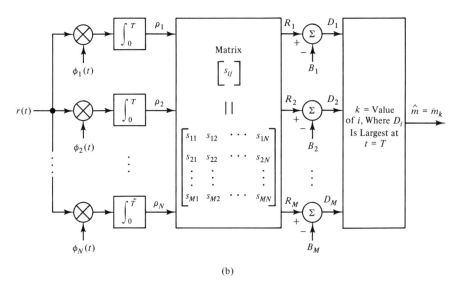

(b)

Figure 6.4-1.   Optimum correlation receivers that (a) use the possible transmitted waveforms and (b) correlate against the orthonormal waveforms.

functions found from the known signals $s_i(t)$ to define the coefficients s_{ij} needed in the matrix (these, of course, define the signal constellation). Both receivers can be simplified somewhat when all possible waveforms $s_i(t)$ have the same energy and are equally probable. Then all biases B_i are the same and the decision rule of (6.4-1) reduces to a test for the largest of the response variables R_i.

Matched Filter Structures

It can be shown that filters with impulse responses, denoted by $h_i(t)$, that are given by

$$h_i(t) = \begin{cases} s_i(T - t), & 0 \le t \le T \\ 0, & \text{elsewhere} \end{cases} \qquad (6.4\text{-}7)$$

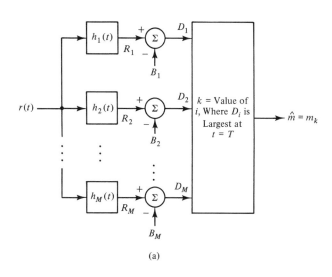

(a)

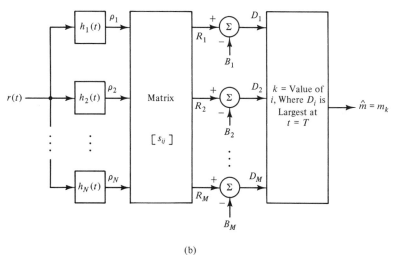

(b)

Figure 6.4-2. Matched filter forms of receivers. (a) Receiver with impulse responses $h_i(t)$ given by (6.4-7) that are equivalent to the receiver of Fig. 6.4-1(a). (b) Receiver with $h_i(t)$ defined by (6.4-8) that is equivalent to that in Fig. 6.4-1(b).

for $i = 1, 2, ..., M$, produce the identical responses to $r(t)$ at the decision time $t = T$ as do the product-integrators in Fig. 6.4-1(a). On replacing these devices by the filters, system performance is unchanged and a matched filter form of receiver is obtained as illustrated in Fig. 6.4-2(a). As the reader may surmise, the name derives from the fact that (6.4-7) defines the white noise matched filter for signal $s_i(t)$. (See (B.9-8).)

In a similar manner, filters defined by impulse responses

$$h_i(t) = \begin{cases} \phi_i(T - t), & 0 \le t \le T \\ 0, & \text{elsewhere} \end{cases} \quad (6.4\text{-}8)$$

can replace the product-integrators in Fig. 6.4-1(b) with no loss in performance. The matched filter receiver is depicted in Fig. 6.4-2(b).

★6.5 *M*-ARY AMPLITUDE SHIFT KEYING

Much of the remainder of this chapter is concerned with specific signal constellations that illustrate the preceding optimum receiver concepts. One of the simplest constellations is associated with *M*-ary amplitude shift keying (*M*-ASK), where a carrier having any one of *M* amplitudes can be sent during a symbol interval.

Signal Constellation

Let the *M* amplitudes A_i be equally probable and uniformly separated by an amount A as defined by

$$A_i = A_1 + (i - 1)A, \qquad i = 1, 2, ..., M. \quad (6.5\text{-}1)$$

The transmitted signals become

$$s_i(t) = A_i \cos(\omega_0 t + \theta_0) = \sqrt{E_i}\, \phi_1(t), \qquad i = 1, 2, ..., M, \quad (6.5\text{-}2)$$

for $0 \le t \le T$, where

$$E_i = \frac{A_i^2 T}{2} \quad (6.5\text{-}3)$$

and only one orthonormal function is required as given by

$$\phi_1(t) = \sqrt{\frac{2}{T}}\cos(\omega_0 t + \theta_0). \quad (6.5\text{-}4)$$

The applicable signal constellation is shown in Fig. 6.5-1. The vector of signal $s_i(t)$ is just a point $s_{i1} = \sqrt{E_i}$ in direction ϕ_1. The distance between points, denoted by d, is found to be

$$d = \sqrt{\frac{A^2 T}{2}}. \quad (6.5\text{-}5)$$

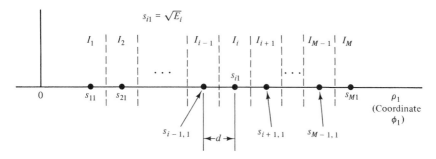

Figure 6.5-1. Signal constellation for M equally spaced M-ASK signals.

Decision region boundaries are located midway between adjacent pairs of points.

System and Its Error Probability

Of the four forms of receiver structure we choose to illustrate only that of Fig. 6.4-2(b) as shown in Fig. 6.5-2. Filter impulse response $h_1(t)$ is given by (6.4-8). Biases B_i are given by (6.4-3) except the term involving $\mathcal{N}_0$ can be omitted because it is the same for all i (when messages are equally probable) and it does not affect the decision rule of (6.4-1).

Average probability of symbol error P_w is given by (6.3-18) using (6.3-15) with only one dimension. After considering property 1 of signal constellations given in Sec. 6.3 and the symmetry of decision regions in Fig.

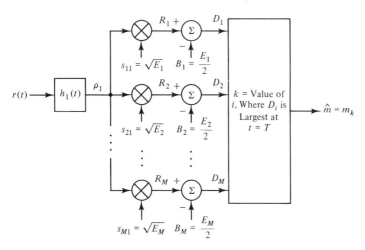

Figure 6.5-2. Optimum M-ASK receiver for waveform set defined by (6.5-1) and (6.5-2).

6.5-1, we see that

$$\int_{I_i} p_r(\rho_1 \mid m_i)\, d\rho_1 = \int_{\sqrt{E_i}-(d/2)}^{\sqrt{E_i}+(d/2)} \frac{1}{\sqrt{\pi \mathcal{N}_0}}\, e^{-(\rho_1 - \sqrt{E_i})^2/\mathcal{N}_0}\, d\rho_1$$

$$= \frac{1}{\sqrt{\pi}} \int_{-d/2\sqrt{\mathcal{N}_0}}^{d/2\sqrt{\mathcal{N}_0}} e^{-\xi^2}\, d\xi \qquad (6.5\text{-}6)$$

$$= 1 - \operatorname{erfc}\left[\frac{d}{2\sqrt{\mathcal{N}_0}}\right], \qquad i = 2, 3, \ldots, M - 1.$$

In a similar manner

$$\int_{I_i} p_r(\rho_1 \mid m_i)\, d\rho_1 = 1 - \frac{1}{2}\operatorname{erfc}\left[\frac{d}{2\sqrt{\mathcal{N}_0}}\right], \qquad i = 1 \text{ or } M. \quad (6.5\text{-}7)$$

After substitution of these two expressions in (6.3-18), we have

$$P_w = \frac{M-1}{M}\operatorname{erfc}\left[\frac{d}{2\sqrt{\mathcal{N}_0}}\right] = \frac{M-1}{M}\operatorname{erfc}\left[\sqrt{\frac{A^2 T}{8\mathcal{N}_0}}\right]. \qquad (6.5\text{-}8)$$

We may compare (6.5-8) with the symbol (word) error probability in *M*-ary PAM given by (4.9-15). By noting that $A^2 T$ is the incremental energy per symbol in the PAM system while it is $A^2 T/2$ in the ASK system, we find the two systems have the same symbol error probability if they both use the same increment of energy per symbol between levels. Since the ASK system can be viewed as a PAM system except with carrier modulation and demodulation operations, we conclude that no performance loss occurs due to the use of a carrier.

Example 6.5-1

As an example of *M*-ASK we compute the average transmitted power for signals defined by (6.5-2) when the *M* messages are equally probable.

$$\text{Average power} = \frac{1}{M}\sum_{i=1}^{M} \frac{A_i^2}{2} = \frac{1}{2M}\sum_{i=1}^{M} [A_1 + (i-1)A]^2.$$

On expanding out the square and using the known sums

$$\sum_{i=1}^{M} i = \frac{M(M+1)}{2}$$

$$\sum_{i=1}^{M} i^2 = \frac{M(M+1)(2M+1)}{6}$$

we get

$$\text{Average power} = \left(\frac{A_1^2}{2}\right) + \left[\frac{A_1 A(M-1)}{2}\right] + \left[\frac{A^2(M-1)(2M-1)}{12}\right].$$

Example 6.5-2

As a second example, suppose we find the amplitude A_1 that minimizes the average transmitted power found in the preceding example. The derivative of the average power is seen to be zero when

$$A_1 = -\frac{A(M-1)}{2}.$$

The second derivative is positive so this value of A_1 does correspond to a minimum. From (6.5-1) we find the other extreme amplitude to be

$$A_M = \frac{A(M-1)}{2}.$$

Thus amplitudes in *M*-ASK should have both positive and negative values, symmetrically displaced about zero, to obtain minimum average transmitted power when messages are equally probable and amplitudes are uniformly separated.

For the binary case, $M = 2$, the amplitudes are simply $-A/2$ and $A/2$. Of course, this corresponds to a carrier being amplitude modulated by a polar waveform format. Thus minimum power binary ASK is the same as binary PSK using a polar data waveform.

★6.6 *M*-ARY PHASE SHIFT KEYING

In *M*-ary PSK (*M*-PSK) the phase of the transmitted signal can take on any one of *M* values in a given symbol interval while amplitude is maintained constant. Thus the signal set $\{s_i(t)\}$ can be defined by

$$s_i(t) = \begin{cases} A\cos(\omega_0 t + \theta_0 + \theta_i), & 0 \le t \le T \\ 0, & \text{elsewhere} \end{cases} \quad (6.6\text{-}1)$$

where

$$\theta_i = \frac{(i-1)2\pi}{M}, \quad i = 1, 2, ..., M. \quad (6.6\text{-}2)$$

Here we assume ω_0 is an integer multiple of $2\pi/T$ and θ_0 is an arbitrary constant phase.

Signal Constellation

By expanding (6.6-1) we have

$$s_i(t) = A\cos(\theta_i)\cos(\omega_0 t + \theta_0) - A\sin(\theta_i)\sin(\omega_0 t + \theta_0) \quad (6.6\text{-}3)$$
$$= s_{i1}\phi_1(t) + s_{i2}\phi_2(t)$$

where

$$\phi_1(t) = \sqrt{\frac{2}{T}}\cos(\omega_0 t + \theta_0) \quad (6.6\text{-}4)$$

$$\phi_2(t) = -\sqrt{\frac{2}{T}}\sin(\omega_0 t + \theta_0) \quad (6.6\text{-}5)$$

are orthonormal functions and

$$s_{i1} = \sqrt{E} \cos(\theta_i) \qquad (6.6\text{-}6)$$

$$s_{i2} = \sqrt{E} \sin(\theta_i) \qquad (6.6\text{-}7)$$

$$\sqrt{E} = \sqrt{\frac{A^2 T}{2}}. \qquad (6.6\text{-}8)$$

The signal constellation applicable to (6.6-3) is shown in Fig. 6.6-1. The special case $M = 8$ is illustrated to demonstrate clearly the decision regions involved. More generally, these wedge-shaped regions are centered on the applicable signal vector, and the wedge's total angle is $2\pi/M$ for the equally probable signals assumed. All signal vectors lay on a circle of radius $\sqrt{E}$ because all signals are assumed to have equal energy over a symbol interval.

Symbol Error Probability

By applying properties 1 and 2 of signal constellations (Sec. 6.3) and noting the symmetry of the decision regions in M-PSK, we find that all integrals in (6.3-18) are equal. Word (symbol) error probability is, therefore,

$$P_w = 1 - \int_{\rho_1 = 0}^{\infty} \int_{\rho_2 = -\rho_1 \tan(\pi/M)}^{\rho_1 \tan(\pi/M)} (\pi \mathcal{N}_0)^{-1} \exp\left[-\frac{(\rho_1 - \sqrt{E})^2}{\mathcal{N}_0} - \frac{\rho_2^2}{\mathcal{N}_0} \right] d\rho_2 \, d\rho_1.$$

$$(6.6\text{-}9)$$

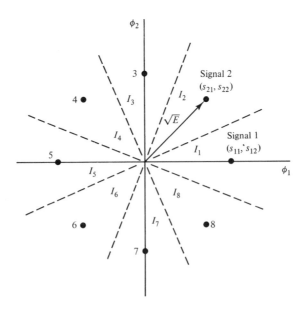

Figure 6.6-1. *M*-ary PSK signal constellation for case $M = 8$.

In writing (6.6-9), use has been made of (6.3-15) with $N = 2$, since only two dimensions are involved and the integrals have been taken over I_1. By simple changes of variables, (6.6-9) can be written as

$$P_w = 1 - \frac{2}{\pi} \int_{u=0}^{\infty} \exp\left[-\left(u - \sqrt{\frac{E}{\mathcal{N}_0}} \right)^2 \right] \int_{\xi=0}^{u \tan(\pi/M)} \exp[-\xi^2]\, d\xi\, du.$$

(6.6-10)

Although the integral over ξ in (6.6-10) can be solved in terms of the error function, the remaining integral apparently has no closed form. It has been shown to have a good approximation [7, 11] for $P_w < 10^{-3}$ and $M > 2$ given by†

$$P_w \approx \text{erfc}\left[\sqrt{\frac{E}{\mathcal{N}_0}}\, \sin\left(\frac{\pi}{M}\right) \right].$$

(6.6-11)

Numerical evaluations of (6.6-10) have also been made. From tabulated data given in Lindsey and Simon [7], we construct the curves of Fig. 6.6-2. These curves are plotted as a function of

$$\varepsilon \triangleq \frac{E}{\mathcal{N}_0 \log_2(M)},$$

(6.6-12)

which is the ratio of average energy per bit of information conveyed in a symbol to $\mathcal{N}_0$. Clearly, as one increases M to transmit more bits of information in a fixed bandwidth (same symbol duration), transmitted power (or energy per symbol) must increase to compensate for the increase in P_w that would otherwise occur.

Example 6.6-1
We compare (6.6-11) for *M*-PSK to (4.9-18) for *M*-PAM when M is large and the same for the two systems. From (6.6-11)

$$P_w \approx \text{erfc}\left[\sqrt{\frac{\pi^2 \log_2(M)}{M^2}\, \varepsilon_{M\text{-PSK}}} \right], \qquad M \gg 1.$$

From (4.9-18)

$$P_w \approx \text{erfc}\left[\sqrt{\frac{3 \log_2(M)}{M^2}\, \varepsilon_{M\text{-PAM}}} \right], \qquad M \gg 1.$$

For the same values of P_w we require

$$\varepsilon_{M\text{-PAM}} = \frac{\pi^2}{3}\, \varepsilon_{M\text{-PSK}} = 3.290\, \varepsilon_{M\text{-PSK}}.$$

Therefore, the *M*-PAM system requires approximately 5.17 dB more average transmitted power than the *M*-PSK system for the same symbol rates, values of M and P_w.

---

† From bounds given in [12] it can be shown that the right side of (6.6-11) is an upper bound on P_w. A lower bound from [12] is $(\frac{1}{2})\text{erfc}[\sqrt{E/\mathcal{N}_0}] < P_w$.

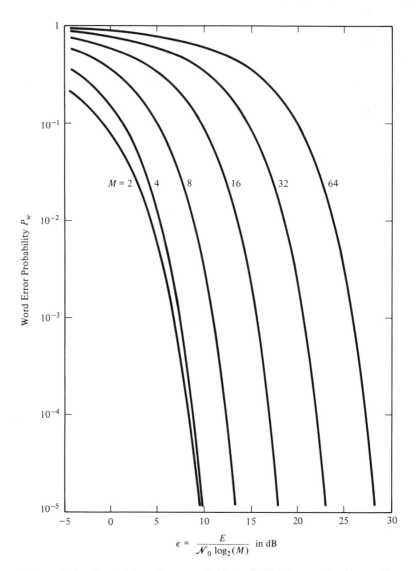

Figure 6.6-2. Symbol (word) error probability of *M*-PSK system having equally probable messages. Curves plotted from data in [7] with permission.

★6.7 ORTHOGONAL SIGNAL SETS

A set of orthogonal signals $\{s_i(t)\}$ can be constructed by choosing orthonormal functions as the signals and assigning only one signal per coordinate in signal space. We have, then, as many signals as there are coordinates and $M = N$. We consider only the case of equally probable messages and equal energy waveforms given by

$$s_i(t) = \begin{cases} \sqrt{E}\ \phi_i(t), & 0 \leq t \leq T \\ 0, & \text{elsewhere} \end{cases} \tag{6.7-1}$$

for $i = 1, 2, \ldots, M = N$. Energy per interval is denoted by E.

Figure 6.7-1 illustrates the orthogonal signal set's constellation for the special case $M = N = 3$.

Decision Rule and System

Because of our assumptions, the biases B_i of (6.4-3) are the same for all messages. They cancel in the decision rule of (6.4-1) so that the optimum decision occurs when the receiver selects message m_k when k corresponds to the largest response variable. That is, k is the value of i for which

$$R_i = \sum_{j=1}^{N} \rho_j s_{ij} \tag{6.7-2}$$

is largest. However, since signal i has a projection only into coordinate i, we have $s_{ij} = 0$ except when $i = j$ where $s_{ii} = \sqrt{E}$. The decision rule reduces to finding which response

$$R_i = \rho_i s_{ii} = \rho_i \sqrt{E} \tag{6.7-3}$$

is largest. Finally, this rule is equivalent to choosing $\hat{m} = m_k$ if

$$\rho_k > \rho_i, \qquad \text{all } i \neq k. \tag{6.7-4}$$

A receiver based on this rule, as derived from Fig. 6.4-2(b), is shown in Fig. 6.7-2.

Error Probability

From symmetry in the signal set, the probability of a correct symbol decision, given a particular message is transmitted, is the same for all

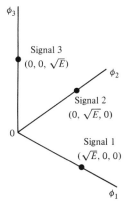

Figure 6.7-1. Signal vectors for $M = N = 3$ orthogonal signals.

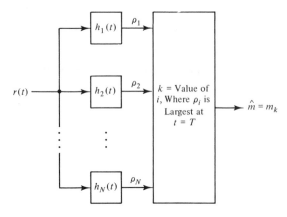

Figure 6.7-2. Optimum receiver for $M = N$ orthogonal, equally probable, equal-energy messages.

messages. This fact means (6.3-17) can be written as

$$P_w = 1 - P(C \mid m_k)$$

$$= 1 - \int_{I_k} \cdots \int p_r(\rho_1, \ldots, \rho_N \mid m_k) \, d\rho_1 \ldots d\rho_N \qquad (6.7\text{-}5)$$

with m_k being the message transmitted and the joint density given by (6.3-15) with $i = k$. Since signal k only has a projection in coordinate k, all $s_{kj} = 0$ except $s_{kk} = \sqrt{E}$ in (6.3-15) and

$$p_r(\rho_1, \ldots, \rho_N \mid m_k) = (\pi \mathcal{N}_0)^{-N/2} e^{-(\rho_k - \sqrt{E})^2/\mathcal{N}_0} \prod_{\substack{i=1 \\ i \neq k}}^{N} \exp\left(-\frac{\rho_i^2}{\mathcal{N}_0}\right). \qquad (6.7\text{-}6)$$

Equation (6.7-4) actually defines the decision region I_k to be used in (6.7-5). For an observed response ρ_k in the correct receiver channel, it says that all other channel responses ρ_i, $i \neq k$, must not exceed ρ_k. Thus (6.7-5) becomes

$$P_w = 1 - \int_{\rho_k = -\infty}^{\infty} \frac{e^{-(\rho_k - \sqrt{E})^2/\mathcal{N}_0}}{\sqrt{\pi \mathcal{N}_0}} \prod_{\substack{i=1 \\ i \neq k}}^{N} \int_{\rho_i = -\infty}^{\rho_k} \frac{\exp(-\rho_i^2/\mathcal{N}_0)}{\sqrt{\pi \mathcal{N}_0}} \, d\rho_i \, d\rho_k$$

$$= 1 - \int_{\rho_k = -\infty}^{\infty} \frac{e^{-(\rho_k - \sqrt{E})^2/\mathcal{N}_0}}{\sqrt{\pi \mathcal{N}_0}} \left[\int_{\rho = -\infty}^{\rho_k} \frac{e^{-\rho^2/\mathcal{N}_0}}{\sqrt{\pi \mathcal{N}_0}} \, d\rho \right]^{N-1} d\rho_k \qquad (6.7\text{-}7)$$

or

$$P_w = 1 - \int_{-\infty}^{\infty} \left\{ 1 - \frac{1}{2} \text{erfc}\left[\frac{\rho}{\sqrt{\mathcal{N}_0}}\right] \right\}^{N-1} \frac{e^{-(\rho - \sqrt{E})^2/\mathcal{N}_0}}{\sqrt{\pi \mathcal{N}_0}} \, d\rho. \qquad (6.7\text{-}8)$$

Although (6.7-8) has no known closed form, it can be upper-bounded by [4, p. 266]†

$$P_w \leq \frac{N-1}{2} \, \text{erfc} \, [\sqrt{E/2\mathcal{N}_0}], \tag{6.7-9}$$

and has been tabulated [7]. Figure 6.7-3 illustrates behavior of (6.7-8) as a function of

$$\varepsilon = \frac{E}{\mathcal{N}_0 \log_2(M)} \tag{6.7-10}$$

using data from [7].

It is interesting to note that the curves of Fig. 6.7-3 imply error-free performance for very large $\log_2(M)$ as long as $\varepsilon > 0.693$ (or -1.59 dB). In practice such an ideal condition is not approached very closely. The limitation does not lie in excess required energy per symbol (see Example 6.7-1), but rather in excessive required bandwidth which increases proportional to M (see Sec. 6.9).

Example 6.7-1

We find the required values of energy to produce $P_w = 2(10^{-4})$ when $\log_2(M) = 1$ (or $M = 2$) and $\log_2(M) = 15$ (or $M = 32{,}768$). From data in [7] or approximately from Fig. 6.7-3, we have $\varepsilon = 12.5$ for $M = 2$ and $\varepsilon = 2.10$ for $M = 2^{15} = 32{,}768$. Thus

$$\frac{\varepsilon \text{ (for } M = 32{,}768)}{\varepsilon \text{ (for } M = 2)} = \frac{2.10}{12.5} = \frac{E \text{ (for } M = 32{,}768) \log_2(2)}{E \text{ (for } M = 2) \log_2(32{,}768)}$$

or

$$\frac{E \text{ (for } M = 32{,}768)}{E \text{ (for } M = 2)} = \frac{2.10(15)}{12.5(1)} = 2.52 \quad \text{(or 4.01 dB)}.$$

The energy required per symbol with $\log_2(M) = 15$ is, therefore, only 2.52 times as great as that needed in a binary system when $P_w = 2(10^{-4})$ and both systems use the same symbol duration.

Simplex Signal Sets

As noted in Sec. 6.3, the origin of signal space can be shifted to a new location to minimize the average energy in the signal set without affecting the symbol error probability when equally probable messages are conveyed over a white Gaussian noise channel. If this signal constellation property is applied to the equal-energy orthogonal signal set, the result is called a *simplex signal set*, which is optimum when energy is constrained [13; 4, p. 260].

Let $(s_{i1}, s_{i2}, \ldots, s_{iM})$ represent the vector of signal $s_i(t)$. Because

---

† The bound becomes increasingly tight as $E/\mathcal{N}_0$ increases for fixed M; for $N = 2$ the bound is the exact result (equality applies) [7, p. 198].

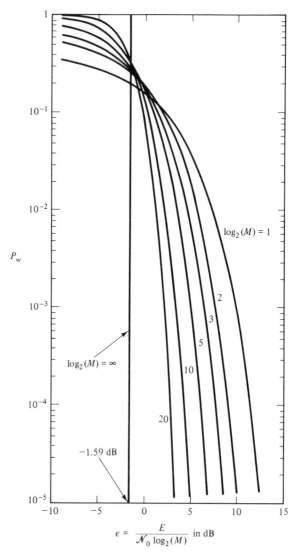

Figure 6.7-3. Symbol error probability for an orthogonal set of equally probable signals. Curves are plotted from data in [7] with permission.

$s_i(t)$ has a projection only into coordinate i, we have

$$s_{ij} = \begin{cases} 0, & j \neq i \\ \sqrt{E}, & j = i, \end{cases} \qquad (6.7\text{-}11)$$

where

$$E = \int_0^T s_i^2(t)\, dt, \qquad \text{all } i \qquad (6.7\text{-}12)$$

is the energy in $s_i(t)$ (same for all i). Thus the vector of $s_i(t)$ is

$$(s_{i1}, s_{i2}, \ldots, s_{ii}, \ldots, s_{iM}) = (0, 0, \ldots, \sqrt{E}, \ldots, 0). \qquad (6.7\text{-}13)$$

The translation to achieve minimum energy is to the constellation's centroid, found by statistically averaging over all signal vectors. This centroid, denoted by the vector $(a_1, a_2, \ldots, a_M)$, is given by (see Prob. 6-33)

$$(a_1, a_2, \ldots, a_M) = \left(\frac{\sqrt{E}}{M}, \frac{\sqrt{E}}{M}, \ldots, \frac{\sqrt{E}}{M} \right). \qquad (6.7\text{-}14)$$

In the new (simplex) coordinates, let the signal vectors have components $s_{ij}^{(s)}$; then

$$s_{ij}^{(s)} = s_{ij} - a_j = \begin{cases} -\sqrt{E}/M, & j \neq i \\ \sqrt{E}\left(1 - \dfrac{1}{M}\right), & j = i, \end{cases} \qquad (6.7\text{-}15)$$

and

$$(s_{i1}^{(s)}, \ldots, s_{ii}^{(s)}, \ldots, s_{iM}^{(s)}) = \left(\frac{-\sqrt{E}}{M}, \ldots, \sqrt{E}\left(1 - \frac{1}{M}\right), \ldots, \frac{-\sqrt{E}}{M} \right). \qquad (6.7\text{-}16)$$

Expressions for the new simplex signals become

$$s_i^{(s)}(t) = \sum_{j=1}^{M} s_{ij}^{(s)} \phi_j(t), \qquad i = 1, 2, \ldots, M. \qquad (6.7\text{-}17)$$

It can be shown (Prob. 6-35) that energies of signals in the simplex set are equal and given by

$$E^{(s)} = E\left(1 - \frac{1}{M}\right). \qquad (6.7\text{-}18)$$

With $M = 2$, 3 and 4, for example, the savings in energy are 3.0, 1.76, and 1.25 dB, respectively; for large M savings become negligible.

An added advantage of the simplex signal set is that its M signals span a space of only $N = M - 1$ dimensions, one less than the original orthogonal set. This fact becomes apparent when we note that the simplex signal vectors result from subtracting the centroid vector (the average over the orthogonal signal vectors) from each orthogonal signal vector. The sum over all simplex signal vectors is therefore the null vector, so any one of the simplex vectors can be expressed as a linear combination of the others, which reduces the number of dimensions by one. An example will help clarify these points.

Example 6.7-2

Consider the binary case where $M = 2$. The applicable signal constellation is shown in Fig. 6.7-4(a). The signals are given by

$$\left. \begin{array}{l} s_1(t) = \sqrt{E}\phi_1(t) \\ s_2(t) = \sqrt{E}\phi_2(t) \end{array} \right\}, \quad 0 \leq t \leq T,$$

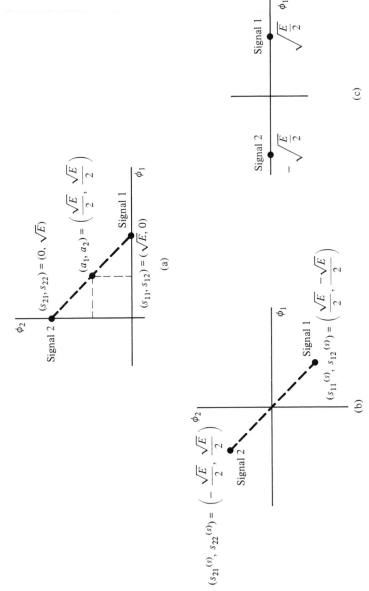

Figure 6.7-4. (a) Signal constellation for a binary orthogonal, equal-energy, equal-probability signal set, (b) the corresponding simplex set, and (c) the constellation equivalent to the simplex set.

so the signal vectors are

$$(s_{11}, s_{12}) = (\sqrt{E}, 0)$$

$$(s_{21}, s_{22}) = (0, \sqrt{E}).$$

The centroid of these vectors is the vector

$$(a_1, a_2) = \left(\frac{\sqrt{E}}{2}, \frac{\sqrt{E}}{2}\right),$$

as shown in the figure.

We shift the origin to the centroid by subtracting the vector (a_1, a_2) from the signal vectors. The new vectors, given by

$$(s_{11}^{(s)}, s_{12}^{(s)}) = \left(\sqrt{E} - \frac{\sqrt{E}}{2}, 0 - \frac{\sqrt{E}}{2}\right) = \left(\frac{\sqrt{E}}{2}, -\frac{\sqrt{E}}{2}\right)$$

$$(s_{21}^{(s)}, s_{22}^{(s)}) = \left(0 - \frac{\sqrt{E}}{2}, \sqrt{E} - \frac{\sqrt{E}}{2}\right) = \left(-\frac{\sqrt{E}}{2}, \frac{\sqrt{E}}{2}\right),$$

are shown in Fig. 6.7-4(b). These vectors, according to (6.7-17), define the simplex signals, which are

$$s_1^{(s)}(t) = \frac{\sqrt{E}}{2}\phi_1(t) - \frac{\sqrt{E}}{2}\phi_2(t) = \frac{\sqrt{E}}{2}[\phi_1(t) - \phi_2(t)]$$

$$s_2^{(s)}(t) = -\frac{\sqrt{E}}{2}\phi_1(t) + \frac{\sqrt{E}}{2}\phi_2(t) = -\frac{\sqrt{E}}{2}[\phi_1(t) - \phi_2(t)].$$

Clearly, the energy in either of these simplex signals is $E/2$, as predicted by (6.7-18). Thus the simplex set uses only half the energy of the orthogonal set.

To show that the simplex set requires only one dimension (one less than the orthogonal set), we observe that the simplex signals can be written as

$$s_1^{(s)}(t) = \sqrt{\frac{E}{2}}\phi_1^{(s)}(t)$$

$$s_2^{(s)}(t) = -\sqrt{\frac{E}{2}}\phi_1^{(s)}(t).$$

Here we define the unit-energy orthogonal function

$$\phi_1^{(s)}(t) = \frac{1}{\sqrt{2}}[\phi_1(t) - \phi_2(t)].$$

These expressions show that the two signals require only one dimension, as shown in Fig. 6.7-4(c).

★6.8 *M-ARY FREQUENCY SHIFT KEYING*

In *M*-ary frequency shift keying (*M*-FSK) the frequency of a carrier is keyed to one of *M* possible frequencies during any symbol interval of duration *T*. Amplitude is usually constant. If we consider only frequencies separated uniformly by an amount $\Delta\omega$, the *M*-FSK signal set can be written as

$$s_i(t) = \begin{cases} A\cos[\omega_0 t + \theta_0 + (i-1)\Delta\omega t], & 0 \le t \le T \\ 0, & \text{elsewhere} \end{cases} \quad (6.8\text{-}1)$$

for $i = 1, 2, ..., M$.

If A and θ_0 are constants, $\Delta\omega$ is an integral multiple of the symbol rate $2\pi/T$, and ω_0 is a multiple of half the symbol rate, it is readily shown that the waveforms

$$\phi_i(t) = \begin{cases} \sqrt{\dfrac{2}{T}}\cos[\omega_0 t + \theta_0 + (i-1)\Delta\omega t], & 0 \le t \le T \\ 0, & \text{elsewhere,} \end{cases} \quad (6.8\text{-}2)$$

are orthonormal on $[0, T]$. (Also see Prob. 6-36.) Thus (6.8-1) can be written as

$$s_i(t) = \sqrt{E}\,\phi_i(t), \quad i = 1, 2, ..., M, \quad (6.8\text{-}3)$$

where

$$E = \frac{A^2 T}{2}. \quad (6.8\text{-}4)$$

From (6.8-3) the signal set is clearly orthogonal because only one signal is assigned to each of the M orthogonal coordinates. One form for the optimum receiver is shown in Fig. 6.7-2; the filters are defined by their impulse responses

$$h_i(t) = s_i(T - t) = \sqrt{E}\,\phi_i(T - t). \quad (6.8\text{-}5)$$

Because the M-FSK signal set is orthogonal, the symbol error probability is given by (6.7-8) for M equally probable signals.

★6.9 QUANTIZED PULSE POSITION MODULATION

Another example of an orthogonal signal set is quantized (M-level) pulse position modulation (M-PPM). PPM is generated by letting the positions of a pulse (relative to a nominal position) vary according to the amplitudes of a message that is quantized to have only M discrete levels. The PPM signal, therefore, has only M discrete time positions that may be occupied. Let the symbol duration T be the interval containing the M positions as shown in Fig. 6.9-1. The pulses

$$\phi_i(t) = \sqrt{\frac{M}{T}}\,\text{rect}\left[\frac{t - \left(i - \frac{1}{2}\right)\left(\frac{T}{M}\right)}{T/M}\right], \quad i = 1, 2, ..., M \quad (6.9\text{-}1)$$

are easily seen to be orthonormal. In terms of the set $\{\phi_i(t)\}$, the available

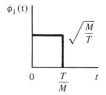

Figure 6.9-1.   Pulse positions of orthonormal pulses in set $\{\phi_i(t)\}$ and the pulse $\phi_1(t)$.

transmitter signals are

$$s_i(t) = A \text{ rect}\left[\frac{t - \left(i - \frac{1}{2}\right)\left(\frac{T}{M}\right)}{T/M}\right] = \sqrt{E}\phi_i(t), \qquad (6.9\text{-}2)$$

where

$$E = \frac{A^2 T}{M} \qquad (6.9\text{-}3)$$

is the energy per pulse.

Symbol error probability for M-PPM is given by (6.7-8) or Fig. 6.7-3 for equally probable message levels where now

$$\varepsilon = \frac{E}{\mathcal{N}_0 \log_2(M)} = \frac{A^2 T}{\mathcal{N}_0 M \log_2(M)}. \qquad (6.9\text{-}4)$$

Interesting comparisons may be made between the orthogonal M-FSK and M-PPM systems. First, as M becomes large the bandwidths of both systems increase but in different ways. Bandwidth in the FSK system increases because M relatively narrowband (order of $1/T$ Hz) waveforms are used that have different spectrum positions (separations about $1/T$ Hz)†. In PPM all transmitted waveforms in the signal set have the *full* bandwidth due to pulses of durations T/M (bandwidth M/T Hz). Thus bandwidth in both cases is about the same and increases linearly with M.

A second comparison of M-PPM with M-FSK considers the pulse amplitudes required to maintain a given performance level (P_w). For equal

† For some values of phase, θ_0, the separation can be as small as $1/2T$ Hz (see Prob. 6-36).

values of M and P_w the two systems require the same values of ε. In the M-FSK system ε is given by

$$\varepsilon = \frac{A^2 T/2}{\mathcal{N}_0 \log_2(M)} \qquad (M\text{-FSK}). \qquad (6.9\text{-}5)$$

It is given by (6.9-4) for M-PPM. On equating the two expressions for the same symbol durations T, we find

$$\frac{A^2}{2}(\text{for } M\text{-FSK}) = \frac{A^2}{M}(\text{for } M\text{-PPM}). \qquad (6.9\text{-}6)$$

In other words, for a given performance in the M-FSK system, the peak pulse power A^2 in M-PPM must *increase* with M to maintain the same performance. This fact places severe constraints on the use of M-PPM for peak-power constrained applications, such as in satellite transponders.

★6.10 BIORTHOGONAL SIGNAL SETS

For a given number of signal space dimensions, N, a simple way to double the number of signals compared to an orthogonal set is to add an additional signal vector in each coordinate that is the negative of the orthogonal signal vector in that coordinate. Signal constellations of this type are called *biorthogonal signal sets*. Figure 6.10-1 depicts a biorthogonal constellation for $N = 2$ dimensions and $M = 4$ signal vectors.

Optimum Receiver

To define the optimum receiver we first define the indexing of signal vectors. We define signals $s_1(t)$ and $s_2(t)$ by vectors $(\sqrt{E}, 0, \ldots, 0)$ and $(-\sqrt{E}, 0, \ldots, 0)$, respectively, in coordinate 1. In coordinate 2 we assign the vector of signal 3 to the positive axis and the vector of signal 4 to the

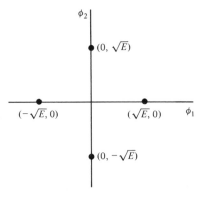

Figure 6.10-1. Signal constellation for biorthogonal signal set when $M = 4$ and $N = 2$.

negative axis. By continuing in this manner, coordinate j has the vector of the odd-numbered signal $s_{2j-1}(t)$ assigned to its positive axis and the vector of the even-numbered signal $s_{2j}(t)$ along the negative axis. With these assignments we have

$$s_{ij} = \begin{cases} \sqrt{E}, & i = 2j - 1, & j = 1, 2, ..., N \\ -\sqrt{E}, & i = 2j, & j = 1, 2, ..., N \\ 0, & \text{otherwise.} \end{cases} \qquad (6.10\text{-}1)$$

The optimum receiver can take on any of the forms of Figures 6.4-1 or 6.4-2. We elect to illustrate only that of Fig. 6.4-1(b).

Because all signals $s_i(t)$ have the same energy E and because all messages have been assumed equally probable, the biases B_i of (6.4-3) are all the same. The receiver can, therefore, be implemented to test for which response variable R_i in Fig. 6.4-1(b) is largest. A further simplification follows the use of (6.10-1). Because every nonzero component s_{ij} has magnitude $\sqrt{E}$, the receiver can simply test normalized response variables $R_i/\sqrt{E}$ for the largest. These considerations lead to the optimum receiver shown in Fig. 6.10-2.

An alternative form of optimum receiver follows the reduction of Fig. 6.4-2(b) in a manner similar to the above. The final receiver is identical to that in Fig. 6.10-2 except the product devices and following integrators are replaced by matched filters defined by (6.4-8).

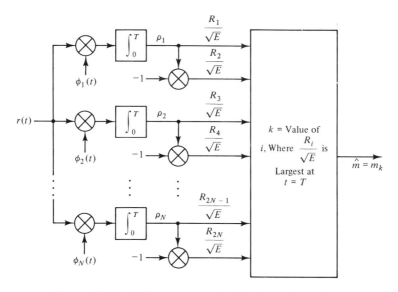

Figure 6.10-2.   Optimum receiver for biorthogonal signal set.

Symbol Error Probability

It can be shown that the symbol error probability P_w is given by [4, 7]

$$P_w = 1 - \int_0^\infty \left\{ 1 - \text{erfc}\left[\frac{\rho}{\sqrt{\mathcal{N}_0}}\right]\right\}^{(M/2)-1} \frac{e^{-(\rho-\sqrt{E})^2/\mathcal{N}_0}}{\sqrt{\pi\mathcal{N}_0}} \, d\rho. \qquad (6.10\text{-}2)$$

No closed form is known for the solution of (6.10-2). However, an upper bound on P_w is known to be [3, 4, 7]†

$$P_w \leq \left(\frac{M-2}{2}\right)\text{erfc}\left[\sqrt{\frac{E}{2\mathcal{N}_0}}\right] + \frac{1}{2}\text{erfc}\left[\sqrt{\frac{E}{\mathcal{N}_0}}\right] \qquad (6.10\text{-}3)$$

and P_w has been numerically evaluated and tabulated [7]. The behavior of P_w is illustrated in Fig. 6.10-3 as plotted from the tabulated data. When these curves for M biorthogonal signals are compared with those of Fig. 6.7-3 for M orthogonal signals, we find nearly the same performance when ε is large and $\log_2(M) \geq 2$, but the biorthogonal signal set requires only half the number of dimensions required by the orthogonal set.

QPSK Example

When $M = 4$ both M-PSK and the biorthogonal signal sets are equivalent to the signal set of QPSK. All three analyses must, therefore, produce the same word error probability. For $M = 4$ minor changes of variables will show that (6.6-10) for M-PSK and (6.10-2) for the biorthogonal sets are the same. These functions are known to have the solution [7]

$$P_w = \text{erfc}\left[\sqrt{\frac{E}{2\mathcal{N}_0}}\right] - \left(\frac{1}{4}\right)\text{erfc}^2\left[\sqrt{\frac{E}{2\mathcal{N}_0}}\right]. \qquad (6.10\text{-}4)$$

To show that (6.10-4) is truly the same as in QPSK, we recall from Chap. 5 that average *bit* error probability was given by (5.4-5). On recognizing that average energy *per symbol* in QPSK is $E = 2A^2 T_b$‡ we write (5.4-5) as

$$P_e = \left(\frac{1}{2}\right)\text{erfc}\left[\sqrt{\frac{E}{2\mathcal{N}_0}}\right]. \qquad (6.10\text{-}5)$$

Now the probability that a word is in error is one minus the probability that both bits on the independent quadrature QPSK channels are correct, which is $(1 - P_e)^2$. Thus

$$P_w = 1 - (1 - P_e)^2 = 2P_e - P_e^2. \qquad (6.10\text{-}6)$$

On substituting (6.10-5), (6.10-6) is seen to equal (6.10-4), as it should.

† The bound is increasingly tight as $E/\mathcal{N}_0$ increases for fixed M.
‡ In Chap. 5 the transmitted waveform's peak amplitude was $\sqrt{2}A$, so peak power is $2A^2/2 = A^2$; energy per symbol of duration $T = 2T_b$ becomes $A^2 2T_b = E$.

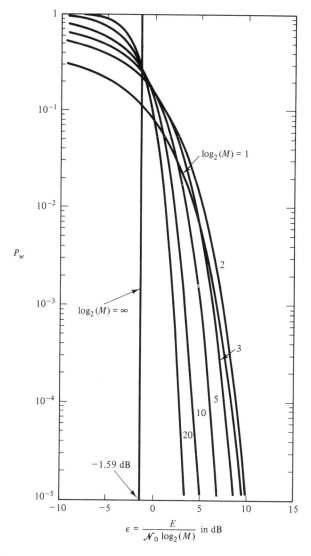

Figure 6.10-3.   Word error probability for optimum system using a biorthogonal signal set for equally probable messages and a white Gaussian noise channel. Curves are plotted from data in Lindsey and Simon [7] with permission.

★6.11 VERTICES OF HYPERCUBE SIGNAL SETS

As a final example of optimum M-ary systems, we consider one in which equally probable signal vectors lie on the corners of a hypercube centered on the origin. Because an N-dimensional hypercube will have 2^N corners, we constrain the problem to the case where $M = 2^N$ messages are used. Fig. 6.11-1 illustrates the geometry involved for $N = 3$ and $M = 8$ signals.

All vectors are equally distant from the origin and will have equal energies per symbol (E). Signal vectors are defined by

$$(s_{i1}, s_{i2}, ..., s_{iN}), \qquad i = 1, 2, ..., M = 2^N \qquad (6.11\text{-}1)$$

where

$$s_{ij} = \pm \sqrt{\frac{E}{N}}, \qquad \text{all } i \text{ and } j. \qquad (6.11\text{-}2)$$

Actual transmitted signals are, of course, given by

$$s_i(t) = \sum_{j=1}^{N} s_{ij}\phi_j(t), \qquad i = 1, 2, ..., M = 2^N. \qquad (6.11\text{-}3)$$

System and Its Error Probability

The optimum system using a vertices of hypercube signal set can employ any one of the optimum receivers of Figs. 6.4-1 or 6.4-2 with the biases B_i neglected (set to zero). The biases can be ignored because of our assumption of equal energy signals and equal probability messages.

Symbol error probability is especially easy to obtain for this signal set. A given message is identified correctly in the receiver if the corresponding signal $s_i(t)$ through its signal vector is correctly identified. The identification amounts to identifying correctly the applicable hypercube corner. Since every corner has a component of its vector in every coordinate, its correct identification amounts to *all* coordinates being correctly demodulated. However, since the noises in all the coordinates are independent with the

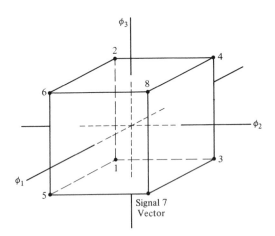

Figure 6.11-1. Signal vectors on vertices of three-dimensional cube (case of $M = 8$, $N = 3$).

same noise powers, the probability of all coordinates being correctly de-modulated is $(1 - p)^N$, where p is the probability that a given coordinate is *not* demodulated correctly. Hence the probability of a symbol (word) error is

$$P_w = 1 - (1 - p)^N. \tag{6.11-4}$$

Probability p is recognized as the average probability of error in a binary decision (one coordinate) involving signals located at $\pm\sqrt{E/N}$. As in polar formatting, this probability readily evaluates to

$$p = \left(\frac{1}{2}\right)\text{erfc}\left[\sqrt{\frac{E}{N\mathcal{N}_0}}\right], \tag{6.11-5}$$

so

$$P_w = 1 - \left\{1 - \left(\frac{1}{2}\right)\text{erfc}\left[\sqrt{\frac{E}{\mathcal{N}_0\log_2(M)}}\right]\right\}^N. \tag{6.11-6}$$

In many practical cases p is small and

$$P_w \approx \left(\frac{N}{2}\right)\text{erfc}\left[\sqrt{\frac{E}{\mathcal{N}_0\log_2(M)}}\right] \tag{6.11-7}$$

for $E/\mathcal{N}_0\log_2(M)$ large.

Polar NRZ Format Example

The vertices of a hypercube signal set has a direct application to the polar-NRZ waveform format. Consider M message levels encoded with N-bit binary codewords, where the polar waveform has rectangular bit pulses of amplitude $\pm A$ and duration T_b. A symbol, or word, has duration

$$T = NT_b \tag{6.11-8}$$

and energy

$$E = A^2T = A^2NT_b. \tag{6.11-9}$$

A codeword waveform, $s_i(t)$, can be expressed as

$$s_i(t) = \sum_{j=1}^{N} \alpha_{ij}\,\text{rect}\left[\frac{t - (j - 0.5)T_b}{T_b}\right], \tag{6.11-10}$$

where

$$\alpha_{ij} = \pm A \tag{6.11-11}$$

with the sign corresponding to data bit j in codeword i. To illustrate these points for the case $M = 8$, $N = 3$, we include Table 6.11-1, where coefficients α_{ij}/A are defined for the eight signals $s_i(t)$.

To express (6.11-10) in terms of signal vectors and orthonormal functions, we write

$$s_i(t) = \sum_{j=1}^{N} s_{ij}\phi_j(t), \tag{6.11-12}$$

TABLE 6.11-1. Codewords, Signal Vectors and Coefficients α_{ij} for $M = 8$ Vectors in $N = 3$ Dimensions

i	Data Codeword	Signal Vector i	α_{ij}/A for j Shown		
			1	2	3
1	**000**	$(-\sqrt{E/N},\ -\sqrt{E/N},\ -\sqrt{E/N})$	-1	-1	-1
2	**001**	$(-\sqrt{E/N},\ -\sqrt{E/N},\ \ \sqrt{E/N})$	-1	-1	1
3	**010**	$(-\sqrt{E/N},\ \ \sqrt{E/N},\ -\sqrt{E/N})$	-1	1	-1
4	**011**	$(-\sqrt{E/N},\ \ \sqrt{E/N},\ \ \sqrt{E/N})$	-1	1	1
5	**100**	$(\ \ \sqrt{E/N},\ -\sqrt{E/N},\ -\sqrt{E/N})$	1	-1	-1
6	**101**	$(\ \ \sqrt{E/N},\ -\sqrt{E/N},\ \ \sqrt{E/N})$	1	-1	1
7	**110**	$(\ \ \sqrt{E/N},\ \ \sqrt{E/N},\ -\sqrt{E/N})$	1	1	-1
8	**111**	$(\ \ \sqrt{E/N},\ \ \sqrt{E/N},\ \ \sqrt{E/N})$	1	1	1

where

$$\phi_j(t) = \left(\frac{1}{\sqrt{T_b}}\right)\text{rect}\left[\frac{t - (j - 0.5)T_b}{T_b}\right], \qquad j = 1, 2, ..., N \qquad (6.11\text{-}13)$$

$$s_{ij} = \pm\sqrt{A^2 T_b} = \pm\sqrt{\frac{E}{N}}, \qquad \text{all } i \text{ and } j. \qquad (6.11\text{-}14)$$

The sign of s_{ij} follows that of α_{ij}. Signal vectors are, of course, given by (6.11-1). Again we refer to Table 6.11-1, this time for construction of example signal vectors in the case $M = 8$, $N = 3$. These vectors correspond to the numbering in Fig. 6.11-1.

The preceding work shows that the polar-NRZ waveform is an example of a vertices of a hypercube signal set. This fact means that (6.11-6) must give the word error probability. Since $E/\mathcal{N}_0 \log_2(M) = E/\mathcal{N}_0 N = A^2 T_b/\mathcal{N}_0$, it is clear that p in (6.11-5) is just P_e, the bit error probability in the polar-NRZ system from (4.4-7). Thus

$$P_w = 1 - [1 - P_e]^N, \qquad (6.11\text{-}15)$$

which was previously found in (4.5-1). There is complete agreement seen between the vector approach taken here and the approach of the earlier work.

6.12 SUMMARY AND DISCUSSION

M-ary systems are the main topics of this chapter. These are systems in which the digital source is allowed to generate any one of M discrete messages in any one symbol interval. For the first time in the book the signal space, or vector, method of representing signals and noise is introduced. The method is a slightly more advanced analysis technique than used elsewhere in the book, but it has far more power. The general M-ary coherent system and the vector representation method are described in Secs. 6.0–6.2.

By using the new analysis method, the optimum M-ary digital system is found. It is defined as the system that produces the largest probability of making a correct decision, in any given symbol interval, as to which symbol is being received at the receiver. The optimum decision rule is developed in Sec. 6.3 that leads to the optimum receiver (Sec. 6.4). The optimum receiver has both a correlation form (Fig. 6.4-1) and a matched filter structure (Fig. 6.4-2). The average probability P_w of making a symbol error is given by (6.3-18) which can only be solved further for specific systems.

The remainder of the chapter explores specific examples of the optimum system. M-ary ASK (M-ASK) is developed in Sec. 6.5 for equally probable messages. The matched filter form of receiver is shown in Fig. 6.5-2. Symbol (word) error probability for the M-ASK system is given by (6.5-8). For the same increment of energy between (equally spaced) message levels, M-PAM, the baseband system of Chap. 4, and M-ASK give the same values of P_w.

M-ary PSK (M-PSK) is a constant amplitude waveform with M possible phases, one of which is used in each symbol interval. A good approximation for P_w is defined by (6.6-11). The exact values of P_w are shown in Fig. 6.6-2. The curves show that significant increases in energy per symbol are needed to maintain P_w at a specified level as M increases. In fact, for fixed P_w and symbol duration, the increase in energy is proportional to $1/\sin^2(\pi/M)$. Thus significant increases in energy (or power) are necessary in M-PSK to maintain a given performance level (value of P_w) as the information rate is increased.

In Sec. 6.7 the orthogonal signal set is introduced. The set defines a class of possible real transmitted waveforms $s_i(t)$, $i = 1, 2, \ldots, M$, that are nonzero only in the symbol interval of duration T that satisfy the orthogonality relation

$$\int_0^T s_i(t)s_j(t)\, dt = \begin{cases} 0, & i \neq j \\ E_i, & i = j. \end{cases} \tag{6.12-1}$$

Here E_i is the energy in $s_i(t)$. In the orthogonal signal set all signals have the same energy $E_i = E$. A system using an orthogonal signal set has the optimum matched filter form of receiver shown in Fig. 6.7-2; its average symbol error probability is given by (6.7-8) which is plotted in Fig. 6.7-3. The curves promise error-free performance as $M \to \infty$. Unfortunately, this limit is difficult to approach in practice because the required channel bandwidth becomes infinite.

An example of an orthogonal signal set is M-ary FSK (M-FSK) discussed in Sec. 6.8. Another example is M-level pulse position modulation (M-PPM) discussed in Sec. 6.9. These two sets of waveforms behave quite differently in a practical sense, but, since they are both orthogonal signal sets, their symbol error probabilities are the same. It is demonstrated in the text, for example, that M-PPM would not be suited to peak-power constrained systems such as a satellite link, whereas M-FSK could be a possible candidate for such a link.

In Sec. 6.10 the biorthogonal signal set is introduced. It is a means of doubling the number of signals in a set by properly combining two orthogonal signal sets. The optimum biorthogonal receiver is shown in Fig. 6.10-2 in its correlation form. Average symbol error probability P_w, as given by (6.10-2), is plotted in Fig. 6.10-3. It is found for large M that the biorthogonal system performs nearly the same as the orthogonal system for small P_w. The QPSK system of Chap. 5 is an example of a biorthogonal signal set when $M = 4$. (Although QPSK was discussed as a *binary* system in Chap. 5, it is in reality a 4-ary system. The earlier binary approach was possible only because of the special parallel structure of the system that made it possible to examine *two* binary paths more or less separately.)

The signal set called vertices of a hypercube derives its name from the geometric appearance of the signal vectors used to define the set. This signal set is important because it defines the N-bit binary code. Of course, it is an M-ary set, and, in the case of the binary code there are $M = 2^N$ possible codewords. Symbol error probability for the vertices of a hypercube set is given by (6.11-6).

REFERENCES

[1] Rice, S. O., Mathematical Analysis of Random Noise, *Bell System Technical Journal*, Vol. 23, July 1944, pp. 282–332, and Vol. 24, January 1945, pp. 46–156. Also reprinted in Wax, N., *Selected Papers on Noise and Stochastic Processes*, Dover Publications, Inc., New York, 1954.

[2] Kotel'nikov, V. A., *The Theory of Optimum Noise Immunity*, McGraw-Hill Book Co., Inc., New York, 1959 (Doctoral dissertation of January 1947, Molotov Energy Institute in Moscow, Russia).

[3] Golomb, S. W. (Editor), *Digital Communications with Space Applications*, Prentice-Hall, Englewood Cliffs, New Jersey, 1964.

[4] Wozencraft, J. M., and Jacobs, I. M., *Principles of Communication Engineering*, John Wiley & Sons, Inc., New York, 1965.

[5] Viterbi, A. J., *Principles of Coherent Communication*, McGraw-Hill Book Co., Inc., New York, 1966.

[6] Sakrison, D. J., *Communication Theory: Transmission of Waveforms and Digital Information*, John Wiley & Sons, Inc., New York, 1968.

[7] Lindsey, W. C., and Simon, M. K., *Telecommunication Systems Engineering*, Prentice-Hall, Inc., Englewood Cliffs, New Jersey, 1973.

[8] Ziemer, R. E., and Tranter, W. H., *Principles of Communications, Systems, Modulation, and Noise*, Houghton Mifflin Co., Boston, Massachusetts, 1976 (see also 2nd ed., 1985).

[9] Korn, G. A., and Korn, T. M., *Mathematical Handbook for Scientists and Engineers*, McGraw-Hill Book Co., Inc., New York, 1961.

[10] Davenport, Jr., W. B., and Root, W. L., *An Introduction to the Theory of Random Signals and Noise*, McGraw-Hill Book Co., Inc., New York, 1958.

[11] Gilbert, E. N., A Comparison of Signaling Alphabets, *Bell System Technical Journal*, Vol. 31, May 1952, pp. 504–522.

[12] Arthurs, E., and Dym, H., On the Optimum Detection of Digital Signals in the Presence of White Gaussian Noise—A Geometric Interpretation and a Study of Three Basic Data Transmission Systems, *IRE Trans. on Communications Systems*, Vol. CS-10, No. 4, December 1962, pp. 336–372.

[13] Landau, H. J., and Slepian, D., On the Optimality of the Regular Simplex Code, *Bell System Technical Journal*, Vol. 45, No. 8, October 1966, pp. 1247–1272.

PROBLEMS

★**6-1.** Extend Example 6.1-1 by proving that the waveforms in the orthonormal set $\{\sqrt{2/T} \cos(n2\pi t/T)\}$ are also orthonormal to the waveforms in the orthonormal set $\{\sqrt{2/T} \sin(n2\pi t/T)\}$.

★**6-2.** Express the Fourier series and coefficient formulas for a periodic signal, as given by (A.4-3) and (A.4-4), in terms of the orthonormal functions of Prob. 6-1.

★**6-3.** (a) Show that the exponential signals

$$\phi_n(t) = (1/\sqrt{T})\exp\left(\frac{-jn2\pi t}{T}\right), \qquad n = 0, \pm 1, \pm 2, \ldots$$

are orthonormal on the interval $[0, T]$.

(b) Write the Fourier series and coefficient formulas for a periodic signal, as given by (A.4-6) and (A.4-7), in terms of the signals $\{\phi_n(t)\}$.

★**6-4.** Define orthonormal functions and coefficients so that a bandlimited function represented by the sampling theorem of (2.1-11) can be written as

$$f(t) = \sum_{k=-\infty}^{\infty} s_k \phi_k(t).$$

(*Hint:* Use results from Example 2.1-1.)

★**6-5.** Given two sinusoids $A\cos(\omega_0 t + \theta_0)$ and $B\sin(\omega_0 t + \theta_0)$, where A, B, and θ_0 are real, but otherwise arbitrary, constants. Find constraints on ω_0 such that the two waveforms are orthogonal on the interval $[0, T]$.

★**6-6.** Define N functions

$$\phi_n(t) = \sqrt{\frac{N}{T}} \, \text{rect} \left\{ \frac{t - [(2n - 1)T/2N]}{T/N} \right\}, \qquad n = 1, 2, \ldots, N$$

orthonormal on $[0, T]$.

(a) For the case $N = 2$, express the four waveforms of a unipolar waveform format of a 2-bit natural binary code in terms of the functions $\phi_n(t)$.

(b) Plot the signal constellation of all waveform vectors.

★6-7. Work Prob. 6-6 except for a polar format.

★6-8. Work Prob. 6-6 except for a 3-bit code.

★6-9. Work Prob. 6-6 except for a 3-bit code and polar format.

★6-10. Given three waveforms as shown in Fig. P6-10. Find orthonormal functions to describe this signal set by use of the Gram-Schmidt procedure. Plot a signal constellation of the signal vectors.

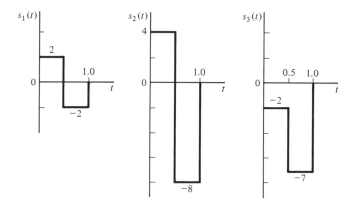

Figure P6-10.

★6-11. (a) Find orthonormal basis functions to describe the waveforms of Fig. P6-11.

(b) Plot the signal constellation.

(c) Find the energies in the signals by direct integration and by use of (6.1-20). Do they both agree?

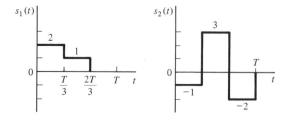

Figure P6-11.

★6-12. Work Prob. 6-11 except for the waveforms of Fig. P6-12.

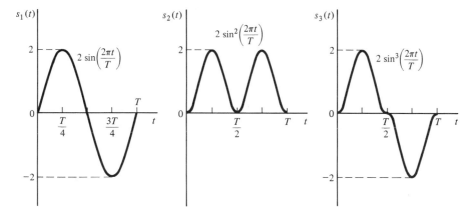

Figure P6-12.

★6-13. Assume (6.2-2) is true and prove that (6.2-3) is valid.

★6-14. From Prob. 6-13 assume (6.2-3) is true and prove that (6.2-4) is true.

★6-15. Prove that (6.2-5) is true if (6.2-1), (6.2-3), and (6.2-4) are true. (*Hint:* Use *Mercer's theorem* [10], one form of which states that

$$R_X(0) = \lim_{N \to \infty} \sum_{n=1}^{N} |\sigma_n|^2 |\psi_n(t)|^2.)$$

★6-16. Use Bayes' theorem to show that (6.3-13) is true and when used in (6.3-12), (6.3-14) results.

★6-17. Justify (6.3-15) and then use it to show that the MAP receiver's decision rule can be written as in (6.3-16).

★6-18. A binary system uses $M = 2$ signals defined on an $N = 2$-dimensional signal space. By solving the MAP receiver decision rule, find exactly what received waveform vector points (ρ_1, ρ_2) fall in decision regions I_1 and I_2. Assume $P_1 = P_2$.

★6-19. Work Prob. 6-18 except assume $P_1 \neq P_2$.

★6-20. Equal-probability binary messages are defined in two-dimensional space by vectors $(-2, 2)$ and $(2, 2)$. If $\mathcal{N}_0 = \frac{4}{9}$ for a white noise channel, define decision regions I_1 and I_2 and use (6.3-18) to compute P_w.

★6-21. Draw a block diagram of an optimum receiver of the form of Fig. 6.4-1(b) for the system of Prob. 6-20. Reduce the receiver to its simplest form.

★6-22. In a particular full duplex modem (separate wires for transmission and reception), M is large while N is small, so the receiver is implemented in the form of Fig. 6.4-1(b). Since the waveforms $\{\phi_j(t)\}$ must be generated for receiver use, draw a block diagram to show how they can also be used to generate the transmitter signals $\{s_i(t)\}$.

★6-23. A transmitter to be used in an M-ASK system has peak and average power limitations of 10 W and 3.5 W, respectively. For $M = 16$ equally probable messages and a constraint on pulse amplitudes of $A_i \geq 0$, all $i = 1, 2, \ldots$, M, select the A_i to best utilize the transmitter. What are the final transmitted peak and average powers? Is the system most affected by peak or average power limitations?

★**6-24.** Work Prob. 6-23 except ignore the constraint $A_i \geq 0$ and let $A_1 = -A_M$.

★**6-25.** An ASK system uses signal amplitudes $A_1 = 0$, $A_2 = 0.5$, $A_3 = 1.0$, and $A_4 = 1.5$ V. If the symbol duration is 10.0 ms, what is the largest allowable channel noise density, $\mathcal{N}_0/2$, that will permit a word error probability $P_w = 5.164(10^{-4})$?

★**6-26.** When the lowest-amplitude signal is zero in M-ASK (when $A_1 = 0$), the decision variable D_1 in Fig. 6.5-2 is identically zero. Give arguments to justify that message 1 will be correctly selected as transmitted ($\hat{m} = m_1$) provided that n_1, the noise on ρ_1, is less than $\sqrt{E_i}/2$ for all $i = 2, 3, \ldots$, M.

★**6-27.** In an M-PSK system $P_w = 10^{-4}$ when $M = 4$.
   (a) What is the required ratio of energy per symbol (E) to $\mathcal{N}_0$?
   (b) If P_w is maintained constant and the same channel is used (same $\mathcal{N}_0$) but M is increased, compute data and plot a curve of the increase in energy E required for $M = 8, 16, 32,$ and 64.

★**6-28.** In a manner similar to Example 6.6-1, compare ε required in an M-PSK system with that of an M-ASK system with $A_1 = -A_M$ to produce the same word error probabilities P_w when M is large.

★**6-29.** An M-ASK system uses $M = 32$ signals, transmits an average power of 140 W when amplitudes are $A_1 = 0$ and $A_i > 0$ for $i > 1$, uses symbols of duration $T = 1000 \ \mu s$, and operates on a channel for which $\mathcal{N}_0 = 9.31(10^{-5})$ W/Hz. An M-PSK system uses the same channel, has the same symbol duration and peak power, and has the same symbol error probability as the M-ASK system.
   (a) Find P_w for the two systems.
   (b) What is the peak transmitted power in the ASK system?
   (c) What is the largest (binary) number of signals (M) that the PSK system can use while satisfying the given constraints?

★**6-30.** A 32-phase M-PSK system transmits symbols of duration 35 μs over a channel with white noise power density $\mathcal{N}_0/2 = 8.89(10^{-7})$ W/Hz. What minimum transmitter average power must the symbol waveforms have if symbol average error probability must not exceed $7.43(10^{-7})$?

★**6-31.** In an M-PSK system $M = 16$. What minimum value of ε is required to guarantee $P_w \leq 3.16(10^{-5})$?

★**6-32.** Consider an orthogonal signal set with $M = 2$ (binary case).
   (a) Define the decision regions I_1 and I_2 when messages are equally probable.
   (b) Use (6.7-5) to find P_w. Compare your result to P_e found in (5.5-11) for coherent binary orthogonal FSK. (*Hint:* Change integration variables by means of a coordinate rotation of $\pi/4$ rad.)

★**6-33.** For an orthogonal signal set corresponding to equally probable messages, determine the mean signal vector and show that it is given by (6.7-14).

★**6-34.** Three signals, each with energy per symbol of 9, belong to an orthogonal signal set.
  (a) Find the vector for the origin shift required to form a simplex set from the orthogonal set.
  (b) Shift the origin and find the vectors of the three signals in the simplex set.
  (c) Write expressions for signals in the simplex set. Show that the third signal can be written in terms of the other two as

$$s_3^{(s)}(t) = a_1 s_1^{(s)}(t) + a_2 s_2^{(s)}(t).$$

  What are a_1 and a_2?

★**6-35.** Show that the energies in the signals of a simplex set are equal and given by (6.7-18).

★**6-36.** If θ_0 equals any integer multiple of $\pi/2$, show that the waveforms of (6.8-2) are orthonormal on $[0, T]$ if ω_0 and $\Delta\omega$ are positive integer multiples of one quarter and one half the symbol rate $2\pi/T$, respectively.

★**6-37.** In a discrete PPM system, 32 pulse positions are possible per symbol interval. If word (symbol) error probability must satisfy $P_w \leq 1.152(10^{-5})$, find the smallest value of ε that is allowed. For the same M and P_w how would peak amplitudes of symbol waveforms of an M-FSK system compare to those of the M-PPM system?

★**6-38.** For equally probable messages draw a block diagram of an optimum receiver for a biorthogonal signal set that is based on reduction of Fig. 6.4-2(a). Reduce the diagram to its simplest possible form.

★**6-39.** Assume the bounding expressions, (6.7-9) and (6.10-3), for P_w in orthogonal and biorthogonal systems are accurate for computations of P_w when $E/\mathcal{N}_0$ is large. Give arguments to support the fact that P_w in the biorthogonal is smaller than that in the orthogonal system for the same $E/\mathcal{N}_0$ and M according to

$$P_w(\text{biorthogonal}) \approx P_w(\text{orthogonal}) - \left(\frac{1}{2}\right)\text{erfc}\left[\sqrt{\frac{E}{2\mathcal{N}_0}}\right].$$

★**6-40.** Show that (6.6-10) and (6.10-2) are the same when $M = 4$.

★**6-41.** A binary system using a polar-NRZ format collects $N = 4$ pulses (bits) over a time $T = 64$ μs and transmits the group as a "word." If word error probability is $P_w = 3.2(10^{-2})$ when pulses of amplitude 2 V are used, what is the channel's noise density, $\mathcal{N}_0/2$?

Appendix A

Review of Deterministic Signals and Networks

A.0 INTRODUCTION

Waveforms may broadly be classified as deterministic or random. A deterministic waveform is one that can be completely described by an equation, a graph, or other definitive manner. A random signal is one in which probability theory is required to describe *characteristics* of the waveform; the signal itself cannot be precisely described. In this appendix we summarize the most important characteristics of deterministic waveforms and linear networks to process them. Random signal theory is reviewed in Appendix B.

A deterministic waveform can also be classified as a *power signal* or an *energy signal*. A power signal has finite power when averaged over all time; it, therefore, also has infinite energy. An energy signal has finite energy but zero power based on averaging over all time. We consider both forms of signal.

A.1 ENERGY SIGNALS

Signals are usually described either in the time domain or in the frequency domain by their *Fourier transform*.

Fourier Transforms

An energy signal $f(t)$ and its Fourier transform $F(\omega)$ are related through the Fourier transform *pair*

$$F(\omega) = \int_{-\infty}^{\infty} f(t)\, e^{-j\omega t}\, dt \qquad (A.1\text{-}1)$$

$$f(t) = \frac{1}{2\pi} \int_{-\infty}^{\infty} F(\omega) e^{j\omega t}\, d\omega. \qquad (A.1\text{-}2)$$

If $f(t)$ is a voltage waveform, $F(\omega)$ is a voltage *density* with the unit V/Hz. It represents the relative voltage amplitudes (complex in general) of the frequencies that constitute $f(t)$. Equation (A.1-2) is called the *inverse Fourier transform*. We sometimes use notation $\mathcal{F}\{\cdot\}$ and $\mathcal{F}^{-1}\{\cdot\}$ to represent taking the direct (Fourier) transform and the inverse transform, respectively. The double arrow is also used to imply a pair. Thus

$$\mathcal{F}^{-1}\{F(\omega)\} = f(t) \longleftrightarrow F(\omega) = \mathcal{F}\{f(t)\}. \qquad (A.1\text{-}3)$$

Under reasonable conditions [1, p. 93] energy signals will always have Fourier transforms. In fact, most waveforms of practical interest, both energy and power, will have transforms.

Properties of Fourier Transforms

We give some very useful properties of Fourier transforms without proofs. Other properties are given in [2]. We assume waveforms $f(t)$ and $f_n(t)$ for $n = 1, 2, \ldots, N$ have Fourier transforms $F(\omega)$ and $F_n(\omega)$ for $n = 1, 2, \ldots, N$, respectively, while ω_0 and t_0 are real constants and α_n for $n = 1, 2, \ldots, N$ are possibly complex constants. The time waveforms may be complex.

The *linearity* property is:

$$f(t) = \sum_{n=1}^{N} \alpha_n f_n(t) \longleftrightarrow \sum_{n=1}^{N} \alpha_n F_n(\omega) = F(\omega). \qquad (A.1\text{-}4)$$

The *time* and *frequency* shifting properties are:

$$f(t - t_0) \longleftrightarrow F(\omega)e^{-j\omega t_0} \qquad (A.1\text{-}5)$$

$$f(t)e^{j\omega_0 t} \longleftrightarrow F(\omega - \omega_0). \qquad (A.1\text{-}6)$$

The *convolution* properties are:

$$f(t) = \int_{-\infty}^{\infty} f_1(\tau)f_2(t - \tau)\, d\tau \longleftrightarrow F_1(\omega)F_2(\omega) = F(\omega) \qquad (A.1\text{-}7)$$

$$f(t) = f_1(t)f_2(t) \longleftrightarrow \frac{1}{2\pi} \int_{-\infty}^{\infty} F_1(\xi)F_2(\omega - \xi)\, d\xi = F(\omega). \qquad (A.1\text{-}8)$$

Thus convolution of two time domain functions corresponds to the product of their two spectrums in the frequency domain. Conversely, the time domain product of two functions corresponds to the convolution of their two spectrums divided by 2π.

The *correlation* properties are:

$$f(t) = \int_{-\infty}^{\infty} f_1^*(\tau)f_2(\tau + t)\, d\tau \longleftrightarrow F_1^*(\omega)F_2(\omega) = F(\omega) \qquad (A.1\text{-}9)$$

$$f(t) = f_1^*(t)f_2(t) \longleftrightarrow \frac{1}{2\pi} \int_{-\infty}^{\infty} F_1^*(\xi)F_2(\xi + \omega)\, d\xi = F(\omega), \qquad (A.1\text{-}10)$$

where the asterisk represents complex conjugation.

Another property can be called *Parseval's theorem* and derives from (A.1-9) with $t = 0$:

$$\int_{-\infty}^{\infty} f_1^*(\tau)f_2(\tau)\, d\tau = \frac{1}{2\pi} \int_{-\infty}^{\infty} F_1^*(\omega)\, F_2(\omega)\, d\omega. \qquad (A.1\text{-}11)$$

Clearly, when $f_1(t) = f_2(t) = f(t)$,

$$\int_{-\infty}^{\infty} |f(t)|^2\, dt = \frac{1}{2\pi} \int_{-\infty}^{\infty} |F(\omega)|^2\, d\omega, \qquad (A.1\text{-}12)$$

which is very important because it shows that the energy in $f(t)$ can be obtained by either a time or frequency domain integration.

From (A.1-1) and (A.1-2) we have the *area* properties:

$$F(0) = \int_{-\infty}^{\infty} f(t)\, dt \qquad (A.1\text{-}13)$$

$$f(0) = \frac{1}{2\pi} \int_{-\infty}^{\infty} F(\omega)\, d\omega. \qquad (A.1\text{-}14)$$

Energy and Energy Density Spectrum

If E denotes the energy in $f(t)$, then from (A.1-12),

$$E = \int_{-\infty}^{\infty} |f(t)|^2\, dt = \frac{1}{2\pi} \int_{-\infty}^{\infty} \mathscr{E}(\omega)\, d\omega \qquad (A.1\text{-}15)$$

where we define

$$\mathscr{E}(\omega) \triangleq |F(\omega)|^2 \qquad (A.1\text{-}16)$$

as the *energy density spectrum*. Clearly the unit of $\mathscr{E}(\omega)$ is energy per hertz if $f(t)$ is a voltage or current waveform.†

† As usual in communication systems theory, we either presume 1-Ω real impedances or assume voltages and currents are in normalized form so that impedances do not show explicitly in energy or power expressions.

A.2 SOME USEFUL ENERGY SIGNALS

Two energy waveforms occur so frequently that they are defined and included here for reference.

Rectangular Function

The *rectangular* (pulse) *function* rect($\cdot$) is defined by

$$f(t) = \text{rect}\left(\frac{t}{\tau}\right) = \begin{cases} 1, & |t| < \tau/2 \\ 0, & |t| > \tau/2 \end{cases} \longleftrightarrow \tau\,\text{Sa}\left(\frac{\omega\tau}{2}\right) = F(\omega) \quad \text{(A.2-1)}$$

where we also define the *sampling function* Sa($\cdot$) by

$$\text{Sa}(x) \triangleq \frac{\sin(x)}{x}. \quad\quad\quad\quad \text{(A.2-2)}$$

The function rect(t/τ) and its spectrum are illustrated in Fig. A.2-1.

Triangular Function

We define the *triangular function* tri($\cdot$) by

$$f(t) = \text{tri}(t/\tau) = \begin{cases} 1 - |t/\tau|, & |t| \leq \tau \\ 0, & |t| > \tau \end{cases} \longleftrightarrow \tau\,\text{Sa}^2\left(\frac{\omega\tau}{2}\right) = F(\omega). \quad \text{(A.2-3)}$$

The function and its spectrum are shown in Fig. A.2-2.

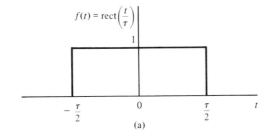

(a)

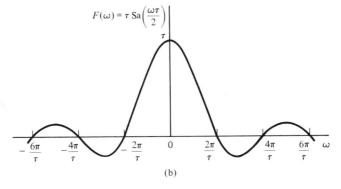

(b)

Figure A.2-1. (a) The rectangular function and (b) its Fourier transform.

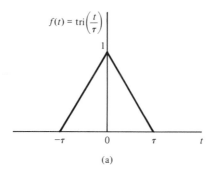

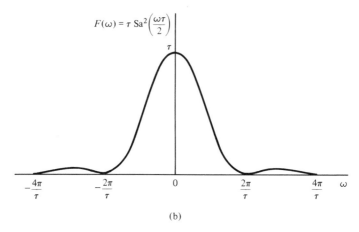

Figure A.2-2. (a) The triangular function and (b) its Fourier transform.

A.3 POWER SIGNALS

As noted earlier, a power signal has a finite power when averaged over all time. In this section we introduce additional details about power signals.

Average Power and Power Density Spectrum

We define

$$\mathscr{A}[\cdot] = \lim_{T \to \infty} \frac{1}{2T} \int_{-T}^{T} [\cdot] \, dt \qquad (A.3\text{-}1)$$

as the infinite-time average of the quantity within the brackets. For a voltage waveform $f(t)$, we have

$$\mathscr{A}[f(t)] = \lim_{T \to \infty} \frac{1}{2T} \int_{-T}^{T} f(t) \, dt, \qquad (A.3\text{-}2)$$

which is the dc component of $f(t)$, or

$$P \triangleq \mathscr{A}[|f(t)|^2] = \lim_{T \to \infty} \frac{1}{2T} \int_{-T}^{T} |f(t)|^2 \, dt, \tag{A.3-3}$$

which is the average power P in $f(t)$. If $f(t)$ is a real waveform, $|f(t)|^2$ is replaced by $f^2(t)$ in (A.3-3).

Average power can also be found from a frequency domain operation. Let $f_T(t)$ represent $f(t)$ truncated to the interval $[-T, T]$ and let its Fourier transform be $F_T(\omega)$ according to

$$f_T(t) = f(t) \, \text{rect}\left(\frac{t}{2T}\right) \longleftrightarrow F_T(\omega). \tag{A.3-4}$$

It can be shown [3, p. 41] that

$$P = \frac{1}{2\pi} \int_{-\infty}^{\infty} \mathscr{S}(\omega) \, d\omega, \tag{A.3-5}$$

where

$$\mathscr{S}(\omega) \triangleq \lim_{T \to \infty} \left\{ \frac{|F_T(\omega)|^2}{2T} \right\} \tag{A.3-6}$$

is called the *power density spectrum* of $f(t)$. The function $\mathscr{S}(\omega)$ has the unit power per hertz.

Time Autocorrelation Function

The time average of the product $f^*(t)f(t + \tau)$, defined by

$$\mathscr{R}(\tau) \triangleq \lim_{T \to \infty} \frac{1}{2T} \int_{-T}^{T} f^*(t)f(t + \tau) \, dt, \tag{A.3-7}$$

is called the *time autocorrelation function* of $f(t)$. It can be shown that $\mathscr{R}(\tau)$ and $\mathscr{S}(\omega)$ are related through the Fourier transform, that is,

$$\mathscr{S}(\omega) = \int_{-\infty}^{\infty} \mathscr{R}(\tau)e^{-j\omega\tau} \, d\tau \tag{A.3-8}$$

$$\mathscr{R}(\tau) = \frac{1}{2\pi} \int_{-\infty}^{\infty} \mathscr{S}(\omega)e^{j\omega\tau} \, d\omega, \tag{A.3-9}$$

so

$$\mathscr{R}(\tau) \longleftrightarrow \mathscr{S}(\omega). \tag{A.3-10}$$

A.4 PERIODIC POWER SIGNALS

The periodic signal is one of the most important types of power signal.

Fourier Series

A periodic signal $f(t)$ with period T_p satisfies the relationship

$$f(t) = f(t + kT_p), \qquad k = \pm 1, \pm 2, \ldots, \qquad (A.4\text{-}1)$$

and can be expressed in terms of an infinite series of time functions known as a *Fourier series*. Two principal forms of Fourier series may be stated. The *trigonometric form* is

$$f(t) = \left(\frac{a_0}{2}\right) + \sum_{n=1}^{\infty} a_n \cos(n\omega_p t) + \sum_{n=1}^{\infty} b_n \sin(n\omega_p t), \qquad (A.4\text{-}2)$$

where the Fourier series coefficients are given by

$$a_n = \left(\frac{2}{T_p}\right) \int_{t_0-(T_p/2)}^{t_0+(T_p/2)} f(t)\cos(n\omega_p t)\, dt, \qquad n = 0, 1, 2, \ldots, \qquad (A.4\text{-}3)$$

$$b_n = \left(\frac{2}{T_p}\right) \int_{t_0-(T_p/2)}^{t_0+(T_p/2)} f(t)\sin(n\omega_p t)\, dt, \qquad n = 1, 2, \ldots, \qquad (A.4\text{-}4)$$

with

$$\omega_p = \frac{2\pi}{T_p} \qquad (A.4\text{-}5)$$

and t_0 is any real constant.

The *complex* or *exponential* Fourier series is

$$f(t) = \sum_{n=-\infty}^{\infty} C_n e^{jn\omega_p t} \qquad (A.4\text{-}6)$$

where

$$C_n = \left(\frac{1}{T_p}\right) \int_{t_0-(T_p/2)}^{t_0+(T_p/2)} f(t) e^{-jn\omega_p t}\, dt, \qquad n = 0, \pm 1, \pm 2, \ldots. \qquad (A.4\text{-}7)$$

Impulse Function

To define and deal with the spectrum (Fourier transform) of a periodic signal it is necessary to introduce the concept of an.impulse function. The *unit impulse function*, sometimes called a *delta function*, denoted by $\delta(t)$, is defined through its integral property

$$\int_{-\infty}^{\infty} \phi(t)\delta(t - t_0)\, dt = \phi(t_0). \qquad (A.4\text{-}8)$$

Here t_0 is a constant and $\phi(t)$ is arbitrary except it is assumed continuous at $t = t_0$. The function $\delta(t - t_0)$ behaves as though it occurs at $t = t_0$, has zero duration, infinite amplitude at $t = t_0$, and area of unity. Other forms of impulses also exist, and discontinuous functions $\phi(t)$ can also be handled [4].

An impulse function $A\delta(t - t_0)$ is a unit impulse (area one) scaled by the amplitude constant A. It is usually shown graphically as a vertical arrow at time $t = t_0$ with amplitude A. Of course, the arrow points to infinity to imply infinite amplitude, whereas the plotted amplitude indicates the scale constant A.

By use of (A.4-8) the Fourier transform of the impulse function is readily established

$$\delta(t) \longleftrightarrow 1 \tag{A.4-9}$$

$$A\delta(t - t_0) \longleftrightarrow A\, e^{-j\omega t_0}. \tag{A.4-10}$$

Other useful properties of impulse functions are, with $\omega_p = 2\pi/T_p$,

$$\sum_{n=-\infty}^{\infty} \delta(t - nT_p) \longleftrightarrow 2\pi \sum_{n=-\infty}^{\infty} \left(\frac{1}{T_p}\right)\delta(\omega - n\omega_p) \tag{A.4-11}$$

$$\sum_{n=-\infty}^{\infty} \delta(t - nT_p) = \left(\frac{1}{T_p}\right) \sum_{n=-\infty}^{\infty} e^{jn\omega_p t}. \tag{A.4-12}$$

Spectrum of a Periodic Signal

By Fourier transforming $f(t)$ of (A.4-6), we obtain the spectrum of a periodic signal

$$F(\omega) = 2\pi \sum_{n=-\infty}^{\infty} C_n\delta(\omega - n\omega_p). \tag{A.4-13}$$

Power Spectrum, Autocorrelation Function, and Power

The power spectrum of the periodic signal is

$$\mathscr{S}(\omega) = 2\pi \sum_{n=-\infty}^{\infty} |C_n|^2\delta(\omega - n\omega_p). \tag{A.4-14}$$

By inverse Fourier transformation the autocorrelation function is

$$\mathscr{R}(\tau) = \sum_{n=-\infty}^{\infty} |C_n|^2 e^{jn\omega_p\tau}. \tag{A.4-15}$$

Average power results from use of (A.4-14) in (A.3-5) or from (A.3-7) with $\tau = 0$:

$$P = \sum_{n=-\infty}^{\infty} |C_n|^2. \tag{A.4-16}$$

A.5   SIGNAL BANDWIDTH AND SPECTRAL EXTENT

Many definitions of bandwidth exist. Generally, bandwidth refers to the band of frequencies that are most important (largest amplitudes) in defining the signal. Spectral extent refers to the band of frequencies outside which the spectral components in a signal may be considered negligible.

Three-dB Bandwidth

The band of positive frequencies over which the magnitude of a signal's spectrum remains above $1/\sqrt{2}$ times its value at a convenient reference frequency, usually located in the band, is called the 3-dB bandwidth. Fig. A.5-1 illustrates 3-dB bandwidth, denoted by W in radians per second, for several spectral shapes. The reference frequency for a baseband (lowpass) signal, as shown in (a), is usually zero, although it could also be another convenient value (1000 Hz for the audio band, for instance). Examples of bandpass signals are shown in (b) and (c). The spectral extent, denoted

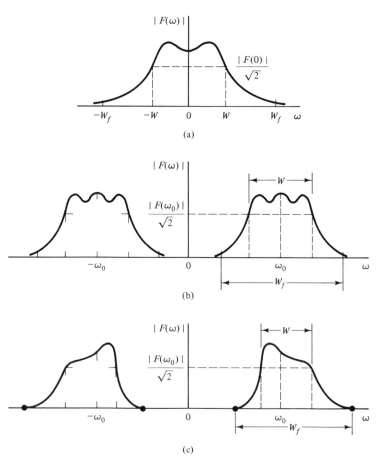

Figure A.5-1. Signal spectrums: (a) Lowpass or baseband, (b) bandpass, and (c) absolutely bandlimited bandpass.

by W_f, is shown in (a) and (b) for typical lowpass and bandpass signals. A waveform having a spectrum that is truly zero outside a band, as shown in (c), is truly bandlimited, and its *absolute bandwidth* is the same as its spectral extent.

In the case where $f(t)$ is periodic, its spectrum's magnitude consists of impulses. Calculation of bandwidth remains the same except the *envelope* of the spectrum's magnitude (locus of tips of positive impulses) is used.

Mean Frequency and RMS Bandwidth

Let $F(\omega)$ be the spectrum of a signal $f(t)$ (lowpass or bandpass). We define *mean frequency* $\overline{\omega}_0$ by

$$\overline{\omega}_0 = \frac{\int_0^\infty \omega |F(\omega)|^2 \, d\omega}{\int_0^\infty |F(\omega)|^2 \, d\omega} \tag{A.5-1}$$

and *rms bandwidth* W_{rms} by

$$W_{\text{rms}}^2 = \frac{\int_0^\infty (\omega - \overline{\omega}_0)^2 |F(\omega)|^2 \, d\omega}{\int_0^\infty |F(\omega)|^2 \, d\omega} \tag{A.5-2}$$

for bandpass signals. For lowpass waveforms, (A.5-2) applies if $\overline{\omega}_0$ is set to zero.

A.6 LINEAR NETWORKS

Impulse Response

A network, having an output port that responds to a signal as excitation at an input port,† is linear if superposition applies. A unit impulse $\delta(t - \tau)$ applied to a linear network causes a response that is characteristic of the network, called its *impulse response* $h(t, \tau)$. The network's response $g(t)$ to an arbitrary input $f(t)$ is given by

$$g(t) = \int_{-\infty}^\infty f(\tau) h(t, \tau) \, d\tau. \tag{A.6-1}$$

If the network is also time invariant such that $h(t, \tau)$ no longer depends *in form* on the time of occurrence of the input impulse, then

$$h(t, \tau) = h(t - \tau) \tag{A.6-2}$$

† We consider only single-input, single-output networks.

and

$$g(t) = \int_{-\infty}^{\infty} f(\tau)h(t - \tau)\, d\tau. \qquad (A.6\text{-}3)$$

This expression applies generally to both energy and power signals.

Transfer Function

By Fourier transformation of (A.6-3) we have

$$G(\omega) = F(\omega)H(\omega), \qquad (A.6\text{-}4)$$

where $G(\omega)$, $F(\omega)$, and $H(\omega)$ are the respective transforms of $g(t)$, $f(t)$, and $h(t)$. $H(\omega)$ is called the *transfer function* of the linear time invariant network. Equation (A.6-4) applies to all signals (energy or power) for which $F(\omega)$ can be defined.

Bandwidth

Bandwidths of a network (3-dB or rms) can be defined in exactly the same way as for waveforms. It is necessary only to use $|H(\omega)|$ in place of $|F(\omega)|$ used earlier in Sec. A.5.

Ideal Networks

It is often desirable to approximate the transfer function of a real filter by an *ideal filter*. We define an ideal filter as one that has unity gain and a perfectly flat (constant) transfer function magnitude at frequencies inside a desired interval and zero response at all other frequencies. This definition is visualized by use of Fig. A.6-1, which illustrates lowpass, highpass, and bandpass ideal filters. In each case the filter may have a linear phase $\theta(\omega) = -\omega\tau$ where τ is the filter's *delay constant*.

Energy and Power Spectrums of Response

If the input $f(t)$ to a linear time invariant network is an energy signal with energy density spectrum $\mathscr{E}_f(\omega) = |F(\omega)|^2$, the energy density of the response $g(t)$, denoted by $\mathscr{E}_g(\omega)$, is

$$\mathscr{E}_g(\omega) = |G(\omega)|^2 = \mathscr{E}_f(\omega)|H(\omega)|^2 \qquad (A.6\text{-}5)$$

from (A.6-4). Total energy E_g in the response becomes

$$E_g = \frac{1}{2\pi} \int_{-\infty}^{\infty} \mathscr{E}_f(\omega)|H(\omega)|^2\, d\omega. \qquad (A.6\text{-}6)$$

If $f(t)$ is a power signal with power density spectrum $\mathscr{S}_f(\omega)$, the power spectrum $\mathscr{S}_g(\omega)$ of $g(t)$ is given by

$$\mathscr{S}_g(\omega) = \mathscr{S}_f(\omega)\, |H(\omega)|^2. \qquad (A.6\text{-}7)$$

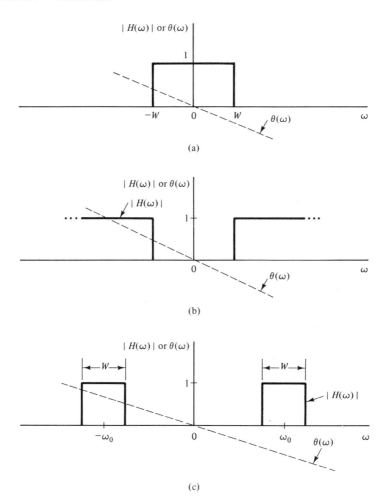

Figure A.6-1.   Ideal filter transfer functions: (a) Lowpass, (b) highpass, and (c) bandpass [3].

Average output power P_g in $g(t)$ becomes

$$P_g = \frac{1}{2\pi} \int_{-\infty}^{\infty} \mathscr{S}_f(\omega)|H(\omega)|^2 \, d\omega. \tag{A.6-8}$$

REFERENCES

[1] Thomas, J. B., *An Introduction to Statistical Communication Theory,* John Wiley & Sons, Inc., New York, 1969.

[2] Peebles, Jr., Peyton Z., *Probability, Random Variables, and Random Signal Principles,* McGraw-Hill Book Co., Inc., New York, 1980.

[3] Peebles, Jr., Peyton Z., *Communication System Principles*, Addison-Wesley Publishing Co., Inc., Reading, Massachusetts, 1976. (Figure A.6-1 has been adapted.)

[4] Korn, G. A. and Korn, T. M., *Mathematical Handbook for Scientists and Engineers*, McGraw-Hill Book Co., Inc., New York, 1961.

PROBLEMS

A-1. Characterize the following signals as either power or energy type.

(a) $f(t) = 3 \cos(10t)$

(b) $f(t) = 1.7 \exp[-6t^2]\cos(3t)$

(c) $f(t) = \dfrac{\cos(6t) + 9 \cos^2(3t)}{1 + 3t^2}$

(d) $f(t) = 3 \exp(-2t)u(t)$

(e) $f(t) = 12 \operatorname{rect}\left(\dfrac{t}{9}\right)$

(f) $f(t) = \sum\limits_{n=-\infty}^{\infty} 3 \operatorname{rect}[4(t - n)]$.

A-2. Find the Fourier transform of the waveforms of Prob. A-1(d).

A-3. Find the Fourier transforms of the waveforms of Prob. A-1(e) and (f).

A-4. Prove (A.1-5) and (A.1-6).

A-5. Prove (A.1-8).

A-6. With α a real number, show that the *scaling property* of Fourier transforms, as given by

$$f(\alpha t) \longleftrightarrow \frac{1}{|\alpha|} F\left(\frac{\omega}{\alpha}\right)$$

with $f(t) \leftrightarrow F(\omega)$, is true.

A-7. For two functions $f(t)$ and $F(\omega)$ that are a Fourier transform pair, show the validity of the *duality property*

$$F(t) \longleftrightarrow 2\pi f(-\omega)$$

of Fourier transforms.

A-8. A complex waveform $f(t)$ has a Fourier transform $F(\omega)$. Find the spectrum of $f^*(t)$ in terms of $F(\omega)$.

A-9. A waveform $f(t)$ is differentiated in time n times. Find the spectrum of $d^n f(t)/dt^n$ in terms of $F(\omega)$, the spectrum of $f(t)$.

A-10. Assume pairs 11 and 20 of Appendix G are true and use them to prove pair 12.

A-11. Use (A.1-9) to find the (auto) correlation function of the rectangular pulse $f(t) = 3 \operatorname{rect}(t/5)$.

A-12. Work Prob. A-11 for the waveform $f(t) = A u(t)\exp(-bt)$, where A and b are positive real constants. Plot your result.

A-13. For the waveform of Prob. A-12: (a) Find the energy density spectrum and the waveform's energy from the energy density. (b) Find the signal's energy from the middle form of (A.1-15) to verify the result of (a).

A-14. Find the spectrum of $\operatorname{rect}(t/\tau)$ and verify that (A.2-1) is valid.

A-15. Work Prob. A-14 for the triangular waveform of (A.2-3).

A-16. Find the Fourier transform of $f(t) = A\mathrm{rect}(t/\tau)\cos(\pi t/\tau)$ and show that it is the sum of two sampling functions by means of the frequency shifting property of Fourier transforms.

A-17. By use of the time average defined in (A.3-1) find the dc component and average power in the signal $f(t) = A\cos^2(\omega_0 t)$, where ω_0 is a constant.

A-18. Work Prob. A-17 for the signal $f(t) = A\cos^4(\omega_0 t)$.

A-19. Use (A.3-7) and (A.3-8) to find the time autocorrelation function and power spectrum of the signal of Prob. A-17.

A-20. Find (a) the trigonometric form and (b) the complex form of the Fourier series of the waveform of Prob. A-1(f).

★A-21. Use Fourier series to prove that

$$\sum_{n=-\infty}^{\infty} \delta(t - nT_p) \longleftrightarrow \frac{2\pi}{T_p} \sum_{n=-\infty}^{\infty} \delta\!\left(\omega - \frac{n2\pi}{T_p}\right),$$

where T_p is a positive constant.

★A-22. Use the result of Prob. A-21 to show that

$$\sum_{n=-\infty}^{\infty} e^{-jn\omega T_p} = \frac{2\pi}{T_p} \sum_{n=-\infty}^{\infty} \delta\!\left(\omega - \frac{n2\pi}{T_p}\right)$$

and

$$\frac{1}{T_p} \sum_{n=-\infty}^{\infty} e^{jn2\pi t/T_p} = \sum_{n=-\infty}^{\infty} \delta(t - nT_p).$$

A-23. If $f(t) \leftrightarrow F(\omega)$, show that the Fourier series coefficients C_n of (A.4-7) that define the complex Fourier series of the periodic signal $f_p(t) = \sum_{n=-\infty}^{\infty} f(t - nT_p)$ are given by

$$C_n = \frac{1}{T_p} F\!\left(\frac{n2\pi}{T_p}\right)$$

where T_p is the period of $f_p(t)$ and $f(t)$ is nonzero only on $(-T_p/2 < t < T_p/2)$.

A-24. Use (A.4-8) to show that $\delta(t - \tau)$ has "area" of one for any value of τ.

A-25. If the argument of a unit impulse function is itself a function $g(t)$, then it can be represented by

$$\delta[g(t)] = \frac{\delta(t - t_0)}{\left|\dfrac{dg(t)}{dt}\right|_{t = t_0}}$$

where t_0 is the value of t for which $g(t) = 0$, that is, t_0 is a root of $g(t) = 0$. Use this result to evaluate $B(t)$ defined by

$$B(t) = \int_{-\infty}^{\infty} \phi(t)\delta(at - b)\, dt$$

where a and b are real constants and $\phi(t)$ is arbitrary except it is continuous at the time of occurrence of the impulse.

A-26. Find the Fourier transform of the signal

$$f(t) = [A + f_m(t)]\cos(\omega_0 t)$$

where A and ω_0 are constants and $f_m(t)$ has the spectrum $F_m(\omega)$.

A-27. Determine the value of k in each of the following signals such that if $\tau \to 0$, then $f(t) \to \delta(t)$ in the limit. (*Hint:* Use tables to find the area in (a), (b), and (f), and assume $\tau > 0$.)

(a) $f(t) = k \exp\left(-\dfrac{t^2}{\tau^2}\right)$

(d) $f(t) = k \exp\left(-\dfrac{|t|}{\tau}\right)$

(b) $f(t) = k \, \mathrm{Sa}\left(\dfrac{t}{\tau}\right)$

(e) $f(t) = k \, \mathrm{rect}\left(\dfrac{t}{\tau}\right)$

(c) $f(t) = k \, \mathrm{tri}\left(\dfrac{t}{\tau}\right)$

(f) $f(t) = k \, \mathrm{Sa}^2\left(\dfrac{t}{\tau}\right)$.

A-28. Evaluate the following integrals:

(a) $\displaystyle\int_{-\infty}^{\infty} \cos(9t)\delta(t - 2)\, dt$

(c) $\displaystyle\int_{-\infty}^{\infty} \delta(t - 3)\exp(-8t^2)\, dt$

(b) $\displaystyle\int_{-\infty}^{\infty} \delta(t + 2)(1 + 4t + 8t^2)\, dt$

(d) $\displaystyle\int_{-\infty}^{\infty} \delta(t)(1 + t^2)^{-1}\, dt$.

A-29. A periodic signal $f(t)$ is constructed by forming replicas of the waveform $f_0(t) = \alpha t^2 \mathrm{rect}[(2t - \tau)/2\tau]$ every nT_p, $n = 0, \pm 1, \pm 2, \ldots$, where α is a constant and $T_p > \tau$. Find (a) the Fourier series and (b) the average power in $f(t)$.

A-30. Find the 3-dB bandwidth of the signal $f(t) = u(t)\exp(-\alpha t)$ where $\alpha > 0$ is a constant.

A-31. Work Prob. A-30 for the signal $f(t) = t^2 u(t)\exp(-\alpha t)$.

A-32. Find the rms bandwidth of the waveform of Prob. A-30. Find the energy in $f(t)$.

A-33. Find the rms bandwidth and energy of the waveform of Prob. A-31.

A-34. A network's transfer function is approximately given by $H(\omega) = 1 + j\omega$. An input baseband signal $f(t)$, having average power P_f and rms bandwidth $W_{f,\mathrm{rms}}$, generates a response $g(t)$. Find the average power P_g in $g(t)$ in terms of P_f and $W_{f,\mathrm{rms}}$.

A-35. A network has transfer function $H(\omega) = 10(1 + j5\omega)^{-1}$. (a) Find the network's impulse response $h(t)$. (b) Use convolution to obtain the network's response to the input signal $f(t) = 12u(t)\exp(-t/10)$.

★A-36. Find (a) the energy density spectrum and (b) the energy in the network's response of Prob. A-35.

Appendix B

Review of Random Signal Theory

B.0 INTRODUCTION

In this appendix we give a brief review of the basic elements of random signal theory that are needed in the main text. We use the words *random signal* to imply either a desired random waveform (usually referred to as the *signal*) or an undesired random waveform such as noise. For additional detail the reader is referred to some of the recent literature [1–9].

Our review begins with the fundamental concept on which all else is based, probability.

B.1 SAMPLE SPACES, EVENTS, AND PROBABILITY

Sample Spaces

An *experiment* for which the *outcomes* of a *trial* are random in their occurrence is called a *random experiment*. An example of a random experiment would be the drawing of a card from a thoroughly shuffled deck of 52 cards. The trial is the actual drawing of a card and the card itself is the outcome; there are 52 possible outcomes in this experiment. The set of all possible outcomes in a given random experiment is called the *sample space* and the outcomes are *elements* of the sample space. In the card-drawing experiment the sample space is called *discrete* because only discrete outcomes are possible.

Different experiments may have different sample spaces. Consider a circular potentiometer having a 360° tap rotation capability that generates any voltage from 0 (at zero-degree position) to, for instance, 15 V (at 360° position). An experiment of randomly placing the tap now produces an outcome (voltage) that can have *any* value of a continuum of values from zero to 15 V. The sample space of this experiment is called *continuous*.

Events

An event is defined as a subset of the sample space. In the card-draw experiment we may be more interested in getting a king than in any one card. Here four elements of the sample space satisfy (make up) the event (subset).

Probability

Probability is a function of the events defined on a sample space. It is a number assigned to each event that defines the relative likelihood that the event can occur. The assignment may be made on the basis of common sense, from the results of measurements, or any reasonable and justifiable basis. For example, it is reasonable in the card-draw experiment that the event "draw a king" will have a probability $\frac{4}{52}$ because there are 4 kings and 52 total cards and all cards are presumed equally likely to be selected.

If events are denoted by capital letters (and elements of the events by lowercase letters) such as A, B, C, and so on, we denote the probability of an event (such as A) by $P(A)$.

Joint Probability

We denote the probability that two events, A and B, will jointly or simultaneously occur on a trial of an experiment by $P(A, B)$. For example, if a box is filled with 90 thoroughly mixed resistors, as shown in Table B.1-1, and if A is the event "draw a 100-Ω resistor" and B is the event "draw a 2-W resistor," then $P(A, B) = \frac{9}{90} = 0.1$. Note that $P(A) = \frac{32}{90}$ and $P(B) = \frac{20}{90}$ in this example.

TABLE B.1-1.   Resistors in a Box by Resistance and Wattage.

Resistance	Wattage			
	$\frac{1}{2}$ W	1 W	2 W	Totals
10 Ω	8	6	5	19
100 Ω	10	13	9	32
1000 Ω	18	15	6	39
Totals	36	34	20	90

For several events A_1, A_2, ..., A_N, their joint probability is denoted by $P(A_1, A_2, ..., A_N)$.

Conditional Probability

Sometimes we are interested in the probability that an event, such as A, occurs, given that some other event, such as B, has occurred. We call this *conditional probability* and denote it by $P(A|B)$. In the example defined in Table B.1-1, we have $P(A|B) = \frac{9}{20}$. Joint and conditional probabilities are related by

$$P(A, B) = P(A|B)P(B), \tag{B.1-1}$$

or, alternatively,

$$P(A, B) = P(B|A)P(A). \tag{B.1-2}$$

By combining (B.1-2) and (B.1-1), we obtain one form of *Bayes' rule*

$$P(A|B)P(B) = P(B|A)P(A). \tag{B.1-3}$$

Statistical Independence

Two events A and B are said to be statistically independent if

$$P(A|B) = P(A), \qquad \text{independent events.} \tag{B.1-4}$$

An equivalent definition is

$$P(A, B) = P(A)P(B), \qquad \text{independent events.} \tag{B.1-5}$$

In the preceding experiment of drawing a resistor from the box of 90 resistors we have $P(A) = \frac{32}{90}$ and $P(A|B) = \frac{9}{20}$ so $P(A|B) \neq P(A)$ which means events A and B are *not* independent.

Multiple events A_1, A_2, ..., A_N are said to be statistically independent if all the conditions

$$P(A_i, A_j) = P(A_i)P(A_j)$$
$$P(A_i, A_j, A_k) = P(A_i)P(A_j)P(A_k) \tag{B.1-6}$$

$$\cdot$$
$$\cdot$$
$$\cdot$$

$$P(A_i, A_j, ..., A_N) = P(A_i)P(A_j) \cdots P(A_N)$$

are satisfied for $1 \leq i < j < k < \cdots \leq N$. There are $2^N - N - 1$ of these conditions.

B.2 RANDOM VARIABLES, DISTRIBUTIONS, AND DENSITIES

In some random experiments outcomes are not numerical, such as in the earlier resistor and card-selection experiments. The concept of a *random variable* allows all random experiments to be represented numerically.

Random Variable

A random variable is defined as a real function of the elements of a sample space. Thus points (elements) of a sample space map into points (numbers) on the real line through the function. Capital letters (W, X, or Y) are used to represent a random variable, and a particular value of the random variable is denoted by lowercase letters (w, x, or y).

Random variables can be discrete, continuous, or a mixture of the two, depending on the sample space and the form of the function defining the random variable.

Distribution Functions

Let A be the set of points in the sample space that maps into the set of real-line points $\{X \leq x\}$. These points correspond to all values of the random variable X that do not exceed an arbitrary number x. The probability of the set A defined on the sample space must equal the probability of the set $\{X \leq x\}$ due to the direct point-to-point mapping involved. This probability is a function of x, which we denote by $P_X(x)$,† that is given by

$$P_X(x) = P\{X \leq x\}. \tag{B.2-1}$$

The function $P_X(x)$ is called the *cumulative probability distribution function* of X. It is a function having the following properties:

$$P_X(-\infty) = 0 \tag{B.2-2a}$$

$$P_X(\infty) = 1.0 \tag{B.2-2b}$$

$$0 \leq P_X(x) \leq 1.0 \tag{B.2-2c}$$

$$P_X(x_1) \leq P_X(x_2) \qquad \text{if } x_1 < x_2 \tag{B.2-2d}$$

$$P\{x_1 < X \leq x_2\} = P_X(x_2) - P_X(x_1). \tag{B.2-2e}$$

For a discrete random variable, $P_X(x)$ will contain steps with amplitudes equal to the probabilities of the possible discrete values of X, denoted by x_i, $i = 1, 2, \ldots, N$:

$$P_X(x) = \sum_{i=1}^{N} P\{X = x_i\} u(x - x_i), \tag{B.2-3}$$

where we define the unit-step function by

$$u(x) = \begin{cases} 1, & x \geq 0 \\ 0, & x < 0. \end{cases} \tag{B.2-4}$$

When more than one random variable is involved, the *joint probability distribution function* is defined by

$$P_{X_1, X_2, \ldots, X_N}(x_1, x_2, \ldots, x_N) = P\{X_1 \leq x_1, X_2 \leq x_2, \ldots, X_N \leq x_N\}. \tag{B.2-5}$$

† The subscript X indicates $P_X(x)$ applies to the random variable X.

Probability Density Functions

The *probability density function,* denoted by $p_X(x)$, is defined as the derivative of the distribution

$$p_X(x) = \frac{dP_X(x)}{dx}. \tag{B.2-6}$$

The properties exhibited by $p_X(x)$ are†

$$0 \leqslant p_X(x) \tag{B.2-7a}$$

$$\int_{-\infty}^{\infty} p_X(x)\, dx = 1 \tag{B.2-7b}$$

$$P_X(x) = \int_{-\infty}^{x} p_X(\xi)\, d\xi \tag{B.2-7c}$$

$$P_X\{x_1 < X \leqslant x_2\} = \int_{x_1}^{x_2} p_X(x)\, dx. \tag{B.2-7d}$$

For a discrete random variable,

$$p_X(x) = \sum_{i=1}^{N} P\{X = x_i\}\delta(x - x_i). \tag{B.2-8}$$

For several random variables the *joint probability density function* is

$$p_{X_1,X_2,\ldots,X_N}(x_1, x_2, \ldots, x_N) = \frac{\partial^N P_{X_1,X_2,\ldots,X_N}(x_1, x_2, \ldots, x_N)}{\partial x_1\, \partial x_2 \cdots \partial x_N}. \tag{B.2-9}$$

Conditional Distribution and Density

By again letting event $A = \{X \leqslant x\}$ and using (B.1-1), we have the probability of $\{X \leqslant x\}$ conditional on some event B having occurred. We call this probability the conditional distribution of X, denoted by $P_X(x|B)$:

$$P_X(x|B) = P\{X \leqslant x|B\} = \frac{P\{X \leqslant x, B\}}{P(B)}. \tag{B.2-10}$$

Conditional density follows the derivative:

$$p_X(x|B) = \frac{dP_X(x|B)}{dx}. \tag{B.2-11}$$

When many random variables, $X_1, X_2, \ldots, X_N$, are to be considered,

---

† Properties (B.2-7a) and (B.2-7b) together are sufficient to guarantee that a function $p_X(x)$ is a valid density function.

we extend (B.2-10) and (B.2-11) to obtain

$$P_{X_1,...,X_N}(x_1, ..., x_N|B) = \frac{P\{X_1 \le x_1, ..., X_N \le x_N, B\}}{P(B)} \quad \text{(B.2-12)}$$

$$p_{X_1,...,X_N}(x_1, ..., x_N|B) = \frac{\partial^N P_{X_1,...,X_N}(x_1, ..., x_N|B)}{\partial x_1 \cdots \partial x_N}. \quad \text{(B.2-13)}$$

In the case where event B in (B.2-10) is defined by a second random variable Y according to $B = \{y - \Delta y < Y \le y + \Delta y\}$, with $\Delta y \to 0$ such that Y approaches a fixed value y, it can be shown that†

$$P_X(x|y) = \lim_{\Delta y \to 0} P\{X \le x|y - \Delta y < Y \le y + \Delta y\}$$

$$= \frac{\displaystyle\int_{-\infty}^{x} p_{X,Y}(\xi, y) \, d\xi}{p_Y(y)} \quad \text{(B.2-14)}$$

$$p_X(x|y) = \frac{p_{X,Y}(x, y)}{p_Y(y)}. \quad \text{(B.2-15)}$$

Statistical Independence

N random variables X_i, $i = 1, 2, ..., N$, are said to be statistically independent if their joint distributions (or densities) satisfy the conditions of (B.1-6). One particular condition,

$$p_{X_1,X_2,...,X_N}(x_1, x_2, ..., x_N) = \prod_{i=1}^{N} p_{X_i}(x_i), \quad \text{(B.2-16)}$$

proves extremely useful in the analysis of many practical systems.

B.3 STATISTICAL AVERAGES

Average of a Function of Random Variables

Let $g(X_1, X_2, ..., X_N)$ be a real function of the N random variables X_i, $i = 1, 2, ..., N$. We define the *statistical average* (also called the *expected value* or *mean value*) of $g(\cdot, \cdots, \cdot)$, denoted either by $\bar{g}$ or $E[g(\cdot, \cdots, \cdot)]$ by

$$\bar{g} = E[g(X_1, X_2, ..., X_N)] \quad \text{(B.3-1)}$$

$$= \int_{-\infty}^{\infty} \cdots \int_{-\infty}^{\infty} g(x_1, x_2, ..., x_N) p_{X_1,X_2,...,X_N}(x_1, x_2, ..., x_N) \, dx_1 \, dx_2 \ldots dx_N.$$

† If Y is a discrete random variable, some care must be exercised in using these results. See [3, pp. 86–89].

Most practical problems require only two random variables, say X and Y. In this special case

$$\bar{g} = E[g(X, Y)] = \int_{-\infty}^{\infty} \int_{-\infty}^{\infty} g(x, y)p_{X,Y}(x, y)\, dx\, dy. \qquad \text{(B.3-2)}$$

Moments

The statistical averages of a number of special functions $g(X, Y)$ are especially important. These averages are called *moments* of the random variables X and Y. For example,

$$\bar{X} = E[X] = \int_{-\infty}^{\infty} xp_X(x)\, dx \qquad \text{(B.3-3)}$$

is called the *mean* or first moment of X (about the origin), and

$$\overline{X^2} = E[X^2] = \int_{-\infty}^{\infty} x^2 p_X(x)\, dx \qquad \text{(B.3-4)}$$

is the *power* in X (second moment about origin). Another moment (about the mean $\bar{X}$) is called the variance of X; it is denoted by σ_X^2 and given by

$$\sigma_X^2 = E[(X - \bar{X})^2] = \int_{-\infty}^{\infty} (x - \bar{X})^2 p_X(x)\, dx. \qquad \text{(B.3-5)}$$

Other important moments are called the *correlation* and the *covariance* of X and Y and are denoted by R_{XY} and C_{XY}, respectively. These quantities are defined by

$$R_{XY} = \overline{XY} = E[XY] = \int_{-\infty}^{\infty} \int_{-\infty}^{\infty} xyp_{X,Y}(x, y)\, dx\, dy \qquad \text{(B.3-6)}$$

$$C_{XY} = \overline{(X - \bar{X})(Y - \bar{Y})} = E[(X - \bar{X})(Y - \bar{Y})]$$

$$= \int_{-\infty}^{\infty} \int_{-\infty}^{\infty} (x - \bar{X})(y - \bar{Y})p_{X,Y}(x, y)\, dx\, dy. \qquad \text{(B.3-7)}$$

Interrelationships of importance are

$$\sigma_X^2 = \overline{X^2} - \bar{X}^2 \qquad \text{(B.3-8)}$$

$$C_{XY} = \overline{XY} - \bar{X}\,\bar{Y} = R_{XY} - \bar{X}\,\bar{Y}. \qquad \text{(B.3-9)}$$

B.4 GAUSSIAN RANDOM VARIABLES

A random variable X is called *Gaussian* if its probability density function is given by

$$p_X(x) = (2\pi\sigma_X^2)^{-1/2} \exp\left[-\frac{(x - \bar{X})^2}{2\sigma_X^2} \right]. \qquad \text{(B.4-1)}$$

This function is plotted in Fig. B.4-1. It is a symmetric function about the mean of X and its spread is proportional to σ_X (called the *standard deviation* of X). The *normalized* Gaussian density results when $\overline{X} = 0$ and $\sigma_X^2 = 1$. Let the notation $p(x)$ denote the normalized case. The *normalized* distribution function is then

$$P(x) = \int_{-\infty}^{x} p(x) \, dx \qquad (B.4\text{-}2)$$

with

$$p(x) = (2\pi)^{-1/2} \exp\left(-\frac{x^2}{2}\right). \qquad (B.4\text{-}3)$$

It is readily shown that the distribution function of the normalized Gaussian random variable is related to $P_X(x)$ by

$$P_X(x) = \int_{-\infty}^{x} p_X(\xi) \, d\xi = P\left(\frac{x - \overline{X}}{\sigma_X}\right). \qquad (B.4\text{-}4)$$

The *error function, complementary error function,* and *Q function* are defined from the normalized Gaussian density by

$$\operatorname{erf}(\beta) \triangleq \frac{2}{\sqrt{\pi}} \int_{0}^{\beta} e^{-\xi^2} \, d\xi, \qquad \operatorname{erf}(-\beta) = -\operatorname{erf}(\beta) \qquad (B.4\text{-}5)$$

$$\operatorname{erfc}(\beta) = 1 - \operatorname{erf}(\beta), \qquad \operatorname{erfc}(-\beta) = 2 - \operatorname{erfc}(\beta) \qquad (B.4\text{-}6)$$

$$Q(\beta) = \frac{1}{\sqrt{2\pi}} \int_{\beta}^{\infty} e^{-\xi^2/2} \, d\xi, \qquad Q(-\beta) = 1 - Q(\beta). \qquad (B.4\text{-}7)$$

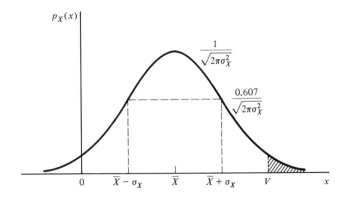

Figure B.4-1.   The Gaussian probability density function.

These are related to the distribution function by

$$P_X(x) = 1 - Q\left(\frac{x - \overline{X}}{\sigma_X}\right) = 1 - \frac{1}{2}\operatorname{erfc}\left(\frac{x - \overline{X}}{\sqrt{2}\,\sigma_X}\right)$$

$$= \frac{1}{2} + \frac{1}{2}\operatorname{erf}\left(\frac{x - \overline{X}}{\sqrt{2}\,\sigma_X}\right). \tag{B.4-8}$$

B.5 RANDOM SIGNALS AND PROCESSES

The purpose of all the preceding theory, of course, is to provide the basis for describing random waveforms. To effectively use the basis, we must now introduce the concept of modeling a real random signal by a real random process.†

Random Process Concept

Consider a real random (noise) waveform, such as shown in Fig. B.5-1 as $x_0(t)$, that might exist at a given point in some system. Another system, built identically to the first, might generate a different random waveform at its similar given point as shown in Fig. B.5-1 as $x_1(t)$. Still other systems, all built identically, would generate other waveforms, such as $x_2(t)$, $x_3(t)$, and so on. In principle we can imagine an infinity of such identical systems, all generating their own waveforms. The collection of these waveforms is called an *ensemble*. The waveforms are different with time, because the thermal agitation of electrons in conductors and semiconductors in the systems that give rise to the noises is different from system to system, even though the systems are built the same.

Another way of viewing the waveforms of Fig. B.5-1 is to *imagine* them to be the possible waveforms that *could have been generated* by *one* network. This approach is especially attractive. The ensemble of all possible waveforms is called a *random process*. The actual random signal is only one ensemble member of the random process of possible signals that, taken together, describe the statistical properties of the signal. For example, at time t_1 the voltage of any one waveform is treated as a specific value of a random variable describing all the waveform values of the members of the process. The mean value of this random variable is a statistical average of all possible voltages that could occur at time t_1; it is called the *ensemble average* at time t_1.

If the random process is denoted by $X(t)$, then the process is interpreted as a random variable X at time t. The probability density is denoted by

---

† We consider only real processes. However, it is also possible to define complex processes [3, 4].

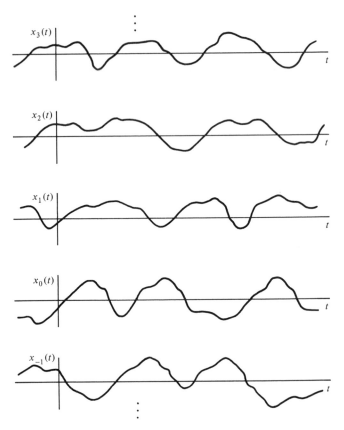

Figure B.5-1. An ensemble of random waveforms $x_i(t)$, $i = 0, \pm 1, \pm 2, \ldots$ that comprise a random process.

$p_X(x; t)$. The average random signal amplitude (dc value) at time t then becomes

$$E[X(t)] = \int_{-\infty}^{\infty} x p_X(x; t)\, dx. \qquad (\text{B.5-1})$$

If $p_X(x; t)$ varies with time, then $E[X(t)]$ may vary with time. By a similar logic we define the *power* in the process by

$$E[X^2(t)] = \int_{-\infty}^{\infty} x^2\, p_X(x; t)\, dx \qquad (\text{B.5-2})$$

which may, in general, vary with time.

Correlation Functions

If random variables defined from the process $X(t)$ at times t_1 and t_2 are $X_1 = X(t_1)$ and $X_2 = X(t_2)$ with joint probability density denoted by

$p_X(x_1, x_2; t_1, t_2)$, the correlation between X_1 and X_2, denoted by $R_{XX}(t_1, t_2)$, is

$$R_{XX}(t_1, t_2) = E[X(t_1)X(t_2)] = \int_{-\infty}^{\infty}\int_{-\infty}^{\infty} x_1 x_2 p_X(x_1, x_2; t_1, t_2)\, dx_1\, dx_2.$$
(B.5-3)

Because the correlation is of two random variables taken from the same process, it is called the *autocorrelation function*. With $t_1 = t$ and $t_2 = t + \tau$, where $\tau = t_2 - t_1$ is the difference in times, we have

$$R_{XX}(t, t + \tau) = E[X(t)X(t + \tau)].$$
(B.5-4)

If two random processes $X(t)$ and $Y(t)$ are considered, the correlation between random variables $X = X(t)$ and $Y = Y(t + \tau)$ is called the *cross-correlation function* of the processes. It is given by

$$R_{XY}(t, t + \tau) = E[X(t)Y(t + \tau)].$$
(B.5-5)

Alternatively,

$$R_{YX}(t, t + \tau) = E[Y(t)X(t + \tau)].$$
(B.5-6)

Stationarity

The mean, power, and autocorrelation functions are all measures of the statistical properties of a process. Many other measures are also possible (higher moments). Broadly speaking, if the process's statistical properties do not vary with time, it is called *stationary*. Although there are many precise definitions of stationarity, one of the most broadly useful is *wide-sense stationarity*. A random process $X(t)$ is wide-sense stationary if two conditions are true:

$$E[X(t)] = \overline{X} = \text{constant}$$
(B.5-7a)

$$E[X(t)X(t + \tau)] = R_{XX}(\tau) \qquad \text{(independent of } t\text{)}.$$
(B.5-7b)

Two processes are *jointly wide-sense stationary* if they are separately wide-sense stationary and their cross correlation functions are independent of absolute time

$$E[X(t)Y(t + \tau)] = R_{XY}(\tau)$$
(B.5-8)

$$E[Y(t)X(t + \tau)] = R_{YX}(\tau).$$
(B.5-9)

An *ergodic process* is a more restricted form of random process. It is one in which the time averages of a single ensemble member of the process are equal to the corresponding statistical averages. If time averages are defined by

$$\mathscr{A}[\cdot] = \lim_{T\to\infty}\frac{1}{2T}\int_{-T}^{T} [\cdot]\, dt$$
(B.5-10)

as usual, then for the ergodic process

$$\overline{X} = E[X(t)] = \mathscr{A}[x(t)] \triangleq \overline{x} \tag{B.5-11}$$

$$\overline{X^2} = E[X^2(t)] = \mathscr{A}[x^2(t)] \triangleq \overline{x^2} \tag{B.5-12}$$

$$R_{XX}(\tau) = E[X(t)X(t + \tau)] = \mathscr{A}[x(t)x(t + \tau)] \triangleq \mathscr{R}_{xx}(\tau), \tag{B.5-13}$$

where we use $\mathscr{R}$ to represent a time-correlation function.

For jointly ergodic processes

$$R_{XY}(\tau) = \mathscr{R}_{xy}(\tau) \triangleq \mathscr{A}[x(t)y(t + \tau)] \tag{B.5-14}$$

$$R_{YX}(\tau) = \mathscr{R}_{yx}(\tau) \triangleq \mathscr{A}[y(t)x(t + \tau)]. \tag{B.5-15}$$

Although statistical averages are most often used in modeling and theoretical analysis of systems, it is the time averages that are most readily measured (approximately) in practice. If the process is ergodic, the two sets of results are equal and the measured results will agree with the theory. In practice, processes are therefore often *assumed* ergodic without actual proof of the assumption.† However, even if the process is not ergodic the validity of the assumption is of little consequence if the theoretical and practical results agree within reasonable bounds, as they often do.

B.6 POWER DENSITY SPECTRUMS

The preceding section discussed ways of describing random processes using the time domain. As with deterministic signals, random signals may be described in the frequency domain. The description involves a power density spectrum instead of a voltage density spectrum, however.

Stationary Processes

If $X_T(t)$ represents a random process $X(t)$ truncated to $-T \le t \le T$ (and zero for $|t| > T$), it can be shown that the process has a *power density spectrum*

$$\mathscr{S}_{XX}(\omega) = \lim_{T \to \infty} \frac{E[|X_T(\omega)|^2]}{2T}, \tag{B.6-1}$$

where $X_T(\omega)$ is the Fourier transform of $X_T(t)$ and the process is assumed to be at least wide-sense stationary.‡ The power spectrum is a real, non-negative, even function of ω for real processes.

For jointly wide-sense stationary processes, $X(t)$ and $Y(t)$, having truncated representations $X_T(t)$ and $Y_T(t)$ with respective Fourier transforms

---

† Proof that a process is ergodic is difficult.

‡ The use of the capital X in both $X_T(\omega)$ and $X_T(t)$ does not imply the same function with different arguments here. It is to be hoped that the reader will be able to resolve this minor clash of notation through context.

$X_T(\omega)$ and $Y_T(\omega)$, *cross-power density spectrums* may be derived as follows:

$$\mathcal{S}_{XY}(\omega) = \lim_{T \to \infty} \frac{E[X_T^*(\omega)Y_T(\omega)]}{2T} \qquad \text{(B.6-2)}$$

$$\mathcal{S}_{YX}(\omega) = \lim_{T \to \infty} \frac{E[Y_T^*(\omega)X_T(\omega)]}{2T}. \qquad \text{(B.6-3)}$$

Power and cross-power density spectrums of jointly wide-sense stationary processes are related to correlation functions as Fourier transform pairs

$$R_{XX}(\tau) \longleftrightarrow \mathcal{S}_{XX}(\omega) \qquad \text{(B.6-4)}$$

$$R_{YY}(\tau) \longleftrightarrow \mathcal{S}_{YY}(\omega) \qquad \text{(B.6-5)}$$

$$R_{XY}(\tau) \longleftrightarrow \mathcal{S}_{XY}(\omega) \qquad \text{(B.6-6)}$$

$$R_{YX}(\tau) \longleftrightarrow \mathcal{S}_{YX}(\omega). \qquad \text{(B.6-7)}$$

Nonstationary Processes

Even if a random process is not stationary, an *average power density spectrum* can be defined as the Fourier transform of the *time average* of the process autocorrelation function

$$\mathcal{S}_{XX}(\omega) = \int_{-\infty}^{\infty} \mathcal{A}[R_{XX}(t, t + \tau)]e^{-j\omega\tau} \, d\tau. \qquad \text{(B.6-8)}$$

Thus

$$\mathcal{A}[R_{XX}(t, t + \tau)] = \frac{1}{2\pi} \int_{-\infty}^{\infty} \mathcal{S}_{XX}(\omega)e^{j\omega\tau} \, d\omega. \qquad \text{(B.6-9)}$$

In the nonstationary case the autocorrelation function cannot be recovered from its average power spectrum; only the *time-averaged* autocorrelation function is recoverable.

In a similar manner, *average cross-power density spectrums* can be defined as the Fourier transforms of the time-averaged cross-correlation functions.

Power

Power in a random process $X(t)$ derives either from its autocorrelation function or its power density spectrum:

$$\overline{X^2} = E[X^2(t)] = R_{XX}(0)$$

$$= \frac{1}{2\pi} \int_{-\infty}^{\infty} \mathcal{S}_{XX}(\omega) \, d\omega. \qquad \text{(B.6-10)}$$

B.7 RANDOM SIGNAL RESPONSE OF NETWORKS

Fundamental Result

A random signal $x(t)$ applied to the input of a single-input, single-output linear time-invariant network will generate an output, or response, given by the convolution of $x(t)$ with the network's impulse response $h(t)$:

$$y(t) = \int_{-\infty}^{\infty} h(\xi)x(t - \xi)\, d\xi$$

$$= \int_{-\infty}^{\infty} x(\xi)h(t - \xi)\, d\xi. \qquad (B.7\text{-}1)$$

The fundamental result (B.7-1) is all that is required to derive all characteristics of the random response $y(t)$.

Since each waveform of a random process $X(t)$ is affected by the network through (B.7-1), the responses correspond to members of a new random process $Y(t)$. As a consequence of this fact we can interpret (B.7-1) as a transformation of one process to another

$$Y(t) = \int_{-\infty}^{\infty} h(\xi)X(t - \xi)\, d\xi$$

$$= \int_{-\infty}^{\infty} X(\xi)h(t - \xi)\, d\xi. \qquad (B.7\text{-}2)$$

Output Correlation Functions

If $X(t)$ is wide-sense stationary and the network is linear and time invariant, it can be shown that the response process $Y(t)$ is wide-sense stationary and $X(t)$ and $Y(t)$ are jointly wide-sense stationary. We state results for only this case.

The mean, power, and autocorrelation function of the response are

$$\overline{Y} = E[Y(t)] = \overline{X} \int_{-\infty}^{\infty} h(\xi)\, d\xi \qquad (B.7\text{-}3)$$

$$\overline{Y^2} = E[Y^2(t)] = \int_{-\infty}^{\infty}\int_{-\infty}^{\infty} R_{XX}(\xi_1 - \xi_2)h(\xi_1)h(\xi_2)\, d\xi_1\, d\xi_2 \qquad (B.7\text{-}4)$$

$$R_{YY}(\tau) = \int_{-\infty}^{\infty}\int_{-\infty}^{\infty} R_{XX}(\tau + \xi_1 - \xi_2)h(\xi_1)h(\xi_2)\, d\xi_1\, d\xi_2. \qquad (B.7\text{-}5)$$

Cross-correlation functions are

$$R_{XY}(\tau) = \int_{-\infty}^{\infty} R_{XX}(\tau - \xi)h(\xi)\,d\xi \qquad \text{(B.7-6)}$$

$$R_{YX}(\tau) = \int_{-\infty}^{\infty} R_{XX}(\tau - \xi)h(-\xi)\,d\xi. \qquad \text{(B.7-7)}$$

Power Density Spectrums

Let a random process $X(t)$, that is at least wide-sense stationary with power density spectrum $\mathcal{S}_{XX}(\omega)$, be applied to the input of a linear time-invariant network having a transfer function $H(\omega)$.† The response random process $Y(t)$ will have a power density spectrum

$$\mathcal{S}_{YY}(\omega) = \mathcal{S}_{XX}(\omega)|H(\omega)|^2. \qquad \text{(B.7-8)}$$

Cross-power density spectrums are related to $\mathcal{S}_{XX}(\omega)$ as follows:

$$\mathcal{S}_{XY}(\omega) = \mathcal{S}_{XX}(\omega)H(\omega) \qquad \text{(B.7-9)}$$

$$\mathcal{S}_{YX}(\omega) = \mathcal{S}_{XX}(\omega)H(-\omega). \qquad \text{(B.7-10)}$$

B.8 BANDPASS RANDOM PROCESSES

Next we assume $X(t)$ is a bandlimited wide-sense stationary, zero-mean real random process having a power density spectrum defined by

$$\mathcal{S}_{XX}(\omega) \begin{cases} \neq 0, & 0 < \omega_1 < |\omega| < \omega_1 + W_f \\ = 0, & \text{elsewhere.} \end{cases} \qquad \text{(B.8-1)}$$

Here ω_1 is a constant and W_f is the spectral extent of the process. In practice, random signals are rarely bandlimited. However, there are always frequencies between which the power spectrum is largest and outside which it is negligible. This means that the assumed power spectrum form is reasonable for most practical problems.

The random process $X(t)$ can be represented in the very useful form

$$X(t) = X_c(t)\cos(\omega_0 t + \theta_0) - X_s(t)\sin(\omega_0 t + \theta_0), \qquad \text{(B.8-2)}$$

where ω_0 and θ_0 are constants; $X_c(t)$ and $X_s(t)$ are lowpass random processes with the following properties [3]:

(1) $X_c(t)$ and $X_s(t)$ are jointly wide-sense stationary   (B.8-3a)

(2) $E[X_c(t)] = 0, \qquad E[X_s(t)] = 0$ (B.8-3b)

(3) $E[X_c^2(t)] = E[x_s^2(t)] = E[X^2(t)]$ (B.8-3c)

---

† We assume the impulse response $h(t) = \mathcal{F}^{-1}[H(\omega)]$ is a real function.

(4) $R_{X_sX_s}(\tau) = R_{X_cX_c}(\tau) = \dfrac{1}{\pi} \displaystyle\int_0^\infty \mathscr{S}_{XX}(\omega)\cos[(\omega - \omega_0)\tau]\,d\omega$ (B.8-3d)

(5) $R_{X_sX_c}(\tau) = -R_{X_cX_s}(\tau) = \dfrac{-1}{\pi} \displaystyle\int_0^\infty \mathscr{S}_{XX}(\omega)\sin[(\omega - \omega_0)\tau]\,d\omega$ (B.8-3e)

(6) $\mathscr{S}_{X_cX_c}(\omega) = \mathscr{S}_{X_sX_s}(\omega) = L_p[\mathscr{S}_{XX}(\omega - \omega_0) + \mathscr{S}_{XX}(\omega + \omega_0)]$ (B.8-3f)

(7) $\mathscr{S}_{X_sX_c}(\omega) = -\mathscr{S}_{X_cX_s}(\omega) = -jL_p[\mathscr{S}_{XX}(\omega - \omega_0) - \mathscr{S}_{XX}(\omega + \omega_0)].$

                                                               (B.8-3g)

Here ω_0 is any convenient frequency in the band $\omega_1 < \omega_0 < \omega_1 + W_f$ and $L_p[\cdot]$ represents taking the lowpass portion of the quantity in brackets.

    One of the most useful applications of (B.8-2) is when $\mathscr{S}_{XX}(\omega)$ is symmetric about ω_0. In this case, $R_{X_cX_s}(\tau) = 0$, $R_{X_sX_c}(\tau) = 0$, $\mathscr{S}_{X_cX_s}(\omega) = 0$ and $\mathscr{S}_{X_sX_c}(\omega) = 0$.

B.9 MATCHED FILTERS

In a number of systems a deterministic signal $f(t)$ is transmitted and received, along with noise $n(t)$ at the receiver. The receiver's problem is often to decide, at some instant in time t_0, whether the signal and noise or just noise is present. Its ability to make this decision is enhanced if the ratio of the signal's power at time t_0 to average noise power is large. In fact, it is possible to select a specific filter, called a *matched filter,* that will make this signal-to-noise ratio *maximum.* We examine such filters in this section.

Colored Noise Case

    Consider an arbitrary linear time-invariant filter with impulse response and transfer function $h(t)$ and $H(\omega)$, respectively. The response $f_o(t)$ generated by the application of $f(t)$ to the filter's input is

$$f_o(t) = \frac{1}{2\pi} \int_{-\infty}^{\infty} F(\omega)H(\omega)e^{j\omega t}\,d\omega \qquad (B.9\text{-}1)$$

where

$$f(t) \longleftrightarrow F(\omega). \qquad (B.9\text{-}2)$$

Average output noise power N_o is

$$N_o = \frac{1}{2\pi} \int_{-\infty}^{\infty} \mathscr{S}_{NN}(\omega)|H(\omega)|^2\,d\omega \qquad (B.9\text{-}3)$$

where $\mathscr{S}_{NN}(\omega)$ is the power density spectrum of the noise $n(t)$ represented by the random process $N(t)$.

    The ratio of signal power at time t_0 to average noise power, which we wish to maximize, is

$$\left(\frac{S_o}{N_o}\right) = \frac{\left|\left(\dfrac{1}{2\pi}\right)\displaystyle\int_{-\infty}^{\infty} F(\omega)H(\omega)e^{j\omega t_0}\, d\omega\right|^2}{\left(\dfrac{1}{2\pi}\right)\displaystyle\int_{-\infty}^{\infty} \mathscr{S}_{NN}(\omega)|H(\omega)|^2\, d\omega}. \tag{B.9-4}$$

By applying *Schwarz's inequality*,† we have

$$\left(\frac{S_o}{N_o}\right) \leq \frac{1}{2\pi}\int_{-\infty}^{\infty} \frac{|F(\omega)|^2}{\mathscr{S}_{NN}(\omega)}\, d\omega. \tag{B.9-5}$$

The maximum (equality) occurs only when we choose the *optimum* transfer function as

$$H_{opt}(\omega) = \frac{1}{2\pi C}\frac{F^*(\omega)}{\mathscr{S}_{NN}(\omega)}e^{-j\omega t_0}. \tag{B.9-6}$$

This filter is called *matched* because its transfer function depends on the signal's spectrum $F(\omega)$. If the signal is changed, the optimum filter must change; it, therefore, must be *matched* to the signal used.

White Noise Case

In the special case of white noise where $\mathscr{S}_{NN}(\omega) = \mathscr{N}_0/2$ (constant) at all frequencies, the optimum filter transfer function is

$$H_{opt}(\omega) = \frac{1}{\pi \mathscr{N}_0 C}F^*(\omega)e^{-j\omega t_0}. \tag{B.9-7}$$

Its impulse response is readily found to be

$$h_{opt}(t) = \frac{1}{\pi \mathscr{N}_0 C}f^*(t_0 - t). \tag{B.9-8}$$

If $f(t)$ is real, this optimum impulse response is a replica of $f(t)$ centered at t_0 but "running backwards" in time.

---

† If $A(\omega)$ and $B(\omega)$ are arbitrary and possibly complex functions of ω, the inequality states (in one of its forms) that

$$\left|\int_{-\infty}^{\infty} A(\omega)B(\omega)\, d\omega\right|^2 \leq \int_{-\infty}^{\infty} |A(\omega)|^2\, d\omega \int_{-\infty}^{\infty} |B(\omega)|^2\, d\omega.$$

The equality holds only if

$$B(\omega) = CA^*(\omega),$$

where C is an arbitrary real constant and * represents complex conjugation.

REFERENCES

[1] Papoulis, A., *Probability, Random Variables, and Stochastic Processes*, 2nd ed., McGraw-Hill Book Co., New York, 1984.

[2] O'Flynn, M., *Probabilities, Random Variables, and Random Processes*, Harper & Row, Publishers, New York, 1982.

[3] Peebles, Jr., Peyton Z., *Probability, Random Variables, and Random Signal Principles*, McGraw-Hill Book Co., New York, 1980.

[4] Miller, K. S., *Complex Stochastic Processes, An Introduction to Theory and Application*, Addison-Wesley Publishing Co., Reading, Massachusetts, 1974.

[5] Melsa, J. L. and Sage, A. P., *An Introduction to Probability and Stochastic Processes*, Prentice-Hall, Inc., Englewood Cliffs, New Jersey, 1973.

[6] Cooper, G. R. and McGillem, C. D., *Probabilistic Methods of Signal and System Analysis*, Holt, Rinehart and Winston, New York, 1971 (see also second edition, 1986).

[7] Davenport, W. B., Jr., *Probability and Random Processes, An Introduction for Applied Scientists and Engineers*, McGraw-Hill Book Co., New York, 1970.

[8] Helstrom, C. W., *Probability and Stochastic Processes for Engineers*, Macmillan Publishing Co., New York, 1984.

[9] Gray, R. M. and Davisson, L. D., *Random Processes: A Mathematical Approach for Engineers*, Prentice-Hall, Inc., Englewood Cliffs, New Jersey, 1986.

PROBLEMS

B-1. In a classroom there are 87 students; 43 have blond hair, 29 have black hair, and 15 have brown hair. The instructor randomly selects a student to answer a question. (a) For this random experiment define a trial, the outcomes, and sample space. (b) What is the probability the selected student's hair is brown? (c) What is the probability the selected student's hair will be blond?

B-2. Two cards are drawn randomly from a thoroughly shuffled ordinary deck of 52 cards. What is the probability that both cards will be aces?

★B-3. Extend Prob. B-2 to the drawing of four cards that are all aces.

B-4. A random experiment consists of drawing a resistor from a box of resistors as defined in Table B.1-1. Define events A = "draw a 10-Ω resistor," B = "draw a 1000-Ω resistor," and C = "draw a 1-W resistor." Find the probabilities: (a) $P(A)$, (b) $P(B)$, (c) $P(C)$, (d) $P(A, B)$, (e) $P(A, C)$, (f) $P(A|B)$, (g) $P(A|C)$, (h) $P(C|A)$, (i) $P(A, B, C)$, and (j) $P(A|B, C)$.

B-5. Are the events A, B, and C of Prob. B-4 statistically independent in any sense, that is, in pairs or as a triple?

B-6. A random experiment consists of throwing *two* dice and observing the sum of the two numbers that show up. (a) Define a sample space for this experiment. (b) Assign probabilities to the elements of the sample space and justify the choices. (c) What are the probabilities of the events A = "sum is 1," B = "sum is 6," and C = "sum is 12"?

B-7. In a box there are 10 green, 8 red, 16 white, 15 blue, and 3 black balls. A ball is randomly selected. (a) What is the probability that the ball is red? (b) What are the probabilities for the other colors?

B-8. A production line manufacturing gunpowder can explode if two control circuits A and B jointly fail. Their probabilities of failure are known to be 0.001 and 0.004, respectively. It is also known that $P(B|A) = 0.006$. (a) What is the probability of the line exploding? (b) What is $P(A|B)$? (c) Are events A and B statistically independent?

B-9. An experiment successfully teaches a rat to select a particular color door to open and receive food. The rat is repeatedly placed before nine doors from which it chooses one. Each door is painted on its left and right halves with one of the three colors red (R), yellow (Y), and blue (B). After many trials it is found that the rat chooses the various doors according to the probabilities of Table PB-9. (a) What is the probability the rat will choose a door with red on it? What are the same probabilities for yellow and blue? To what color was the rat taught to respond? (b) Find the probabilities that the rat chooses pairs of colors. That is, find probabilities $P(R, Y)$, $P(R, B)$, and $P(Y, B)$. What color does the rat seem to prefer most as a second color choice?

TABLE PB-9.   Probabilities of Door Selections by a Rat.

		Door Right Side Color		
		R	Y	B
Door Left Side Color	R	0.05	0.12	0.10
	Y	0.15	0.20	0.08
	B	0.13	0.10	0.07

B-10. Define equations that must be true if three events A_1, A_2, and A_3 are to be statistically independent.

B-11. The sample space, denoted by the set S, in a random experiment is $S = \{1, 2, 3.5, 6\}$. For each random variable X defined, find the set of values X may take on. Here s_i, $i = 1, 2, 3, 4$ are the elements of S, referred to generally by s.

(a) $X = 6s$

(b) $X = 12s^2 - 2s$

(c) $X = \dfrac{(s^3 + 3)}{5}$

(d) $X = 3 \exp\left(\dfrac{-s}{2}\right)$

(e) $X = \cos\left(\dfrac{\pi s}{2}\right)$

(f) $X = \left[1 + \left(\dfrac{s^2}{6}\right)\right]^{-1}$.

B-12. A sample space contains all numbers s defined by $2 \leq s \leq 5$. Rework Prob. B-11 using this new sample space.

B-13. A random variable X is discrete and has the possible values of the set $\{1, 3, 5, 9, 13\}$. The respective probabilities of the values form the set $\{0.05, 0.15, 0.25, 0.40, 0.15\}$. (a) Plot the distribution function of X. (b) Plot the density function. (c) Find the probability $P\{X \leqslant 7.0\}$.

B-14. The random variable of Prob. B-13 is converted to a new random variable Y according to $Y = 2X - 6$. (a) Plot the distribution function of Y. (b) Plot the density function of Y.

★B-15. A random variable X is transformed to a new random variable Y through a transformation $Y = T(X)$. Let $T(\cdot)$ be either a monotonically increasing or decreasing continuous function and let X be a continuous random variable with probability density function $p_X(x)$. Show that the density function $p_Y(y)$ is given by

$$p_Y(y) = p_X(x) \left| \frac{dx}{dy} \right|,$$

where $x = T^{-1}(y)$ and $T^{-1}(\cdot)$ is the inverse of $T(\cdot)$.

B-16. If a continuous random variable X is transformed to a new random variable $Y = aX + b$, a and b being real constants, use the results of Prob. B-15 to show that the density of Y is

$$p_Y(y) = p_X\left(\frac{y - b}{a}\right) \frac{1}{|a|}.$$

B-17. A random variable X is called *uniformly distributed* on (a, b), with $b > a$, if

$$p_X(x) = \begin{cases} \dfrac{1}{b - a}, & a < x < b \\ 0, & \text{elsewhere.} \end{cases}$$

(a) Find all moments, denoted by m_n, defined by

$$m_n = E[X^n], \qquad n = 0, 1, 2, \dots.$$

(b) Find the variance of X.

B-18. The uniform random variable of Prob. B-17 is transformed to a new random variable $Y = CX^3$ where $C > 0$ is a constant. (a) Find and plot the density function of Y. (b) For $a > 0$ find and plot the distribution function of Y.

B-19. (a) Find the distribution function of the uniformly distributed random variable of Prob. B-17. (b) Let $x_1 = 0.9a + 0.1b$ and $x_2 = 0.2a + 0.8b$. Find $P\{x_1 < X \leqslant x_2\}$.

B-20. Determine which of the following functions are not valid density functions and state reasons why.

(a) $q(x) = (\frac{1}{8})u(x)\exp(-x/6)$
(b) $q(x) = u(x + 6) - u(x - 10)$
(c) $q(x) = [u(x + 1) - u(x - 1)](\frac{1}{2})\cos(\pi x)$.

B-21. Work Prob. B-20 for the given functions.

(a) $q(x) = -2u(x)\exp(-x^2/4)$
(b) $q(x) = (2\pi)^{-1/2}\exp(-x^2)$
(c) $q(x) = 12(1 + x^2)^{-2}$.

B-22. The *Laplace* density as defined by

$$p_X(x) = ae^{-|x-m|/b}$$

is a valid density function for arbitrary real constants m and b if a is chosen properly. Find: (a) the proper value of a, (b) the mean, and (c) the variance of the random variable X.

B-23. The *exponential* random variable X has a density

$$p_X(x) = \left(\frac{1}{b}\right)u(x)\exp\left(-\frac{x}{b}\right)$$

for $b > 0$ a real constant. Find: (a) the mean, (b) the variance, (c) the distribution function of X, and (d) the probability that the random variable can have values larger than its mean.

B-24. The *Rayleigh* random variable X is defined by the density

$$p_X(x) = \left(\frac{2x}{b}\right)u(x)\exp\left(-\frac{x^2}{b}\right)$$

for $b > 0$ a real constant. Find: (a) the mean, (b) the variance, (c) the distribution function of X, (d) the *mode* of X, defined as the value of X at which $p_X(x)$ reaches a maximum, and (e) the probability that X can have values exceeding the mode.

B-25. A Rayleigh random variable X has $b = 4$ (see Prob. B-24). Find $P\{X \leq 3\}$.

B-26. Two statistically independent Rayleigh random variables X and Y are defined as in Prob. B-24 with b equal to 2 and 4, respectively. Find $P\{1.5 < X, 1.0 < Y \leq 2.0\}$.

B-27. The joint density function of random variables X and Y is

$$p_{X,Y}(x, y) = \left(\frac{1}{12}\right)\text{rect}\left(\frac{x}{3}\right)\text{rect}\left(\frac{y}{4}\right).$$

(a) Find $P\{Y \leq X/2\}$. (b) Find $P\{Y \leq 2X\}$.

★**B-28.** The density function of a single random variable can be obtained by integrating out all other random variables in a joint density function. The resulting density is called the *marginal density function*. Thus for two random variables X and Y,

$$p_X(x) = \int_{-\infty}^{\infty} p_{X,Y}(x, y)\, dy$$

$$p_Y(y) = \int_{-\infty}^{\infty} p_{X,Y}(x, y)\, dx.$$

Use these results to find the marginal densities of X and Y when

$$p_{X,Y}(x, y) = u(x)u(y)x\exp(-x - xy).$$

By application of (B.2-15), also find the conditional densities $p_X(x|y)$ and $p_Y(y|x)$.

B-29. Find (a) the mean of X, $E[X]$, (b) the mean of Y, $E[Y]$, and (c) the correlation of X and Y, $R_{XY} = E[XY]$ for random variables X and Y defined by the joint density in Prob. B-28.

B-30. A Gaussian random variable X has mean $\overline{X} = 2$ and variance $\sigma_X^2 = 1.44$. Find $P\{0.32 < X \leqslant 2.84\}$.

B-31. What is the probability that the magnitude of a zero-mean Gaussian random variable can have values larger than twice its standard deviation?

B-32. A zero-mean Gaussian random variable X with variance 4 is changed to a new variable Y according to $Y = 8 + (X/2)$. (a) Find and plot the density function of Y. (b) What are the mean and variance of Y? (c) Find $E[X^2]$ and $E[Y^2]$.

★B-33. Let X and Y be statistically independent random variables with respective densities $p_X(x)$ and $p_Y(y)$. Show that the density of the sum $W = X + Y$ is given by

$$p_W(w) = \int_{-\infty}^{\infty} p_X(w - y)p_Y(y)\, dy.$$

(*Hint:* Equate probabilities $P_W(w) = P\{W \leqslant w\} = P\{X + Y \leqslant w\}$.)

B-34. Show that the first right-hand-side equation in (B.4-8) is true.

B-35. Show that the random process $X(t) = A\cos(\omega_X t + \phi_X + \Theta_X)$ is wide-sense stationary, where A, ϕ_X, and ω_X are real constants; Θ_X is a random variable uniform on $[-\pi, \pi]$.

B-36. Work Prob. B-35 except for the process $Y(t) = B\cos(\omega_Y t + \phi_Y + \Theta_X)$.

B-37. (a) Find the cross-correlation function $E[X(t)Y(t + \tau)]$ of the random processes defined in Probs. B-35 and B-36. (b) What conditions can be placed on the various constants such that $X(t)$ and $Y(t)$ are jointly wide-sense stationary?

B-38. Show that (B.5-11) and (B.5-13) are true for the random process of Prob. B-35.

B-39. Find the autocorrelation function and power spectrum of the process of Prob. B-35.

B-40. The power spectrum of a process $X(t)$ is

$$\mathcal{S}_{XX}(\omega) = \frac{8}{1 + (\omega^2/100)}.$$

(a) Use (B.6-10) to find the power in the process. (b) Find the process's autocorrelation function. (c) Verify the power in $X(t)$ by the relationship $E[X^2(t)] = R_{XX}(0)$.

B-41. Find the power in processes having the following power spectrums:

(a) $\mathcal{S}_{XX}(\omega) = \dfrac{12}{(10^4 + \omega^2)^2}$

(b) $\mathcal{S}_{XX}(\omega) = \dfrac{100\,\omega^2}{(100 + \omega^2)^2}$

(c) $\mathcal{S}_{XX}(\omega) = 8\exp\left(\dfrac{-\omega^2}{1000}\right)$

(d) $\mathcal{S}_{XX}(\omega) = \dfrac{\omega^2/10}{(10 + \omega^2)^3}.$

B-42. Assume the cross-power spectrum of two processes is defined by

$$\mathscr{S}_{XY}(\omega) = \begin{cases} a + \dfrac{j\omega}{W}, & -W < \omega < W \\ 0, & \text{elsewhere,} \end{cases}$$

where $W > 0$ and a is a real constant. Find the crosscorrelation function $R_{XY}(\tau)$.

B-43. Define a process $X(t)$ as in Prob. B-35 except let Θ_X be uniform on $[0, \pi/2]$; it becomes a nonstationary process. (a) Find the autocorrelation function of $X(t)$. (b) Find the time average of the autocorrelation function. (c) Find the power spectrum of $X(t)$.

B-44. White noise with autocorrelation function $(\mathscr{N}_0/2)\delta(\tau)$ is applied to a lowpass filter for which $H(\omega) = [1 + j(\omega/10)]^{-1}$. (a) Find the filter's impulse response. (b) Find the autocorrelation function of the output noise process $Y(t)$ by use of (B.7-5). (c) Use (B.7-6) to find $R_{XY}(\tau)$.

B-45. Show that (B.7-3) is equivalent to

$$\overline{Y} = \overline{X}H(0)$$

where $h(t) \leftrightarrow H(\omega)$.

B-46. Begin with (B.7-2) and derive (B.7-5) and (B.7-6).

B-47. Show that (B.7-5) can be written in the form

$$R_{YY}(\tau) = \int_{-\infty}^{\infty} R_{XY}(\tau + \xi)h(\xi)\,d\xi$$

$$= \int_{-\infty}^{\infty} R_{YX}(\tau - \xi)h(\xi)\,d\xi.$$

B-48. By Fourier transformation of $R_{YY}(\tau)$, using (B.7-5), show that (B.7-8) is true.

B-49. Fourier transform the crosscorrelation functions of (B.7-6) and (B.7-7) to show that (B.7-9) and (B.7-10) result.

B-50. If $H(\omega)$ is the transfer function of a network, its *noise bandwidth* W_N (rad/s) is defined by

$$W_N = \frac{\displaystyle\int_0^{\infty} |H(\omega)|^2\,d\omega}{|H(\omega_0)|^2}$$

where ω_0 is the midband frequency (rad/s). For a lowpass network $\omega_0 = 0$. Find the noise bandwidth for the network for which

$$H(\omega) = \frac{a}{[1 + (\omega/W)^2]^2}$$

with a and $W > 0$ constants.

B-51. If white noise is applied to a linear network, show that the output noise power is the same as if the network were replaced by an ideal filter (rectangularly shaped $|H(\omega)|$) with the same midband gain and bandwidth equal to the network's noise bandwidth.

B-52. The power spectrum of a bandpass random process $X(t)$ is

$$\mathscr{S}_{XX}(\omega) = P[u(\omega - \omega_0 + W_L) - u(\omega - \omega_0 - W_H)$$
$$+ u(\omega + \omega_0 + W_H) - u(\omega + \omega_0 - W_L)],$$

where P, ω_0, W_L and W_H are constants, $W_L + W_H = W$, and $\omega_0 > W$.
(a) Plot $\mathscr{S}_{XX}(\omega)$. (b) Find and plot the power and cross-power density spectrums
of $X_c(t)$ and $X_s(t)$ used in the representation of (B.8-2). (c) What is the largest
bandwidth that $X_c(t)$ and $X_s(t)$ may have if W is to be kept constant and
what must W_L and W_H be for this bandwidth to occur? (d) Repeat part (c)
to find the *smallest* bandwidth.

B-53. Find the crosscorrelation function $R_{X_cX_s}(\tau)$ for the process defined in Prob.
B-52.

★B-54. Let a wide-sense stationary random process $X(t)$ be applied to one side of
a product device. The other input to the product is the signal $A\cos(\omega_0 t + \theta_0)$ where A, ω_0 and θ_0 are constants. Show that the power spectrum of the
output $Y(t)$ of the product device is

$$\mathscr{S}_{YY}(\omega) = \frac{A^2}{4}[\mathscr{S}_{XX}(\omega - \omega_0) + \mathscr{S}_{XX}(\omega + \omega_0)]$$

where $\mathscr{S}_{XX}(\omega)$ is the power spectrum of $X(t)$.

B-55. Derive (B.9-5) from (B.9-4) using Schwarz's inequality.

B-56. A rectangular pulse of duration τ_0 defined by

$$f(t) = A \operatorname{rect}\left[\frac{t - (\tau_0/2)}{\tau_0}\right]$$

is applied, along with white noise of power density $\mathscr{N}_0/2$ to a matched filter.
(a) Find $H_{opt}(\omega)$ if $H_{opt}(0) = 1$ is required. (b) Find $h_{opt}(t)$.

B-57. Work Prob. B-56 for the triangular pulse

$$f(t) = A \operatorname{tri}\left[\frac{(t - \tau_0)}{\tau_0}\right].$$

B-58. For a signal $f(t)$ in white noise with power density $\mathscr{N}_0/2$, show that (B.9-5)
can also be written as

$$\left(\frac{S_o}{N_o}\right) \le \frac{2}{\mathscr{N}_0} \int_{-\infty}^{\infty} |f(t)|^2 \, dt.$$

B-59. Show that (B.9-8) is true.

Appendix C

Trigonometric Identities

$$\cos(x \pm y) = \cos(x)\cos(y) \mp \sin(x)\sin(y) \qquad \text{(C-1)}$$

$$\sin(x \pm y) = \sin(x)\cos(y) \pm \cos(x)\sin(y) \qquad \text{(C-2)}$$

$$\cos\left(x \pm \frac{\pi}{2}\right) = \mp \sin(x) \qquad \text{(C-3)}$$

$$\sin\left(x \pm \frac{\pi}{2}\right) = \pm \cos(x) \qquad \text{(C-4)}$$

$$\cos(2x) = \cos^2(x) - \sin^2(x) \qquad \text{(C-5)}$$

$$\sin(2x) = 2\sin(x)\cos(x) \qquad \text{(C-6)}$$

$$2\cos(x)\cos(y) = \cos(x - y) + \cos(x + y) \qquad \text{(C-7)}$$

$$2\sin(x)\sin(y) = \cos(x - y) - \cos(x + y) \qquad \text{(C-8)}$$

$$2\sin(x)\cos(y) = \sin(x - y) + \sin(x + y) \qquad \text{(C-9)}$$

$$2\cos^2(x) = 1 + \cos(2x) \qquad \text{(C-10)}$$

$$2\sin^2(x) = 1 - \cos(2x) \qquad \text{(C-11)}$$

$$4\cos^3(x) = 3\cos(x) + \cos(3x) \qquad \text{(C-12)}$$

$$4\sin^3(x) = 3\sin(x) - \sin(3x) \qquad \text{(C-13)}$$

$$8\cos^4(x) = 3 + 4\cos(2x) + \cos(4x) \qquad \text{(C-14)}$$

$$8\sin^4(x) = 3 - 4\cos(2x) + \cos(4x) \qquad \text{(C-15)}$$

$$A \cos(x) - B \sin(x) = R \cos(x + \theta) \qquad \text{(C-16)}$$

where

$$R = \sqrt{A^2 + B^2} \qquad \text{(C-17)}$$

$$\theta = \tan^{-1}(B/A) \qquad \text{(C-18)}$$

$$A = R \cos(\theta) \qquad \text{(C-19)}$$

$$B = R \sin(\theta) \qquad \text{(C-20)}$$

$$2 \cos(x) = e^{jx} + e^{-jx} \qquad \text{(C-21)}$$

$$2j \sin(x) = e^{jx} - e^{-jx} \qquad \text{(C-22)}$$

Appendix D

Useful Integrals and Series

INDEFINITE INTEGRALS

Rational Algebraic Functions

$$\int x^n \, dx = \frac{x^{n+1}}{n+1}, \qquad 0 \le n \tag{D-1}$$

$$\int \frac{dx}{x} = \ln|x| \tag{D-2}$$

$$\int \frac{dx}{x^n} = \frac{-1}{(n-1)x^{n-1}}, \qquad 1 < n \tag{D-3}$$

$$\int (a + bx)^n \, dx = \frac{(a + bx)^{n+1}}{b(n+1)}, \qquad 0 \le n \tag{D-4}$$

$$\int \frac{dx}{a + bx} = \frac{1}{b} \ln|a + bx| \tag{D-5}$$

$$\int \frac{dx}{(a + bx)^n} = \frac{-1}{(n-1)b(a + bx)^{n-1}}, \qquad 1 < n \tag{D-6}$$

$$\int \frac{dx}{a^2 + b^2 x^2} = \frac{1}{ab} \tan^{-1}\left(\frac{bx}{a}\right) \tag{D-7}$$

$$\int \frac{dx}{(a^2 + x^2)^2} = \frac{x}{2a^2(a^2 + x^2)} + \frac{1}{2a^3} \tan^{-1}\left(\frac{x}{a}\right)$$ (D-8)

$$\int \frac{dx}{(a^2 + x^2)^3} = \frac{x}{4a^2(a^2 + x^2)^2} + \frac{3x}{8a^4(a^2 + x^2)} + \frac{3}{8a^5} \tan^{-1}\left(\frac{x}{a}\right)$$ (D-9)

$$\int \frac{x\, dx}{a^2 + x^2} = \frac{1}{2} \ln(a^2 + x^2)$$ (D-10)

$$\int \frac{x^2\, dx}{a^2 + x^2} = x - a \tan^{-1}\left(\frac{x}{a}\right)$$ (D-11)

$$\int \frac{dx}{a^4 + x^4} = \frac{1}{4a^3 \sqrt{2}} \ln\left(\frac{x^2 + ax\sqrt{2} + a^2}{x^2 - ax\sqrt{2} + a^2}\right) + \frac{1}{2a^3 \sqrt{2}} \tan^{-1}\left(\frac{ax\sqrt{2}}{a^2 - x^2}\right)$$ (D-12)

$$\int \frac{x^2\, dx}{a^4 + x^4} = -\frac{1}{4a\sqrt{2}} \ln\left(\frac{x^2 + ax\sqrt{2} + a^2}{x^2 - ax\sqrt{2} + a^2}\right) + \frac{1}{2a\sqrt{2}} \tan^{-1}\left(\frac{ax\sqrt{2}}{a^2 - x^2}\right)$$ (D-13)

Trigonometric Functions

$$\int \cos(x)\, dx = \sin(x)$$ (D-14)

$$\int x \cos(x)\, dx = \cos(x) + x \sin(x)$$ (D-15)

$$\int x^2 \cos(x)\, dx = 2x \cos(x) + (x^2 - 2)\sin(x)$$ (D-16)

$$\int \sin(x)\, dx = -\cos(x)$$ (D-17)

$$\int x \sin(x)\, dx = \sin(x) - x \cos(x)$$ (D-18)

$$\int x^2 \sin(x)\, dx = 2x \sin(x) - (x^2 - 2)\cos(x)$$ (D-19)

Exponential Functions

$$\int e^{ax}\, dx = \frac{e^{ax}}{a}$$ (D-20)

$$\int xe^{ax}\, dx = e^{ax}\left[\frac{x}{a} - \frac{1}{a^2}\right]$$ (D-21)

$$\int x^2 e^{ax}\, dx = e^{ax}\left[\frac{x^2}{a} - \frac{2x}{a^2} + \frac{2}{a^3}\right]$$ (D-22)

DEFINITE INTEGRALS

$$\int_0^\infty e^{-a^2x^2}\, dx = \sqrt{\pi}/2a, \qquad 0 < a \tag{D-23}$$

$$\int_0^\infty x^2 e^{-x^2}\, dx = \sqrt{\pi}/4 \tag{D-24}$$

$$\int_0^\infty \mathrm{Sa}(x)\, dx = \int_0^\infty \frac{\sin(x)}{x}\, dx = \frac{\pi}{2} \tag{D-25}$$

$$\int_0^\infty \mathrm{Sa}^2(x)\, dx = \pi/2 \tag{D-26}$$

FINITE SERIES

$$\sum_{n=1}^{N} n = \frac{N(N+1)}{2} \tag{D-27}$$

$$\sum_{n=1}^{N} n^2 = \frac{N(N+1)(2N+1)}{6} \tag{D-28}$$

$$\sum_{n=1}^{N} n^3 = \frac{N^2(N+1)^2}{4} \tag{D-29}$$

$$\sum_{n=0}^{N} x^n = \frac{x^{N+1} - 1}{x - 1} \tag{D-30}$$

Appendix E

Gaussian (Normal) Probability Density and Distribution Functions

For zero mean and unit variance:

$$p(x) = \frac{1}{\sqrt{2\pi}} e^{-x^2/2} \tag{E-1}$$

$$P(x) = \frac{1}{\sqrt{2\pi}} \int_{-\infty}^{x} e^{-u^2/2} \, du \tag{E-2a}$$

$$P(-x) = 1 - P(x) \tag{E-2b}$$

TABLE E-1. The Gaussian or Normal Density and Distribution.

x	$p(x)$	$P(x)$	x	$p(x)$	$P(x)$
0.00	0.3989	0.5000	1.40	0.1497	0.9192
0.05	0.3984	0.5199	1.50	0.1295	0.9332
0.10	0.3970	0.5398	1.60	0.1109	0.9452
0.15	0.3945	0.5596	1.70	0.0940	0.9554
0.20	0.3910	0.5793	1.80	0.0790	0.9641
0.25	0.3867	0.5987	1.90	0.0656	0.9713
0.30	0.3814	0.6179	2.00	0.0540	0.9772
0.35	0.3752	0.6368	2.10	0.0440	0.9821
0.40	0.3683	0.6554	2.20	0.0355	0.9861
0.45	0.3605	0.6736	2.30	0.0283	0.9893
0.50	0.3521	0.6915	2.40	0.0224	0.9918
0.55	0.3429	0.7088	2.50	0.0175	0.9938
0.60	0.3332	0.7257	2.60	0.0136	0.9953
0.65	0.3230	0.7422	2.70	0.0104	0.9965
0.70	0.3123	0.7580	2.80	0.0079	0.9974
0.75	0.3011	0.7734	2.90	0.0060	0.9981
0.80	0.2897	0.7881	3.00	0.0044	0.9987
0.85	0.2780	0.8023	3.20	0.0024	0.9993
0.90	0.2661	0.8159	3.40	0.0012	0.9997
0.95	0.2541	0.8289	3.60	0.0006	0.9998
1.00	0.2420	0.8413	3.80	0.0003	0.9999
1.10	0.2179	0.8643	4.00	0.0001	1.0000
1.20	0.1942	0.8849	4.50	0.0	1.0000
1.30	0.1714	0.9032			

SOURCE: Peebles, Jr., p. 2, *Communication System Principles*, Addison-Wesley Publishing Co., Inc., Reading, Massachusetts, 1976.

For arbitrary mean m and variance σ^2:

$$p_X(x) = \frac{1}{\sqrt{2\pi\sigma^2}} e^{-(x-m)^2/2\sigma^2} \tag{E-3}$$

$$P_X(x) = P\left(\frac{x-m}{\sigma}\right) \tag{E-4}$$

Appendix F

Table of Error Functions

The complementary error function is defined by

$$\text{erfc}(x) = 1 - \text{erf}(x) \qquad\qquad \text{(F-1)}$$

where $\text{erf}(x)$ is the error function given by

$$\text{erf}(x) = \frac{2}{\sqrt{\pi}} \int_0^x e^{-\xi^2}\, d\xi. \qquad\qquad \text{(F-2)}$$

For negative x

$$\text{erf}(-x) = -\text{erf}(x) \qquad\qquad \text{(F-3)}$$

$$\text{erfc}(-x) = 1 + \text{erf}(x) = 2 - \text{erfc}(x). \qquad\qquad \text{(F-4)}$$

For $x \geqslant 2$ the approximation

$$\text{erfc}(x) \approx \frac{e^{-x^2}}{x\sqrt{\pi}}, \qquad x \gg 1 \qquad\qquad \text{(F-5)}$$

has 10.5% or less error. Equations (F-1), (F-2), and (F-5) are tabulated.

x	erf (x)	erfc (x)	$\dfrac{e^{-x^2}}{x\sqrt{\pi}}$
0.0	0.0	1.0	
0.1	0.1125	0.8875	
0.2	0.2227	0.7773	
0.3	0.3286	0.6714	
0.4	0.4284	0.5716	
0.5	0.5205	0.4795	
0.6	0.6039	0.3961	
0.7	0.6778	0.3222	
0.8	0.7421	0.2579	
0.9	0.7969	0.2031	
1.0	0.8427	0.1573	0.2076
1.1	0.8802	0.1198	0.1529
1.2	0.9103	$0.8969 \ (10^{-1})$	0.1114
1.3	0.9340	$0.6599 \ (10^{-1})$	$0.8008 \ (10^{-1})$
1.4	0.9523	$0.4771 \ (10^{-1})$	$0.5676 \ (10^{-1})$
1.5	0.9661	$0.3389 \ (10^{-1})$	$0.3964 \ (10^{-1})$
1.6	0.9763	$0.2365 \ (10^{-1})$	$0.2726 \ (10^{-1})$
1.7	0.9838	$0.1621 \ (10^{-1})$	$0.1844 \ (10^{-1})$
1.8	0.9891	$0.1091 \ (10^{-1})$	$0.1228 \ (10^{-1})$
1.9	0.9928	$0.7210 \ (10^{-2})$	$0.8033 \ (10^{-2})$
2.0	0.9953	$0.4678 \ (10^{-2})$	$0.5167 \ (10^{-2})$
2.1	0.9970	$0.2979 \ (10^{-2})$	$0.3266 \ (10^{-2})$
2.2	0.9981	$0.1863 \ (10^{-2})$	$0.2028 \ (10^{-2})$
2.3	0.9989	$0.1143 \ (10^{-2})$	$0.1237 \ (10^{-2})$
2.4	0.9993	$0.6885 \ (10^{-3})$	$0.7408 \ (10^{-3})$
2.5	0.9996	$0.4070 \ (10^{-3})$	$0.4357 \ (10^{-3})$
2.6	0.9998	$0.2360 \ (10^{-3})$	$0.2515 \ (10^{-3})$
2.7	0.9999	$0.1343 \ (10^{-3})$	$0.1426 \ (10^{-3})$
2.8	0.9999	$0.7501 \ (10^{-4})$	$0.7932 \ (10^{-4})$
2.9	1.0000	$0.4110 \ (10^{-4})$	$0.4331 \ (10^{-4})$
3.0	1.0000	$0.2209 \ (10^{-4})$	$0.2321 \ (10^{-4})$
3.5	1.0000	$0.7431 \ (10^{-6})$	$0.7713 \ (10^{-6})$
4.0	1.0000	$0.1542 \ (10^{-7})$	$0.1587 \ (10^{-7})$
5.0	1.0000	$0.1564 \ (10^{-11})$	$0.1567 \ (10^{-11})$

Appendix G

Table of Useful Fourier Transform Pairs

In the accompanying table the functions $f(\cdot)$, $f_n(\cdot)$, $F_n(\cdot)$ and $F(\cdot)$ may be complex. Constants t_0, ω_0, a, τ, W, and σ are real, whereas α_n may be complex. Special functions are defined as follows:

$$u(\xi) = \begin{cases} 1, & \xi > 0 \\ 0, & \xi < 0 \end{cases}$$

$$\text{rect}(\xi) = \begin{cases} 1, & |\xi| < 1/2 \\ 0, & |\xi| > 1/2 \end{cases}$$

$$\text{tri}(\xi) = \begin{cases} 1 - |\xi|, & |\xi| < 1 \\ 0, & |\xi| > 1 \end{cases}$$

$$\text{Sa}(\xi) = \xi^{-1}\sin(\xi)$$

$$\text{sgn}(\xi) = \begin{cases} 1, & \xi > 0 \\ -1, & \xi < 0. \end{cases}$$

Pair Number	$f(t)$	$F(\omega)$	Notes
1	$f_n(t)$	$F_n(\omega)$	
2	$\displaystyle\sum_{n=1}^{N} \alpha_n f_n(t)$	$\displaystyle\sum_{n=1}^{N} \alpha_n F_n(\omega)$	
3	$f(t - t_0)$	$F(\omega)e^{-j\omega t_0}$	
4	$f(t)e^{j\omega_0 t}$	$F(\omega - \omega_0)$	
5	$f(at)$	$\dfrac{1}{|a|} F\left(\dfrac{\omega}{a}\right)$	

6	$F(t)$	$2\pi f(-\omega)$	
7	$\dfrac{d^n f(t)}{dt^n}$	$(j\omega)^n F(\omega)$	
8	$(-jt)^n f(t)$	$\dfrac{d^n F(\omega)}{d\omega^n}$	
9	$\displaystyle\int_{-\infty}^{t} f(\tau)\,d\tau$	$\dfrac{F(\omega)}{j\omega} + \pi F(0)\delta(\omega)$	
10	$\delta(t)$	1	
11	1	$2\pi\delta(\omega)$	
12	$u(t)$	$\pi\delta(\omega) + \left(\dfrac{1}{j\omega}\right)$	
13	$\dfrac{1}{2}\delta(t) - \dfrac{1}{j2\pi t}$	$u(\omega)$	
14	$\text{rect}\left(\dfrac{t}{\tau}\right)$	$\tau\,\text{Sa}\left(\dfrac{\omega\tau}{2}\right)$	$\tau > 0$
15	$\dfrac{W}{\pi}\text{Sa}(Wt)$	$\text{rect}\left(\dfrac{\omega}{2W}\right)$	$W > 0$
16	$\text{tri}\left(\dfrac{t}{\tau}\right)$	$\tau\,\text{Sa}^2\left(\dfrac{\omega\tau}{2}\right)$	$\tau > 0$
17	$\dfrac{W}{\pi}\text{Sa}^2(Wt)$	$\text{tri}\left(\dfrac{\omega}{2W}\right)$	$W > 0$
18	$e^{j\omega_0 t}$	$2\pi\delta(\omega - \omega_0)$	
19	$\delta(t - t_0)$	$e^{-j\omega t_0}$	
20	$\text{sgn}(t)$	$\dfrac{2}{j\omega}$	
21	$\dfrac{j}{\pi t}$	$\text{sgn}(\omega)$	
22	$\cos(\omega_0 t)$	$\pi[\delta(\omega - \omega_0) + \delta(\omega + \omega_0)]$	
23	$\sin(\omega_0 t)$	$-j\pi[\delta(\omega - \omega_0) - \delta(\omega + \omega_0)]$	
24	$u(t)\,e^{-at}$	$\dfrac{1}{a + j\omega}$	$a > 0$
25	$u(t)te^{-at}$	$\dfrac{1}{(a + j\omega)^2}$	$a > 0$
26	$u(t)t^2 e^{-at}$	$\dfrac{2}{(a + j\omega)^3}$	$a > 0$
27	$u(t)t^3 e^{-at}$	$\dfrac{6}{(a + j\omega)^4}$	$a > 0$
28	$e^{-a|t|}$	$\dfrac{2a}{a^2 + \omega^2}$	$a > 0$
29	$e^{-t^2/(2\sigma^2)}$	$\sigma\sqrt{2\pi}\,e^{-\sigma^2\omega^2/2}$	$\sigma > 0$

Index